Chirality: Physical Chemistry

ACS SYMPOSIUM SERIES **810**

Chirality: Physical Chemistry

Janice M. Hicks, Editor

Georgetown University

American Chemical Society, Washington, DC

Library of Congress Cataloging-in-Publication Data

Chirality: physical chemistry / Janice M. Hicks, editor.

　　p. cm.—(ACS symposium series ; 810)

Includes bibliographical references and index.

　ISBN 0–8412–3737–9

　　1. Chirality—Congresses. 2. Chemistry, Physical and theoretical— Congresses.

　I. Hicks, Janice M. II. Series.

QD481.C555　2002
541.2′25—dc21　　　　　　　　　　　　　　　　2001053393

The paper used in this publication meets the minimum requirements of American National Standard for Information Sciences—Permanence of Paper for Printed Library Materials, ANSI Z39.48–1984.

Distributed by Oxford University Press

The cover is reprinted with permission. Copyright 2000 Corel Corporation and Corel Corporation Ltd.

PRINTED IN THE UNITED STATES OF AMERICA

Foreword

The ACS Symposium Series was first published in 1974 to provide a mechanism for publishing symposia quickly in book form. The purpose of the series is to publish timely, comprehensive books developed from ACS sponsored symposia based on current scientific research. Occasionally, books are developed from symposia sponsored by other organizations when the topic is of keen interest to the chemistry audience.

Before agreeing to publish a book, the proposed table of contents is reviewed for appropriate and comprehensive coverage and for interest to the audience. Some papers may be excluded to better focus the book; others may be added to provide comprehensiveness. When appropriate, overview or introductory chapters are added. Drafts of chapters are peer-reviewed prior to final acceptance or rejection, and manuscripts are prepared in camera-ready format.

As a rule, only original research papers and original review papers are included in the volumes. Verbatim reproductions of previously published papers are not accepted.

ACS Books Department

Contents

Advances in Chiral Spectroscopy

New Approaches in Chiral Recognition

Surface Chirality and Chiral Nanostructures

Indexes

Preface

Chirality is a compelling topic with classic unanswered questions and great research problems still not resolved. This book evolved from the first symposium on the Physical Chemistry of Chirality held in San Francisco, California at the National Meeting of the American Chemical Society (ACS) at the Renaissance Hotel in March 2000. The range of topics explored is broad, relating to structure, recognition, and optical response of chiral systems through various new instrumental and theoretical approaches. My intention for the symposium, and now for the book, is that those studying interesting new chiral systems, such as aggregates and surfaces, will interact with those at the forefront of developing new methods for studying chirality both experimentally and theoretically, in order to stimulate new research directions. Because of an explosion of new papers in this area since the time of the meeting, I have tried to update the references in the Overview Chapter as well as possible.

I thank George Schatz for suggesting this symposium and for inviting me to lead it. The job has been a pleasure and honor, especially given the wonderful group this meeting drew together. I thank the sponsors of the meeting: OLIS, Inc. and the ACS Petroleum Research Fund. I also thank Carolyn Dozier and Emily Dozier for their support during the preparation of this book.

Janice M. Hicks*
Department of Chemistry
Georgetown University
37th and O Streets, NW
Washington, DC 20057-1227

*Current address: National Science Foundation, 4201 Wilson Boulevard, Room 1054, Arlington, VA 22230 (phone: 703/292-4956; fax: 703/292-9037; email: jhicks@nsf.gov)

Introduction

Chapter 1

The Physical Chemistry of Chirality

Janice M. Hicks[1]

Department of Chemistry, Georgetown University, Washington DC 20057
[1]Current address: National Science Foundation, 4201 Wilson Boulevard, Room 1055, Arlington VA 22230

New advances in experimental and theoretical physical chemistry have made possible many innovations in the study of chiral molecules, nano-structures and surfaces of solids. This introduction reviews some of the background and current highlights of the topic for a general chemistry audience.

Chirality is a geometrical concept that has captivated chemists since the time of Louis Pasteur. In 1848 at the age of 26, Pasteur discovered that solutions of crystals that he separated into right-hemihedral and left-hemihedral shapes could rotate linearly polarized light in a right or left-handed sense corresponding to the crystal shapes *(1)*. Lord Kelvin, in his *Baltimore Lectures on Molecular Dynamics and the Wave Theory of Light*, called this notion "chirality" (from the Greek kheir, for hand), and defined it in this manner: "I call any geometric figure, or group of points, chiral, and say it has chirality, if its image in a plane mirror, ideally realized, cannot be brought to coincide with itself" *(2)*. Kelvin's definition applies to any object, but of interest here is when the objects are molecules, supramolecular structures or surfaces of solids.

The concept of chirality in chemistry might be merely an academic one if it were not for the stunning fact that most of biochemistry is chiral *(3)*. Proteins, DNA, amino acids, sugars and many natural products such as steroids, hormones, and pheromones possess chirality. Indeed, classes of molecules such as amino acids are largely found to be homochiral (in the L form for amino acids, where L and D refer to levo (left) and dextro (right) molecular configurations. This system is giving way to another using S (sinister, left) and R (rectus, right)). The exact origin of homochirality is one of the great, unanswered questions in evolutionary science *(4) (5)*. For chiral chemicals sampled from extraterrestrial environments or meteorites, detection of a preferred chiral sense could, some argue, be associated with life.

In fact, it is usually the case that natural product molecules contain many stereogenic centers each, giving rise to a large number of possible absolute structures. Stereochemical assignment has proven to be a major challenge *(6)* and some of the motivation for developing physical tools originates with this issue.

Because most biological receptors and membranes are chiral, many drugs, herbicides, pesticides and other biological agents must themselves possess chirality for binding and action to occur. Most medicines such as ibuprofen, and other bioactive compounds, such as nicotine are therefore chiral. Synthetic processes ordinarily produce a 50:50 (racemic) mixture of left-handed and right-handed molecules (so-called enantiomers), and often the two enantiomers behave differently in a biological system. Ropivacaine was the first chirally pure local anesthetic, offered as the S(-) enantiomer, and it has lower cardiotoxicity than the racemic mixture *(7)*. Positron emission tomography scans of Ritalin in the brain reveal that the D form concentrates in the striatum, whereas its mirror-image form distributes nonspecifically over the brain *(8)*. The list of molecules having differing biological activities for right and left-handed forms is extensive *(9)*. (The famous case of the sedative thalidomide is controversial *(10)*.) Chiral purity has become, then, a major concern in the pharmaceutical industry, and chemists are striving to develop methods to 1. separate enantiomers in batch by chromatography (increasingly, by simulated moving bed chromatography) or through the use of membranes and 2. design syntheses, perhaps using biocatalysts, that enable just one enantiomer to form. The 6 billion dollar market for single enantiomers is expected to grow rapidly. Another new effort involves environmental science. Concern about the chirality of pollutants in the environment, such as herbicides and pesticides, is increasing since this information is vital to assessing bioavailability, toxicity and transport *(11)*. Reliable sensors capable of determining concentrations of enantiomers are needed.

Much has been written about chirality in molecules *(12)*, separations *(13)*, and optical effects of chiral molecules *(14) (15)*. This book focuses on the new work of physical chemists applying advanced methods in experimental and theoretical chemistry and physics to problems in chirality. Issues of absolute molecular structure, dynamics, reactivity, energetics and interaction of chiral

4

matter with light have been illuminated with these recent advances. Chiral molecules have been studied in the gas phase, in liquids, on surfaces, and in structured materials.

Some of the newest innovations in the physical chemistry of chirality involve spectroscopy. Historically, early physicists and physical chemists explained Pasteur's discovery achieving the earliest knowledge about the interaction of light with chiral media. This work culminates in Rosenfeld's initial explanation of the quantum origins of optical activity in 1928 *(16)* and Tinoco's more complete theory in 1960 *(17)*. Recently, new instruments such as the Vibrational Circular Dichroism Spectrometer, new methods such as nonlinear optics and *ab initio* calculations, and new light sources such as synchroton radiation have allowed a resurgence of productivity in the area of chiroptical spectroscopy.

A second new area for physical chemistry is the study of chiral surfaces. Chiral surfaces could potentially be used to catalyze the synthesis of optically pure samples, with the benefit of easily retrieving the catalyst afterwards. Fundamental studies of chiral surfaces underlie this effort to design heterogeneous chiral catalysts. Understanding molecule/surface interactions is also important for the improvement of chiral separations via chromatography. A prevailing model of retention of analytes in chiral stationary and mobile phases involves a "three point recognition" process *(18)*. If an analyte contacts the target at at least 3 of its geometric points, the chirality of the analyte can be recognized and one enantiomer retained in favor of the other (the basis for separation). Physical models of the molecular process, and data to support them, will allow advancements to occur in this technology.

One new approach to the study of surface chirality is second-order nonlinear optical spectroscopy methods such as Second Harmonic Generation (SHG), where the chiroptical effects are large and originate selectively from the interface where symmetry is broken *(19) (20)*. New kinds of information can be obtained, for example, the spectrum of a unique chiral parameter characteristic of the molecule (χ_{xyz}) can be measured *(21)*. Enantiomeric excess at the interface can be measured *(22)*. Absolute orientation of the molecules at the surface (pointed up or down) can be obtained if their chirality is known *(19)*. It is possible to probe structural transitions in proteins as they adsorb to surfaces *(23)*. Another major surface science technique, scanning tunneling microscopy (STM) is also being applied to determine chirality on surfaces in a direct manner by molecular scale imaging.

Finally, physical chemists as well as synthetic chemists are participating in creating new chiral nanometer-sized particles. Chiral concepts could be of use in designing molecular motors, e.g. propellers *(24)*, and molecular electronics, one of the goals of nanotechnology.

Spectroscopy

Certainly spectroscopy is a dominant tool for the study of chiral molecules, having set the course from Pasteur's work onward. Interestingly, the light/matter interactions in linear spectroscopy are weak, as we shall see. Nevertheless, due to excellent polarization optics, measurements of optical activity are routine even in the undergraduate organic chemistry course.

Electronic Spectroscopy

There are two most common types of chiroptical measurements: optical activity (also called polarimetry) is the rotation of linearly polarized light by chiral media. This is a scattering method, most commonly performed at the Na line at 589 nm (which is not usually in resonance with a molecular energy transition). When conducted as a function of wavelength (through a resonance), this is called optical rotatory dispersion (ORD). The rotation angle, δ, is proportional to the pathlength of the cell containing the sample, and the difference in the refractive index of the sample for left and right circularly polarized light (n_l - n_r). This latter quantity is typically around 10^{-6}, but with sufficient pathlength, several degrees rotation result. Feynman presents a satisfying physical explanation of this rotation *(25)*.

The other most common linear spectroscopic method is circular dichroism spectroscopy (CD), which is an absorption method and is performed at wavelengths resonant with a molecular energy transition. CD is measured as the difference between the absorption by a sample of left versus right circularly polarized light ($\Delta\varepsilon=\varepsilon_l$ - ε_r, where ε is the molar extinction coefficient). For a certain handedness of the molecule and resonance, left or right circularly polarized light will be absorbed more strongly, thus $\varepsilon_l \neq \varepsilon_r$ for the solution containing these molecules. A typical $\Delta\varepsilon$ / ε is about 10/10,000 or 0.1%.

The strengths of ORD and CD are related to the Rotatory Power R_{ab} = $\mu_{ab}\cdot m_{ba}$, where a and b are molecular states, μ is the electric dipole transition moment and **m** is the magnetic dipole transition moment *(16)*. For ORD and CD to occur then, μ and **m** must have components parallel to each other. This does not occur for symmetric molecules, and even in chiral molecules, these effects are small. Enantiomers have opposite signs of R and thus their ORD and CD spectra are opposite signs from each other. In real molecules, one must sum over all the states (not just a and b).

Modern ORD and CD spectrometers function in the ultraviolet and visible regions of the spectrum, and therefore measure electronic transitions. They are averaging. The methods are inherently sensitive to chirality, unlike other powerful structural tools such as Nuclear Magnetic Resonance (NMR), which requires chiral shift agents to determine chirality (26). The timescales of the ORD and CD processes are very fast, associated with the time for light to scatter or be absorbed (10^{-14} to 10^{-15} s). This is much faster than NMR. A tremendous literature on ORD and CD exists, with a great deal of biological work. Secondary structures of proteins in solution can be differentiated using signature CD spectra in the 190 - 240 nm region, associated with the n->π* and π -> π * transitions of the amide backbone (27). Using lasers, time-resolved CD measurements have been made on ultrafast timescales for the detailed kinetic study of protein folding, for example (28).

One of the drawbacks of ORD and CD is that the electronic transitions on which the effects are based tend to be broad and sometimes overlapping, making interpretation difficult. CD and ORD are of little use in studying molecules that do not have accessible chromophores. Transmissivity of samples at wavelengths below 190 nm is a challenge. Computation of spectra from first principles is arduous for both ORD and CD because of the difficulty of getting accurate electronic states from the calculations. Beratan et al. in chapter 8 show some recent results dealing with semi-empirical approaches to predicting molar optical rotations (α) at a single wavelength for large chiral molecules containing several chiral centers. For large molecules (greater than about one hundred atoms), *ab initio* methods are not yet of sufficient accuracy. Small molecules have been studied with success (29) (30).

Vibrational Spectroscopy

When CD is conducted in the infrared region of the spectrum, vibrational transitions are excited in the molecule, and vibrational CD (VCD) is achieved. Despite the fact that vibrational CD is several orders of magnitude smaller than electronic CD (due in part to the low frequency of the optical transitions), VCD is now fast becoming a useful method for absolute molecular structure determination in solution when chiral species are involved. VCD was first implemented in 1974 by Holzwarth et al. *(31)* and Nafie et al. *(32)*. Raman optical activity (ROA) involves the difference in Raman scattering intensity from chiral samples for left versus right circularly polarized light, and gives complementary vibrational information. ROA was first implemented by Barron et al. in 1973 *(33)*.

With new instrumental advances as described by Nafie in chapter 6, VCD can be measured in the region > 700 cm^{-1} with typical resolutions of 1 to 5 cm^{-1}, free of baseline artifacts. Raman optical activity captures 80 - 1700 cm^{-1}. Near-IR VCD has recently been studied by Abbate *(34)*. Both methods contain a great deal more stereochemical information than electronic CD, and further, much more than in regular IR spectroscopy, because IR spectra often are identical for different conformations of flexible molecules. In chapter 2, Stephens describes how Density Functional Theory calculations combined with the spectral data yield detailed information on the molecular conformations present in solution. One challenge to theorists is the inclusion of solvent molecules, which may affect conformations *(35)*. Freedman et al. (chapter 5) describe a method for plotting vibrational transition current density that helps to visualize the origin of VCD in molecules.

ROA is advantageous over VCD for the study of proteins and viruses, since water absorptions can be avoided. This work is described by Barron in chapter 3. One frontier according to this author is the application of ROA to the study of RNA/protein complexes. Using ^{13}C labels, Keiderling et al. use VCD to study peptide dynamics, probing specific sites in the peptide as it unfolds. In work described in chapter 4, these authors outline a modified theoretical approach for these large molecules, which are too large for *ab initio* work. In chapter 7, Polavarapu reports on recent VCD studies of the peptide gramicidin in various organic solvents and ion environments. The peptide has many conformations due to alternating D and L amino acids (rare in nature, as mentioned previously).

What are some of the challenges of VCD and ROA? One of the drawbacks to VCD is the length of time required to obtain a spectrum (about one hour). This precludes faster time-resolved work with the present instrument. The signals are stronger in ROA because a shorter wavelength excitation is used (often the 514 nm line of the Ar ion laser). In both VCD and ROA, the spectra may be difficult to interpret without extensive calculations. Because of signal to noise problems, fairly concentrated samples are required.

8

X-Ray Spectroscopy

The availability of left and right circularly polarized synchrotron radiation providing bright X-rays has made possible the first natural circular dichroism studies in the X-ray region in 1998 by Alagna et al. *(36)*. Peacock et al. (chapter 12) and Stewart et al. (chapter 13) address the instrumental and theoretical aspects of the phenomena. The advantage of the method is that the local chirality around a given atom can be probed. One example given is the study of the environment of Nd^{+3} ion in a chiral organometallic complex. The method should be readily applicable to biological samples already being studied by EXAFS and XANES.

Nonlinear Optical Spectroscopy

The discussion thus far has centered on spectroscopy where the response of the molecule is linear in the photon flux. With lasers, one can achieve light fluxes of sufficient intensities to permit the simultaneous interaction of two or more photons with a molecule or material, nonlinear optical (NLO) spectroscopy *(37)*. When two photons combine in a scattering process to form one photon, this is termed second order. For two initial photons of equal energy, the output light frequency is doubled, and the process is called Second Harmonic Generation (SHG). The microscopic property of the molecule that governs this NLO response is called the hyperpolarizability, β, a third rank tensor. By symmetry, SHG is forbidden in all centrosymmetric materials, including liquids. SHG is a useful process for changing the colors of laser beams, and also for opto-electronic applications. A more general second order NLO case is the adding of two photons of arbitrary energy, e.g. one IR photon plus one visible photon to form another visible photon of higher energy. This is termed Sum Frequency Generation (SFG). With a few exceptions, the consequences of chirality of the NLO media in these processes have only recently been studied *(38) (39)*. Because NLO are governed by high rank tensors (e.g. $\chi^{(2)}$, the macroscopic third rank tensor governing SHG and SFG), the information content in the signal is different, and in some cases, more powerful, than in linear spectroscopy.

One of the first such studies in 1966 by Rentzepis et al. focused on SFG from a chiral liquid *(40)*. SFG should be observed from a solution of chiral molecules (but not SHG), it was argued, based on symmetry. This early paper reported such a signal from an optically pure arabinose solution. Recently, several groups have been inspired by this work to pursue it further. A spectroscopic method that could selectively examine only the chiral species in a solution above a zero background of the other achiral solutes could be of potential use. Buckingham et al. further develop the theory for this work in chapter 9. Albrecht et al. examine characteristics of SFG from chiral solutions, concentrating on factors (such as phase matching) that could make the signal difficult to observe (chapter 10). Both of these groups did not observe signals in preliminary attempts, and the early observations were deemed artifactual *(41)*. Kulakov et al. have recently reported the observation of SFG from a solution of limonene *(42)*. This signal was enhanced by an IR resonance. It was three orders of magnitude smaller in intensity than expected. Messmer et al. have observed chiral SFG from monolayers of lauryl leucine, and the signal intensities were much weaker than the usual surface SFG, which is electric dipole-allowed by symmetry-breaking by the surface *(43)*. Shen et al. observed visible-visible SFG from a concentrated R-binaphthol solution with the sum frequency resonant with an electronic transition in the molecule *(44)*. The topic of SFG from chiral liquid samples will no doubt be further scrutinized.

In addition to extensive work performed on SHG with respect to chiral surfaces, where the surface sensitivity of SHG is exploited *(39)* as mentioned previously, other motivations prompt attention to SHG and chirality. Harris et al. point out that with a timed-sequence of pulses, it could be possible to study chiral vibrations in achiral molecules with SHG *(45)*.

A laser scanning microscope based on SHG has been reported, and images of living cells obtained. The chirality of the staining dyes plays a significant role in contrast generation *(46)*. Because SHG is a two photon process, higher spatial resolution is possible (since the signal is proportional to the laser intensity squared, the beam diameter is narrowed upon signal generation). Photobleaching and phototoxicity can also be minimized using SHG. In a similar application, SFG has been used in a Near Field Scanning Microscope to image a zinc selenide disk *(47)*.

There is intense effort to improve materials used for frequency doubling *(48)*, and the inherent asymmetry of chiral molecules may be of some advantage. In chaper 11, Persoons et al. describe their studies on supramolecular long chain-terminated helicenes, where chiral purity results in SHG signal enhancements of three orders of magnitude. This group is interested in designing molecules that will have large magnetic dipole contributions to hyperpolarizability. They also seek possible applications of chiral NLO materials.

Chiral Vapors

We move now to questions of molecular structure and recognition involving chirality. The weak interactions involved in chiral discrimination such as were discussed for the "three point model" *(18)* lend themselves to gas phase techniques, where complexes can be stabilized in cold molecular beams. Further, quantitative data obtained in this manner are of great value to theorists as they attempt to model chiral molecular interactions. Zehnacker et al. in chapter 16 present their work on complexation of chiral molecules in molecular beams using high resolution electronic spectroscopy as a sensitive probe of interactions. A fluorescent chiral chromophore 2-naphthyl-1-ethanol is used as one partner in the complex. The spectroscopic properties of this selectand are modified by complexation with the chiral guest, and the modification is different for the RR and SR pairs. Heterochiral complexes are found in this work to have longer lifetimes. Precise binding energy measurements can be made. Vaccaro et al. report on a very sensitive cavity-ring-down apparatus for the measurement of optical rotations by chiral molecules in the gas phase *(49)*.

Several laboratories are now utilizing similar gas phase host/guest chemistry as a mass spectrometric analytical method to detect enantiomeric excess *(50-52)*. Electrospray ionization or fast-atom bombardment methods are used to introduce the samples to the spectrometer. Enantiomeric impurities of a few percent in picomolar concentrations of analyte can be detected. Because mass spectrometry is not sensitive to chirality a priori, chiral complexing agents are employed. The mass spectrometric methods take advantage of very high throughput capabilities and extraordinary sensitivity. Miniature instruments are under consideration for remote sampling, including extraterrestial environments.

Using ion trap tandem mass spectrometry, Cooks et al. discovered a magic number effect in the clustering of homochiral serine *(53)*. A singly protonated serine octamer forms under positive ion electrospray conditions. Calculations show that heterochiral octamers are less stable than the homochiral counterpart. The work has implications for the origin of homochirality of proteins.

Chiral Surfaces

Chiral surfaces are increasingly of interest for applications such as stereoselective chemical synthesis, surface modification in microelectronics and sensors, separation of chiral compounds, protein adsorption and crystal growth. However, we lack a mechanistic understanding of how these surfaces work. Lipkowitz addresses the questions: Where in or around the chiral sites does the analyte prefer to bind? Which intermolecular forces are responsible for complexation, and which are responsible for chiral discrimination? How chiral does a molecule have to be for enantioselectivity to occur? *(54-55)* His computational work addresses permethyl-β-cyclodextrin,, which is a chiral stationary phase for gas chromatography, and he finds, for example, that the complexation is due to van der Waals forces whereas the chiral recognition is due to dispersion forces. He also finds that most of the enantiodifferentiation occurs at sites inside the cyclodextrin ring, even though the outside also possesses chirality.

For the systematic experimental study of chiral surfaces, there are two approaches: first, one can adsorb chiral molecules on flat surfaces. Issues regarding conformation of adsorbed molecules are important for reactivity and recognition processes. In chapter 14, Hermann et al. provide an extensive review of the recent use of scanning probe microscopy (SPM) for analyzing order phenomena in chiral molecular films, including liquid crystals, and for imaging chiral molecules on surfaces. Using scanning tunneling microscopy (STM) Wolkow et al. studied chiral alkenes on Si(100) and were able to determine the absolute configuration (R or S) for each of the chiral centers formed on chemisorption (chapter 20). Flynn et al. report in chapter 15 of their STM studies on racemic 2-bromohexadecanoic acid physisorbed at a liquid/solid interface. In the presence of achiral hexadecanoic acid, the chiral species spontaneously segregate into enantiomerically-pure domains on a graphite surface. The authors present evidence that the absolute chirality of individual molecules can be directly determined from STM images.

Ernst et al. adsorbed heptahelicene on a Ni (111) surface in ultrahigh vacuum *(56)*. The two dimensional ordered structure of intact molecules was studied using Low Energy Electron Diffraction (LEED) and STM, and were found to be close packed. The organic layer was then subjected to metal vapor deposition in an attempt to prepare a metal surface with preferred handedness. This first attempt proved to be unsuccessful *(55)*. A similar approach was used by Lorenzo et al. in creating a chiral catalyst mimic by the adsorption of (R, R)-tartartic acid molecules on Cu (110) surfaces *(57)*.

A second approach to the experimental study of chiral surfaces is to utilize chiral solid surfaces occurring naturally. Quartz was used in one of the first demonstrations of catalytic stereospecificity *(58)*. Clay surfaces such as montmorillonites can be chiral. These natural chiral surfaces have been examined as possible catalysts of reactions such as RNA synthesis, and thus have been associated with the origin of life *(59)*. Solid D- or L- $NaOCl_3$ has recently been shown to catalyze the enantioselective addition of a zinc compound (di-isoproplyzinc) to a substituted aldehyde (2-(tert-butylethynyl)pyrimidine-5-carbaldehyde) *(60)*.

The theme of spontaneous resolution of racemates at interfaces is continued in the work of Lahav in chapter 17. This work addresses the question of how chiral bias can be propagated in a system. Grazing incidence X-ray diffraction studies are used to provide direct information on the structure and dynamics of monolayers on a liquid surface. From slow evaporation of a solution containing glycine and a racemic mixture of α-amino acids, an enantiomeric excess of the chiral amino acids is found at the air/water interface. The crystalline glycine face exposed to the solution is of one handedness, the air side another. Molecules from solution adsorb enantioselectively to the organic solid, leaving behind an enrichment in the solution of the other enantiomer. This spontaneous generation of chirality followed by amplification is thought to offer an explanation for homochirality of oligopeptides from hydrophobic α-amino acids by their polymerization at the interface. In chapter 21, Kondepudi et al. offer another chiral resolution and amplification scheme as they examine the crystallization phenomena of stirred solutions of $NaOCl_3$, which are apparently autocatalytic.

Orme et al. have observed chiral morphologies on the surface of an achiral solid, calcite, through the selective binding of D- or L- aspartic acid to the surface of the crystal. Atomic force microscopy images show both growth hillocks and dissolution pit geometries that are chiral, macroscopic chirality traced directly from molecular chirality (61).

Kink sites on high Miller index surfaces of clean metal single crystals are chiral when the step lengths or step faces on either side of the kink are unequal (62). An example is Ag(643). Chapter 18 by Attard extends this definition to include cases where the step lengths or faces may be equal. Monte Carlo simulations by Sholl et al. predicted that binding energies of chiral hydrocarbons on chiral Pt surfaces should vary with enantiomer by measurable amounts (63). Chapters 18 by Attard et al. and 19 by Gellman et al. provide the first experimental examples of this type of enantioselectivity in chemical reactions on kinked metal surfaces. The method appears quite promising.

Chiral Nanostructures

Nanometer-scale chiral supramolecular structures have been studied including calixerane-based helical tubular structures (64), lipid tubules (65), polymers (66) and vacuum deposited inorganic chiral materials (67). Tiny nanoclusters of gold containing between 20 and 40 atoms encapsulated by a common biomolecule (glutathione) display distinctly chiral properties (68). Many of these superstructures are shown to exhibit large optical activity. In Chapter 22, Thomas et al. use AFM and optical microscopy to study the growth of giant phospholipid tubule formation. These novel new materials are potential candidates for applications such as sensors, separations, and optical devices. The phenomena of enhanced chiroptical signals in extended structures presumably is due to the coupling of chromophores, and is also not yet well understood.

Single wall carbon nanotubes are among perhaps the most promising candidates for molecular electronics: the use of individual molecules as functional devices *(69)*. The nanotubes are either one dimensional metals or semiconductors depending on their diameter and chirality.

Nanosecond switching times are predicted for molecular chiroptical dipole switches, where a combination of light and electric field simultaneously reverse both the chirality and dipole direction. Information stored in an array of these molecules can be read nondestructively with circularly polarized light *(70)*. The applications include optical memory*(71)*.

Photochemical work will be spurred by the recent advance by Rikken et al. who were able to direct a chemical reaction enantiospecifically using a magnetic field in combination with unpolarized light *(72)*. Enantiomeric excess of the product was shown to depend linearly on the magnetic field strength. This so-called magnetochiral anisotropy, though a small effect here, may present a model for the origin of the homochirality of life.

Conclusion

There is a great deal of activity in the study of chirality within the subdiscipline of physical chemistry, and even more when one includes the physical approaches of analytical, materials and nano-chemistry. It is important that scientists studying interesting new chiral systems, such as aggregates and surfaces, interact with those at the forefront of developing new methods for studying chirality experimentally and theoretically. Some frontiers of the next decade might be to:

- Image and monitor significant chiral molecules (such as proteins) at interfaces or in cells to the point of being able to recognize their structures and to measure their kinetics *in situ*. Some approaches to this include SHG imaging, SFG in the near field as well as tip-modified AFM.
- Theoretically predict chiroptical values including spectra. Model parameters concerning molecular and solid chiral interactions, leading to an understanding of chiral recognition.
- Produce chirally pure drugs and other biological products as needed. This might be accomplished through use of heterogeneous catalysis. Separate enantiomers of species as desired without the use of environmentally destructive solvents.
- Produce reliable molecular motors for the ultimate miniaturization of electronics. Some of these might employ ideas from the unique features of chiral molecules.

With the arrival of chiral-sensitive, rugged and transportable instruments, it may be possible that extraterrestial environments will be explored for evidence of life, revealing exciting clues about the origin of life.

References

1. Pasteur, L. *Ann. Chim. Phys.* **1848**, *III*, 24, 442.
2. Lord Kelvin *Baltimore Lectures (1884) on Molecular Dynamics and the Wave Theory of Light;* Clay and Sons: London, 1904, p. 449.
3. Stryer, L. *Biochemistry,* 4th ed., W. H. Freeman and Co.: New York, 1995.
4. Siegel, J. S. *Chirality* **1998**, *10*, 24-27
5. Jacques, J. *The Molecule and Its Double*; McGraw Hill: New York, 1993.
6. Eliel, E. L.; Wilen, S. H. *Stereochemistry of Organic Compounds*, John Wiley: New York, 1993.
7. deJong, Rudolph H. *Local Anesthetics*; Mosby: St. Louis, MO, 1994.
8. Ding, Y.-S., Fowler, J. S., Volkow, N. D., Dewey, S. L., Wang, G.-J., Logan, J., Gatley, S. J., Pappas, N. *Psychopharmacology* **1997**, *131*, 71-78.
9. Crossley, R. *Chirality and the Biological Activity of Drugs*; CRC Press: Boca Ratan Florida, 1995.
10. Eriksson T, Bjorkman S, Roth B, Hoglund P. *J. Pharm. Pharmacology* **2000**, *52*, 807-817
11. Garrison, A. W. In *Encyclopedia of Analytical Chemistry*; Meyers, R. A., Ed.; John Wiley and Sons: New York, 2000 pp. 6147-6158.
12. Nakanishi, K.; Berova, N.; and Woody, R. W. *Circular Dichroism: Principles and Applications;* VCH Publishers: New York, 1994.
13. Stalcup, A. M. In *Encyclopedia of Separation Technology* John Wiley: New York, 1997.
14. Charney, E. *The Molecular Basis of Optical Activity*; Robert E. Krieger Publishing Co.: Malabar FL, 1985.
15. *Vibrational Optical Activity: From Fundamentals to Biological Applications*; Faraday Discussions No. 99; The Royal Society of Chemistry: London, 1994.
16. Rosenfeld, L. *Z. Phys.* **1928**, *52*, 161.
17. Tinoco, Jr., I. *J. Chem. Phys.* **1960**, *33*, 1332.
18. Pirkle, W. H. and Pochapsky T. C. *Chem. Rev.,* **1989**, *89*, 347.
19. Byers, J. D.; Yee, H. I.; Petralli-Mallow, T.; Hicks, J. M. *Phys. Rev. B* **1994** *49*, 14643-14647.
20. Kauranen, M.; Verbiest, T.; Maki, J. J.; Persoons, A. *J. Chem. Phys.* **1994** *101* 8193-8199.
21. Byers, J. D.; Hicks, J. M. *Chem. Phys. Lett.* **1994** *231*, 216.
22. Byers, J. D.; Yee, H. I.; Hicks, J. M. *J. Chem. Phys.* **1994**, *101*, 6233.
23. Hicks, J. M.; Petralli-Mallow, T. *Appl. Phys B* **1999** *68*, 589-593.
24. Vacek, J.; Michl, J. *Proc. Nat. Acad. Sci.* **2001** *98*, 5481-5486.
25. Feynman, R. P.; Leighton, R. B.; Sands, M.; *The Feynman Lectures on Physics;* Addison Wesley: Reading, MA, 1963, Vol. 1, p. 33-6.
26. Wenzel, T. J. In *Encyclopedia of Spectroscopy and Spectrometry;* Academic Press: New York, 2000, Vol. 1, pp. 411-421.

27. Creighton, T. E. *Proteins: Structures and Molecular Properties;* W. H. Freeman and Co.: New York, 1993, p. 191.
28. Chen, E. F., Wood, M. J.; Fink, A. L.; Kliger, D. S. *Biochem.* **1998**, *37*, 5589-5598.
29. Stephens, P. J.; Devlin, F. J.; Cheeseman, J. R.; Frisch, M. J. *J. Phys. Chem. A* **2001**, *105*, 5356-5371.
30. Furche, F.; Ahlrichs, R.; Wachsman, C.; Weber, E.; Sobanski, A.; Vogtle F.; Grimme, S. *J. Am. Chem. Soc.* **2000**, 122, 1717-1724.
31. Holzwarth, G.; Hsu, E. C.; Mosher, H. S.; Faulker, T. R.; Moscowitz, A. *J. Am. Chem. Soc.* **1974**, 96, 251.
32. Nafie, L. A.; Cheng, J. C.; Stephens, P. J. *J. Am. Chem. Soc.* **1975**, *97*, 3842.
33. Barron, L.D.; Bogard, M. P.; Buckingham, A. D. *J. Am. Chem. Soc.* **1973**, 95, 603.
34. Longhi, G.; Boiadjiev, S.; Lightner, D. A.; Bertucci, C.; Salavadori, P.; Abbate, S. *Enantiomer* **1998,** *3,* 337-347.
35. Knapp-Mohammady, M.; Jalkanen, K. J.; Nardi, F.; Wade, R. C.; Suhai, S. *Chem. Phys.* **1999**, *240* 63-77.
36. Alagna, L.; Prosperi, T.; Turchini, S.; Goulon, J.; Rogalev, A.; Goulon-Ginet, C.; Natoli, C. R.; Peacock, R. D.; Stewart, B. *Phys. Rev. Lett.* **1998**, *80*, 4799-4802.
37. Shen, Y. R. *The Principles of Nonlinear Optics*; Wiley: New York, 1984.
38. for a review up to 1994, see Petralli-Mallow, T.; Wong, T. M.; Byers, J. D.; Yee, H. I.; Hicks, J. M. *J. Phys. Chem.* **1993**, *97*, 1383.
39. Hicks, J. M.; Petralli-Mallow, T.; Byers, J. D. *Faraday Discussions No. 99*; The Royal Society of Chemistry: London, 1994, p.341-357.
40. Rentzepis, P. M.; Giordmaine, J. A.; Wecht, K. W. *Phys. Rev. Lett.* **1966**, *16*, 792-794.
41. Fisher, P.; Wiersma, D. S.; Rhigini, R.; Champagne, B.; Buckingham, A. D. *Phys. Rev. Lett.* **2000**, *85*, 4253-4256.
42. Kulakov, T.; Belkin, M.; Yan, L.; Ernst, K.-H.; Shen, Y. R., *Phys. Rev. Lett.*, **2000**, *85*, 4474-4477.
43. Messmer, M. C.; Pizzolatto, R. L., Wolf, L. K., Yang, Y. J. *J. Abst. Pap. Am. Chem.* S. 219: 190-PHYS, Part 2, MAR 26, 2000.
44. Belkin, M. A., Han. S. H., Wei, X., Shen, Y. R. *Phys. Rev. Lett.* , **2001**, *87*, 113001.
45. Harris, RA . *Abst. Pap. Am. Chem.* S. 219: 190-PHYS, Part 2, MAR 26, 2000.
46. Campagnola, P. J.; Wei, M., Lewis, A.; Loew, L. M.; *Biophys. J.* **1999**, *77*, 3341-3349.
47. Schaller, R. D.; Saykally, R. J. *Langmuir*, **2001**, *17*, 2055-2058.
48. Prasad, P. N.; Williams, D. J. *Introduction to Nonlinear Optical Effects in Molecules and Polymers*; John Wiley and Sons: New York NY 1991.

16

49. Muller, T.; Wiberg, K. B.; Vaccaro, P. H. *J. Phys. Chem. A* **2000**, *104*, 5959.
50. Liang, Y. J.; Bradshaw, J. S.; Izatt, R. M.; Pope, R. M.; Dearden, D. V. *Int. J. Mass. Spec.* **1999**, 187, 977-988.
51. Grigorean, G.; Lebrilla C. B. *Anal. Chem.* **2001**, *73*, 1684-1691
52. Tao, W. A.; Gozzo, F. C.; Cooks, R. G. *Anal. Chem.* **2001**, *73*, 1692-1698.
53. Cooks, R. G.; Zhang, D.; Kim, K. J.; Gozzo, F. C.; Eberlin, M. N. *Anal. Chem.* **2001**, *73*, 3646-3655.
54. Lipkowitz K. B.; Pearl, G.; Coner, R.; Peterson, M. A. *J. Am. Chem. Soc.* **1997,** *119*, 600-610.
55. Lipkowitz, K. B.; Coner, R.; Peterson, M. A. *J. Am. Chem. Soc.* **1997**, *119*, 11269-11276.
56. Ernst, K. H.; Bohringer, M.; McFadden, C. F.; Hug, P.; Muller, U.; Ellerbeck, E. *Nanotechnology* **1999**, *10*, 1-7.
57. Lorenzo, M. O.; Baddeley, C. J.; Muryn, C.; Raval, R. *Nature* **2000**, *404*, 376-379.
58. Schwab, G. M. and Rudolph, L. *Naturwiss* **1932**, *20*, 363.
59. Joshi, P. C.; Pitsch, S.; Ferris, J. P. *Chem. Commun.* **2000**, *24*, 2497-2498.
60. Sato, I.; Kadowaki, K.; Soai, K. *Angew Chemie* **2000**, *112,* 4087.
61. Orme, C. A., Noy, A., Wierzbicki, A., McBride, M. T., Grantham, M., Teng, H. H., Dove, P. M., DeYoreo, J. J. *Nature* **2001**, *411*, 775-779.
62. McFadden, C. F., Cremer, P. S.; Gellman, A. J. *Langmuir* **1996**, *12*, 2483.
63. Sholl, D. S. *Langmuir* **1998**, *14*, 862-867.
64. Orr, G. W.; Barbour, L. J.; Atwood, J. L. *Science* **1999**, *285*, 1049.
65. Spector, M. S.; Selinger, J. V.; Singh, A.; Rodriguez, J. M.; Price, R. R.; Schnur, J. M. *Langmuir* **1998**, *14*, 3493-3500.
66. Srinivasrao, M. *Curr. Op. Coll. Int. Sci.* **1999**, *4*, 147-152.
67. Hodgkinson, I., Wu, Q.H. *Adv. Mat.* **2001**, *13*, 889.
68. Schaaff, T. G.; Whetten, R. L. *J. Phys. Chem. B* **2000**, *104*, 2630-2641.
69. Yao, Z.; Postma, H.W.C.; Balents, L.; Dekker, C. *Nature* **1999**, *402*, 273.
70. Hutchinson, K. A.; Parakka, J. P.; Kesler, B. S., Schumaker, R. *Photonics West Conf.* Jan. 28, 2000, San Jose CA.
71. Dantas, S. D.; Barone, P. M. V. B.; Braga, S. F.; Galvao, D. S. *Synthetic Metals* **2001**, *116*, 275-279.
72. Rikken, G. L. J. A., Raupach, E. *Nature* **2000**, *405*, 932-935.

Vibrations of Chiral Molecules

Chapter 2

Determination of the Structures of Chiral Molecules Using Vibrational Circular Dichroism Spectroscopy

P. J. Stephens, F. J. Devlin, and A. Aamouche

Department of Chemistry, University of Southern California,
Los Angeles, CA 90089–0482

INTRODUCTION

Chiral molecules exhibit Vibrational Circular Dichroism (VCD) [1]. The VCD spectrum of a chiral molecule is dependent on its structure. It is extremely sensitive to the Absolute Configuration: enantiomeric molecules exhibit mirror-image VCD spectra. In the case of flexible molecules, it is also dependent on the conformation. Thus, in principle, VCD spectroscopy permits both the determination of the Absolute Configuration and the Conformational Analysis of chiral molecules. To date, VCD has not been widely used. Until recently, it has not been practicable to reliably extract structural information from VCD spectra. Recently, this situation has been radically altered by the development of a methodology permitting the prediction of VCD spectra with very high reliability [2]. This methodology is based on Density Functional Theory (DFT) [3] and uses Gauge-Invariant Atomic Orbitals (GIAOs) [4]. We refer to it here as the DFT/GIAO methodology. The accuracy of this methodology, when state-of-the-art density functionals and suitable basis sets are used, is remarkable [5]. It permits, for the first time, the routine application of VCD spectroscopy to structural chemistry.

In this presentation, we describe the DFT/GIAO methodology and illustrate its utilisation via recent applications.

THEORY

The vibrational unpolarised absorption and circular dichroism spectra of a molecule in dilute solution are given by [6]:

$$\bar{\varepsilon}(v) = \frac{8\pi^3 N v}{(2.303)3000 h c} \sum_{g,k} \alpha_g D_{gk} f_{gk}(v_{gk}, v) \tag{1}$$

$$\Delta\varepsilon(v) = \frac{32\pi^3 N v}{(2.303)3000 h c} \sum_{g,k} \alpha_g R_{gk} f_{gk}(v_{gk}, v) \tag{2}$$

where

$$D_{gk} = |\langle g | \vec{\mu}_{el} | k \rangle|^2 \tag{3}$$

$$R_{gk} = \text{Im}[\langle g | \vec{\mu}_{el} | k \rangle \cdot \langle k | \vec{\mu}_{mag} | g \rangle] \tag{4}$$

α_g is the fractional population of state g. $f_{gk}(v_{gk}, v)$ is the normalized lineshape of the excitation $g \rightarrow k$ at frequency v_{gk}. $\vec{\mu}_{el}$ and $\vec{\mu}_{mag}$ are the electric dipole and magnetic dipole moment operators respectively. For fundamental vibrational transitions of the ith normal mode, within the Harmonic Approximation (HA) [7]

$$\langle 0 | (\mu_{el})_\beta | 1 \rangle_i = \left(\frac{\hbar}{4\pi v_i}\right)^{1/2} \sum_{\lambda\alpha} S_{\lambda\alpha,i} P_{\alpha\beta}^\lambda \tag{5}$$

$$\langle 0 | (\mu_{mag})_\beta | 1 \rangle_i = (4\pi\hbar \, v_i)^{1/2} \sum_{\lambda\alpha} S_{\lambda\alpha,i} M_{\alpha\beta}^\lambda \tag{6}$$

where

$$P_{\alpha\beta}^\lambda = E_{\alpha\beta}^\lambda + N_{\alpha\beta}^\lambda$$

$$E_{\alpha\beta}^\lambda = \left(\frac{\partial}{\partial X_{\lambda\alpha}} \langle \psi_G | \left(\mu_{el}^e\right)_\beta | \psi_G \rangle\right)_0$$

$$N_{\alpha\beta}^\lambda = (Z\lambda e) \, \delta_{\alpha\beta}$$

$$M_{\alpha\beta}^{\lambda} = E_{\alpha\beta}^{\lambda} + J_{\alpha\beta}^{\lambda} \qquad (7)$$

$$I_{\alpha\beta}^{\lambda} = \left\langle \left(\frac{\partial\psi_G}{\partial X_{\lambda\alpha}}\right)_0 \middle| \left(\frac{\partial\psi_G}{\partial H_\beta}\right)_0 \right\rangle$$

$$J_{\alpha\beta}^{\lambda} = \frac{i}{4\hbar c} \sum_{\gamma} (Z_\lambda e) R_{\lambda\gamma}^{0} \, \varepsilon_{\alpha\beta\gamma}$$

The $S_{\lambda\alpha,i}$ matrix interrelates Cartesian displacement coordinates $X_{\lambda\alpha}$ and normal coordinates Q_i:

$$X_{\lambda\alpha} = \sum_{i} S_{\lambda\alpha,i} Q_i \qquad (8)$$

Normal mode frequencies are v_i. $S_{\lambda\alpha,i}$, Q_i and v_i are obtained by diagonalizing the mass-weighted Cartesian Harmonic Force Field (HFF): $\partial^2 W_G/\partial X_{\lambda\alpha}\partial X_{\lambda'\alpha'}$. $P_{\alpha\beta}^{\lambda}$ and $M_{\alpha\beta}^{\lambda}$ are the Atomic Polar Tensors (APTs) [8,9] and the Atomic Axial Tensors (AATs) [7,9] respectively. Ψ_G is the wavefunction of the ground electronic state. $\partial\Psi_G/\partial H_\beta$ is the derivative of Ψ_G in the presence of the perturbation $-(\mu_{mag}^{e})_\beta H_\beta$; this derivative is well-known in the theory of magnetic properties [4b].

HFFs, APTs and AATs are all expressed in terms of derivatives with respect to $X_{\lambda\alpha}$ and H_β. They are most accurately and efficiently calculated using ab initio Analytical Derivative (AD) methods [10]. Rapid convergence to the complete basis set limit requires "perturbation-dependent (PD) basis sets". In the case of derivatives with respect to $X_{\lambda\alpha}$ the use of nuclear-position dependent basis sets is universal [11]. In the case of derivatives with respect to H_β GIAOs [4] are the standard choice. The first calculations of vibrational rotational strengths based on HFFs, APTs and AATs calculated using ab initio AD methods and PD basis sets were carried out using the Hartree-Fock/Self-Consistent Field (HF/SCF) and Multi-Configuration Self-Consistent Field (MCSCF) ab initio methodologies [12]. The former does not include correlation. The latter includes correlation, but only to a limited extent; it is also very much more laborious computationally. DFT permits the inclusion of correlation, in principle (given an exact functional) completely, and its computational cost is only a little heavier than HF/SCF. The calculation of DFT HFFs and APTs using AD methods and PD basis sets was implemented in the early 1990s [13]. The calculation of DFT AATs using AD methods and PD basis sets was reported in 1996 [2]. As of 1998, DFT HFFs, APTs and AATs became available in the widely distributed program, GAUSSIAN [14].

DFT calculations require the choice of a functional. At this time, the most accurate functionals are the "hybrid" functionals introduced by Becke [15]. By now, many functionals of this type have been developed. The most commonly used, probably, is B3LYP [16]. The original Becke functional is B3PW91 [15,16].

The accuracy of VCD spectra predicted using the DFT/GIAO methodology can be assessed by comparison to experiment for small rigid molecules, where the complications of conformational flexibility are absent and vibrational analysis is straightforward. Methyl oxirane (propylene oxide), **1**, is such a molecule. Thorough studies of the mid-IR VCD spectrum of **1** have recently been carried out [17]. The experimental spectrum of a CCl_4 solution of **1** is shown in Figure 1, together with DFT/GIAO predictions obtained using the B3LYP functional and a range of basis sets. The largest basis sets (TZ2P and cc-pVTZ) give very similar results. As the basis set size diminishes the spectrum changes; the change is modest at the 6-31G* level and drastic at the 3-21G level. Spectra predicted using a range of functionals at the cc-pVTZ basis set level are shown in Figure 2. The hybrid functionals B3LYP and B3PW91 give very similar results. The "non-local" functional BLYP gives a spectrum similar to B3LYP. The "local" functional LSDA, however, gives a substantially different spectrum. The spectra predicted using hybrid functionals and large basis sets are in the best agreement with experiment, showing that (as expected) they are the most accurate.

Quantitative evaluation of DFT/GIAO calculations can be carried out by comparing predicted vibrational rotational strengths to values obtained from the experimental spectra via Lorentzian fitting. Results for a variety of functionals and basis sets are given in Table I. The most accurate calculations yield average (absolute) deviations of calculated rotational strengths from experimental values in the range 3-4 x 10^{-44} esu^2cm^2. Experimental errors average 1-2 x 10^{-44} esu^2cm^2. Thus, calculational error for the most accurate calculations is approaching experimental error. The calculations do not include either anharmonicity or solvent effects. In the case of the mid-IR spectrum of **1** in CCl_4 it is clear that their contributions are small.

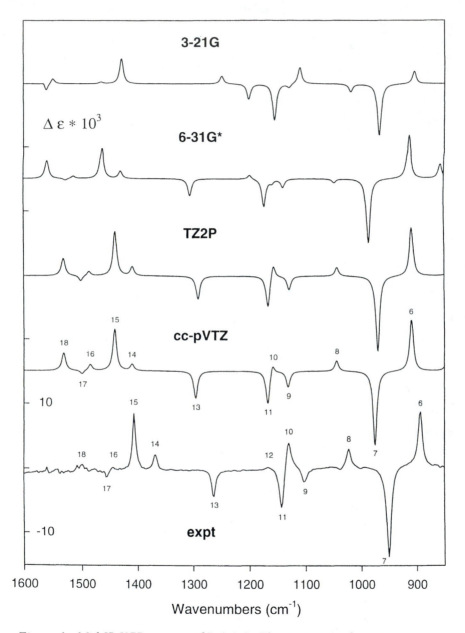

*Figure 1. Mid-IR VCD spectra of R-(+)-**1**. The experimental spectrum is in CCl₄ solution. DFT/GIAO spectra are calculated using the B3LYP functional and a range of basis sets. Band shapes are Lorentzian (γ= 4.0 cm⁻¹). Fundamentals are numbered.*

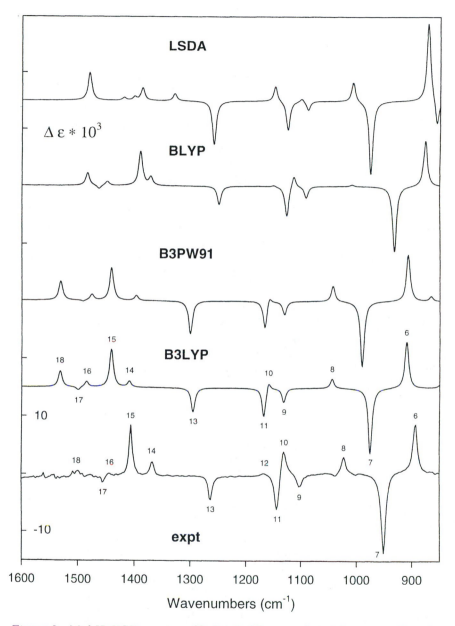

Figure 2. Mid-IR VCD spectra of R-(+)-1. The experimental spectrum is as in Figure 1. DFT/GIAO spectra are calculated using the cc-pVTZ basis set and a range of functionals. Band shapes are Lorentzian (γ= 4.0 cm⁻¹). Fundamentals are numbered.

APPLICATIONS

1. Conformational Analysis

Different conformations of a molecule exhibit different VCD spectra. Consequently, the experimental conformation(s) of a flexible molecule can be elucidated from its VCD spectrum. We have recently studied the molecule 3-methylcyclohexanone, **2** [*18*]. Two low-energy conformations, **e** and **a**, are expected in which the 3-methyl group is respectively equatorial and axial. Mid-IR B3PW91/TZ2P VCD spectra of **e** and **a** are shown in Figure 3, together with the experimental VCD spectrum of a CCl_4 solution of **2**. Comparison of calculated and experimental spectra shows immediately that **e** is the dominant conformation present in the solution: the spectra of **e** and **a** are very different; only that of **e** resembles the experimental spectrum. Careful examination of the experimental spectrum, however, reveals a number of bands which cannot be assigned to conformation **e** but which can be assigned to conformation **a**. These bands are indicated in Figure 3. Comparison of the intensities of the bands assigned to either conformation **e** or **a** to predicted intensities is shown in Figure 4. The ratio of the slopes of the straight line fits to the plots for **a** and **e** equals the conformational equilibrium constant, K_c for the equilibrium **e** $\Leftrightarrow$ **a**. In turn, this yields a conformational free energy difference, $\Delta G_c = 1.23$ kcal/mole. This value is similar to those obtained previously for solutions of **2**, Table II. The B3PW91/TZ2P energies of **e** and **a** differ by 1.47 kcal/mole. The difference from ΔG_c can be attributed to errors in the DFT calculations, solvent effects and entropy.

The DFT/GIAO methodology has been applied to the Conformational Analysis of a number of molecules. These include 1,5-dimethyl-6,8-dioxabicylo [3.2.1] octane [*19*], methyl lactate [*20*], dimethyl tartrate [*20*], a hexahydro-methyl-naphthalenone [*21*], a tetrahydro-methyl-naphthalenedione [*21*] and a bis-oxazoline [*22*].

2. Determination of Absolute Configuration (AC)

Enantiomeric forms of a molecule exhibit mirror-image spectra. Consequently, the AC of a chiral molecule can be elucidated from its VCD spectrum. The protocol is very simple when the molecule is conformationally rigid. The experimental VCD spectrum is compared to the spectra predicted for the two enantiomers. The AC of the experimental sample is that of the enantiomer whose VCD spectrum matches the observed spectrum in sign. The AC of Tröger's Base, **3**, was recently studied using VCD [*23*] since electronic CD (ECD) [*24*] and X-ray crystallography [25] studies had resulted in opposite ACs. The B3PW91 spectrum of (R,R)-**3** is compared to the experimental

Table I. Accuracy of Calculated DFT/GIAO Rotational Strengths of Methyl Oxirane **1**[a] [*17*]

Basis Set	B3LYP Error[b]	Functional	cc-pVTZ Error[b]
3-21G	10.5	LSDA	6.7
6-31G*	6.5	BLYP	3.9
6-31G**	6.2	BH and H	9.2
cc-pVDZ	6.4	BH and HLYP	6.0
TZ2P	3.9	B3LYP	3.9
cc-pVTZ	3.9	B3PW91	4.3
VD3P	3.4	B3P86	4.3
cc-pVQZ	3.7	PBE1PBE	4.7

[a] Rotational strengths in 10^{-44} esu^2 cm^2.

[b] Error is average absolute deviation of calculated and experimental rotational strengths for fundamentals 6-18. Experimental rotational strengths were obtained by Lorentzian fitting to the VCD spectrum of **1** in CCl$_4$ solution.

Table II. Experimental Conformational Free Energy Difference of **2**[a][*18*]

ΔG_c	Method	Solvent
0.50	electronic CD	ether/isopentane/ethanol
0.75	electronic CD	methylcyclohexane
1.1	NMR	CDCl$_3$[b]
1.62	REMPI	gas phase
1.23	VCD	CCl$_4$

[a] Energy difference of **a** and **e**; in kcal/mol.

[b] **2** is complexed with Yb(fod)$_3$ shift reagent.

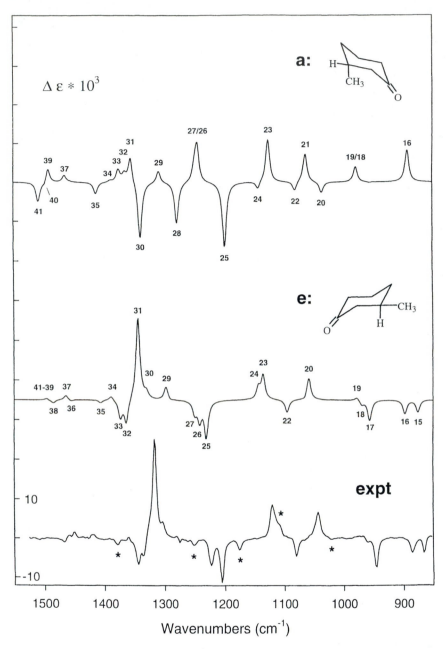

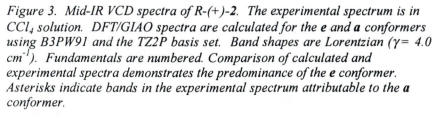

Figure 3. Mid-IR VCD spectra of R-(+)-2. The experimental spectrum is in CCl₄ solution. DFT/GIAO spectra are calculated for the e and a conformers using B3PW91 and the TZ2P basis set. Band shapes are Lorentzian (γ= 4.0 cm⁻¹). Fundamentals are numbered. Comparison of calculated and experimental spectra demonstrates the predominance of the e conformer. Asterisks indicate bands in the experimental spectrum attributable to the a conformer.

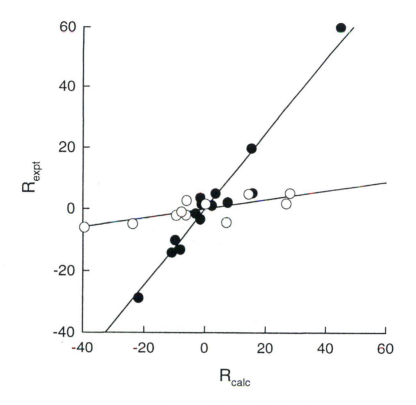

Figure 4. Comparison of calculated B3PW91/TZ2P and experimental rotational strengths for R-(+)-2. • *and o denote fundamentals of e and a, respectively. R is in 10^{-44} esu^2 cm^2.*

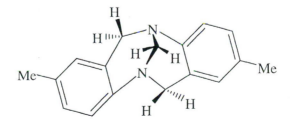

(R,R)-3

spectrum of a CCl_4 solution of (-)-**3** in Figure 5. It is obvious that the AC is (R,R) - (-) / (S,S) - (+), in agreement with the X-ray crystallography result and in disagreement with the ECD result.

When the molecule is conformationally flexible, the VCD spectra of all populated conformations must be calculated and the population-weighted composite spectrum compared to experiment. An example is provided by the methyl ester of spiropentylcarboxylic acid, **4**, whose AC was recently determined using VCD [26]. B3LYP/TZ2P and B3PW91/TZ2P calculations predict two low energy conformations of **4**: I and II. In both, the C(=O)OC atoms of the ester group are coplanar; this plane is perpendicular to that of the adjacent cyclopropyl ring. In I, the C=O group points over the cyclopropyl ring; II is obtained by a 180° rotation of the ester group. II is predicted to be ~1.0 kcal/mole higher in energy than I. B3LYP/TZ2P VCD spectra of R-I and R-II, the population-weighted composite spectrum and the experimental spectrum of a CCl_4 solution of (+)-**4**, multiplied by -1, are shown in Figure 6. The assignment of the experimental spectrum is indicated in Figure 6. The spectrum is very similar to that of conformation I, as expected given its preponderance. However, bands originating in II are clearly observed. The predicted composite spectrum for R-**4** and the experimental spectrum of (+)-**4**, multiplied by -1, are in excellent agreement. The AC can be unambiguously assigned as R- (-) / S- (+).

The DFT/GIAO methodology has also been applied to the determination of the AC of desflurane [27] and 1,2,2,2-tetrafluoroethyl methyl ether [28].

CONCLUSIONS

The development of the DFT/GIAO methodology, together with the availability of commercial VCD instrumentation, now permits the routine use of VCD in elucidating the stereochemistry of chiral molecules. Improvements in density functionals, the inclusion of anharmonicity and the incorporation of solvent effects will further increase the accuracy of the DFT/GIAO methodology, and further enhance its utility. Such developments can be anticipated in the very near future.

ACKNOWLEDGMENTS

We are grateful to Drs. M.J. Frisch and J.R. Cheeseman, Gaussian Inc., for their continual assistance, to Elf/Sanofi for a post doctoral fellowship (to A. Aamouche) and to NIH and NSF for financial support.

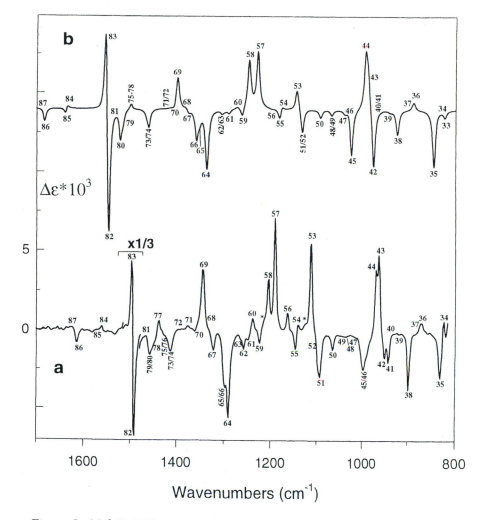

*Figure 5. Mid-IR VCD spectra of **3**. The experimental spectrum, **a**, is of the (-) isomer and is in CCl₄ solution. The DFT/GIAO spectrum, **b**, is calculated for the (R,R) enantiomer using B3PW91 and the 6-31G* basis set. Band shapes are Lorentzian (γ= 4.0 cm⁻¹). Fundamentals are numbered. Asterisks indicate non-fundamentals.*

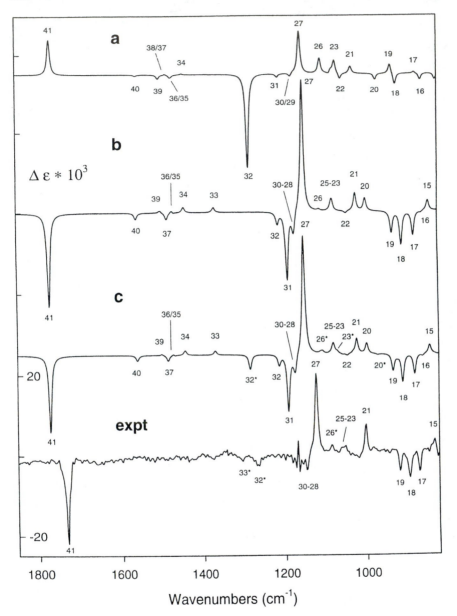

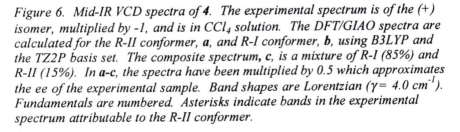

*Figure 6. Mid-IR VCD spectra of **4**. The experimental spectrum is of the (+) isomer, multiplied by -1, and is in CCl₄ solution. The DFT/GIAO spectra are calculated for the R-II conformer, **a**, and R-I conformer, **b**, using B3LYP and the TZ2P basis set. The composite spectrum, **c**, is a mixture of R-I (85%) and R-II (15%). In **a-c**, the spectra have been multiplied by 0.5 which approximates the ee of the experimental sample. Band shapes are Lorentzian (γ = 4.0 cm⁻¹). Fundamentals are numbered. Asterisks indicate bands in the experimental spectrum attributable to the R-II conformer.*

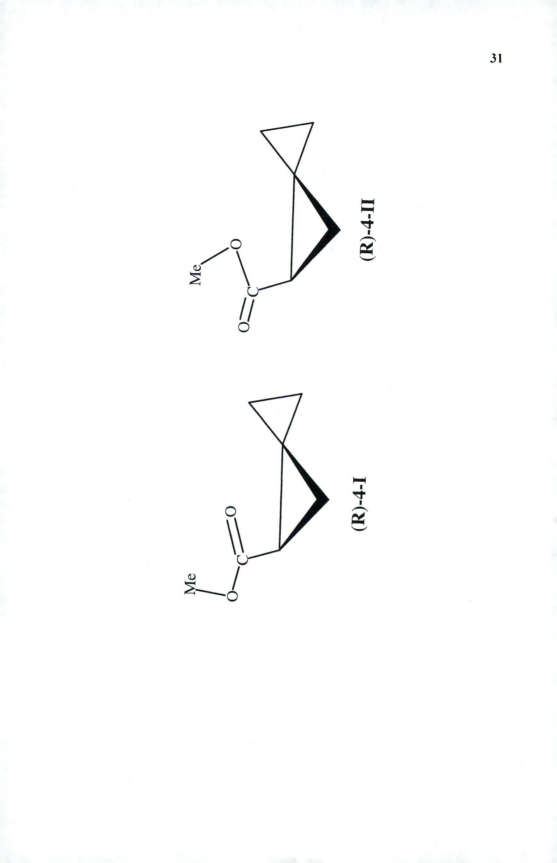

(R)-4-II

(R)-4-I

References

1. (a) Stephens, P. J.; Lowe, M. A. *Ann. Rev. Phys. Chem.* **1985**, 36, 213-241.
   (b) Nafie, L. A. *Ann. Rev. Phys. Chem.* **1997**, 48, 357-386.
2. Cheeseman, J. R; Frisch, M. J.; Devlin, F. J; Stephens P. J. *Chem. Phys. Lett.* **1996**, 252, 211-220.
3. Laird, B. B.; Ross, R. B.; Ziegler, T., Eds.; *Chemical Applications of Density Functional Theory*; American Chemical Society Symposium Series 629; ACS 1996.
4. (a) London, F. *J. Phys. Radium* **1937**, 8, 397.
   (b) Ditchfield, R. *Mol. Phys.* **1974**, 27, 789-807.
5. (a) Stephens, P. J.; Ashvar, C. S.; Devlin, F. J.; Cheeseman, J. R.; Frisch, M. J. *Mol. Phys.* **1996**, 89, 579-594.
   (b) Devlin, F. J.; Stephens, P. J.; Cheeseman, J. R.; Frisch, M. J. *J. Phys. Chem.* **1997**, 101, 6322-6333.
   (c) Devlin, F. J.; Stephens, P. J.; Cheeseman, J. R.; Frisch, M. J. *J. Phys. Chem.* **1997**, 101, 9912-9924.
   (d) Ashvar, C. S.; Devlin, F. J.; Stephens, P. J. *J. Am.Chem. Soc.* **1999**, 121, 2836-2849.
6. Schellman, J. A. *Chem. Revs.* **1975**, 75, 323-331
7. (a) Stephens, P. J. *J. Phys. Chem.* **1985**, 89, 748-752.
   (b) Stephens, P. J. *J. Phys. Chem.* **1987**, 91, 1712-1715.
8. Person, W. B.; Newton, J. H. *J. Chem. Phys.* **1974**, 61, 1040-1049.
9. Stephens, P. J.; Jalkanen, K. J.; Amos, R. D.; Lazzeretti, P.; Zanasi, R. *J. Phys. Chem.* **1990**, 94, 1811-1830.
10. (a) Amos, R. D. *Adv. Chem. Phys.* **1987**, 67, 99-153.
    (b) Pulay, P. *Adv. Chem. Phys.* **1987**, 69, 241.
    (c) Yamaguchi, Y.; Osamura, Y.; Goddard, J.D.; Schaeffer, H. F. *A New Dimension to Quantum Chemistry: Analytic Derivative Methods in Ab Initio Quantum Chemistry*; OUP 1994.
11. Hehre, W. J.; Schleyer, P. R.; Radom, L.; Pople, J. A. *Ab Initio Molecular Orbital Theory;* Wiley: New York, 1986.
12. Bak, K. L.; Jørgensen, P.; Helgaker, T.; Ruud, K.; Jensen, H. J. A. *J. Chem. Phys.* **1993**, 98, 8873-8887; **1994**, 100, 6620-6627.
13. (a) Johnson, B. G.; Frisch, M. J. *Chem. Phys. Lett.* **1993**, 216, 133-140.
    (b) Johnson, B. G.; Frisch, M. J. *J. Chem. Phys.* **1994**, 100, 7429-7442.
14. Frisch, M. J. et al. *Gaussian 98*, Gaussian Inc., Pittsburgh, PA.
15. Becke, A. D. *J. Chem. Phys.* **1993**, 98, 1372-1377; 5648-5652.
16. Stephens, P. J.; Devlin, F. J.; Chabalowski, C. F.; Frisch, M. J. *J. Phys. Chem.* **1994**, 98, 11623-11627.
17. Devlin, F. J.; Stephens, P. J. (to be published).
18. Devlin, F. J.; Stephens, P. J. *J. Am. Chem. Soc.* **1999**, 121, 7413-7414.
19. Ashvar, C. S.; Stephens, P. J.; Eggimann, T.; Wieser, H. *Tetrahedron*: *Asymmetry* **1998**, 9, 1107-1110.
20. Stephens, P. J.; Devlin, F. J. *Chirality* **2000**, 12, 172-179.

21. Aamouche, A.; Devlin, F .J.; Stephens, P. J. *Conformations of Chiral Molecules in Solution: Ab Initio Vibrational Absorption and Circular Dichroism Studies of 4,4a,5,6,7,8-Hexahydro-4a-Methyl 3(3H)-Naphthalenone and 3,4,8,8a-Tetrahydro-8a-Methyl-1,6(2H,7H)-Naphthalenedione, J. Am. Chem. Soc.* **2000,** in press.

22. Aamouche, A.; Devlin, F. J.; Stephens, P. J. *Molecular Structure in Solution: An Ab Initio Vibrational Spectroscopy Study of the* 8a-*Conformations of the Chiral Bis-Oxazoline: 2,2′-Methylenebi*s[3a, Dihydro-8H-Indeno[1,2-d]Oxazole], (to be published).

23. (a) Aamouche, A.; Devlin, F. J.; Stephens, P. J. *J. Chem. Soc. Chem. Comm.* **1999,** 361-362.
    (b) Aamouche, A.; Devlin, F. J.; Stephens, P. J. *J. Am. Chem. Soc.* **2000,** 122, 2346-2354.

24. Mason, S. F.; Vane, G. W.; Schofield, K.; Wells, R. J.; Whitehurst, J. S. *J. Chem. Soc. B* **1967,** 553-556.

25. Wilen, S. H.; Qi, J. Z.; Williard, P. G. *J. Org. Chem.* **1991,** 56, 485-487.

26. Devlin, F. J.; Stephens, P. J.; Cheeseman, J. R.; Frisch, M. J.; Oesterle, C.; Wiberg, K. B. (to be published).

27. Polavarapu, P. L.; Zhao, C.; Cholli, A. L.; Vernice, G. G. *J. Phys. Chem. B* **1999,** 103, 6127-6132.

28. Polavarapu, P. L.; Zhao, C.; Ramig, K. *Tetrahedron: Asymmetry* **1999,** 10, 1099-1106.

Chapter 3

New Insight into Solution Structure and Dynamics of Proteins, Nucleic Acids, and Viruses from Raman Optical Activity

L. D. Barron[1], E. W. Blanch[1], A. F. Bell[1], C. D. Syme[1], L. Hecht[1], and L. A. Day[2]

[1]Department of Chemistry, University of Glasgow, Glasgow G12 8QQ, United Kingdom
[2]Public Health Research Institute, 455 First Avenue, New York, NY 10016

Raman optical activity measures vibrational optical activity by means of a small difference in the intensity of Raman scattering from chiral molecules in right and left circularly polarized incident light. The sensitivity of ROA to chirality makes it an incisive probe of biomolecular structure and dynamics in aqueous solution. This article reviews the basic theory and instrumentation of ROA, and describes recent results which illustrate how ROA provides new insight into current biomedical problems including protein misfolding and disease, and virus structure at the molecular level.

Introduction

Studies of the structure and dynamics of biomolecules in aqueous solution remains at the forefront of biomedical science. The potential value of vibrational spectroscopy, both infrared and Raman, has been greatly enhanced in this area by the addition of the new dimension of optical activity *(1-3)*, which confers an exquisite sensitivity to the absolute stereochemistry and conformation of chiral molecules.

The word 'chiral', meaning handed, was first introduced into science by Lord Kelvin *(4)*, Professor of Natural Philosophy in the University of Glasgow. Phenomena which are sensitive to molecular chirality include optical rotation, and ultraviolet circular dichroism (UVCD) where left and right circularly polarized UV light is absorbed slightly differently. Such 'chiroptical' techniques have a special sensitivity to the three dimensional structure of a chiral molecule and are widely used in the conformational analysis of biomolecules. The importance of the newer vibrational optical acivity methods is that they are sensitive to chirality associated with all 3N−6 fundamental vibrational transitions, where N is the number of atoms, and therefore have the potential to provide much more stereochemical information than UVCD which measures optical activity associated with the far fewer accessible electronic transitions.

Vibrational optical activity in typical chiral molecules in the liquid phase was first observed by Barron *et al.* in 1973 *(5)* using a Raman optical activity technique in which, as depicted in Figure 1, a small difference in the intensity of Raman scattering is measured using right- and left-circularly polarized incident light. Until recently, however, lack of sensitivity restricted ROA to favourable samples such as small chiral organic molecules. But major advances in instrumentation have, in recent years, rendered biomolecules in aqueous solution accessible to ROA studies *(6)*. Conventional Raman spectroscopy has a number of favourable characteristics which have led to many applications in biochemistry. In particular, the complete vibrational spectrum from ~100 to 4000 cm^{-1} is accessible on one simple instrument, and both H_2O and D_2O are excellent solvents for Raman studies. Since ROA is sensitive to chirality, it is able to build on these advantages by adding to Raman spectroscopy an extra sensitivity to the three-dimensional structure, which opens a new window on biomolecular problems *(7)*.

Theory of ROA

The fundamental scattering mechanism responsible for ROA was discovered by Atkins and Barron in 1969 *(8)*, who showed that interference between waves scattered via the polarizability and optical activity tensors of a chiral molecule yields a dependence of the scattered intensity on the degree of circular

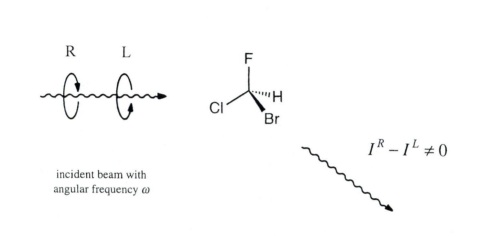

R   L

incident beam with
angular frequency ω

$I^R - I^L \neq 0$

Raman component of scattered beam
with angular frequency $\omega - \omega_\nu$

Figure 1. The basic ROA experiment measures a small difference in the intensity of Raman scattering from chiral molecules in right- and left-circularly polarized incident light.

polarization of the incident light and to a circular component in the scattered light. Barron and Buckingham (9) subsequently developed a more complete version of the theory and introduced the following definition of the dimensionless circular intensity difference (CID)

$$\Delta = \left(I^R - I^L\right)\!\big/\!\left(I^R + I^L\right)$$

as an appropriate experimental quantity, where I^R and I^L are the scattered intensities in right- and left-circularly polarized incident light. In terms of the electric dipole–electric dipole molecular polarizability tensor $\alpha_{\alpha\beta}$ and the electric dipole–magnetic dipole and electric dipole–electric quadrupole optical activity tensors $G'_{\alpha\beta}$ and $A_{\alpha\beta\gamma}$ (1,2,9), the CIDs for *forward* (0°) and *backward* (180°) scattering from an isotropic sample are as follows:

$$\Delta(0°) = 4\big[45\alpha G' + \beta(G')^2 - \beta(A)^2\big]\big/c\big[45\alpha^2 + 7\beta(\alpha)^2\big]$$

$$\Delta(180°) = 24\big[\beta(G')^2 + \tfrac{1}{3}\beta(A)^2\big]\big/c\big[45\alpha^2 + 7\beta(\alpha)^2\big]$$

where the isotropic invariants are defined as

$$\alpha = \tfrac{1}{3}\alpha_{\alpha\alpha}, \quad G' = \tfrac{1}{3}G'_{\alpha\alpha}$$

and the anisotropic invariants as

$$\beta(\alpha)^2 = \tfrac{1}{2}\big(3\alpha_{\alpha\beta}\alpha_{\alpha\beta} - \alpha_{\alpha\alpha}\alpha_{\beta\beta}\big), \quad \beta(G')^2 = \tfrac{1}{2}\big(3\alpha_{\alpha\beta}G'_{\alpha\beta} - \alpha_{\alpha\alpha}G'_{\beta\beta}\big),$$

$$\beta(A)^2 = \tfrac{1}{2}\omega\alpha_{\alpha\beta}\varepsilon_{\alpha\gamma\delta}A_{\gamma\delta\beta}$$

We are using a Cartesian tensor notation in which a repeated Greek suffix denotes summation over the three components, and $\varepsilon_{\alpha\beta\gamma}$ is the third-rank unit antisymmetric tensor. These results apply specifically to Rayleigh scattering. For Raman scattering the same basic CID expressions apply but with the molecular property tensors replaced by corresponding vibrational Raman transition tensors.

Using a bond polarizability theory of ROA for the case of a molecule composed entirely of idealized axially symmetric bonds, the relationships

$\beta(G')^2 = \beta(A)^2$ and $\alpha G' = 0$ are found *(1)*. Within this model, therefore, isotropic scattering makes zero contribution to the ROA intensity which is generated exclusively by anisotropic scattering, and the forward and backward CIDs reduce to

$$\Delta(0°) = 0, \quad \Delta(180°) = 32\beta(G')^2/c\left[45\alpha^2 + 7\beta(\alpha)^2\right]$$

Hence unlike the conventional Raman intensity, which is the same in the forward and backward directions, the ROA is maximized in backscattering and is zero in forward scattering. These considerations lead to the important conclusion that *backscattering boosts the ROA signal relative to the background Raman intensity and is therefore the best experimental strategy for most ROA studies of biomolecules in aqueous solution.*

The normal modes of vibration of biomolecules can be highly complex, with contributions from local vibrational coordinates in both the backbone and sidechains. The advantage of ROA is that it is able to cut through the complexity of the corresponding vibrational spectra since the largest ROA signals are often associated with the vibrational coordinates which sample the most rigid and chiral parts of the structure. These are usually within the backbone and often give rise to ROA band patterns characteristic of the backbone conformation. Polypeptides in the standard conformations defined by characteristic Ramachandran ϕ,ψ angles found in secondary, loop and turn structures within native proteins are particularly favourable in this respect since signals from the peptide backbone usually dominate the ROA spectrum, unlike the parent conventional Raman spectrum in which bands from the amino acid side chains often obscure the peptide backbone bands.

Since the time scale of the Raman scattering event ($\sim 3.3 \times 10^{-14}$ s for a vibration with wavenumber shift 1000 cm^{-1} excited in the visible) is much shorter than that of the fastest conformational fluctuations in biomolecules, the ROA spectrum is a superposition of 'snapshot' spectra from all the distinct chiral conformers present in the sample. This, together with the dependence of ROA on chirality, leads to an enhanced sensitivity to the dynamic aspects of biomolecular structure. For example, the ROA from enantiomeric structures, such as chiral two-group units with torsion angles of approximately equal magnitude but opposite sign, will tend to cancel as the mobile structure explores the set of accessible conformations. In contrast, observables that are 'blind' to chirality, such as conventional Raman band intensities, tend to be additive and hence much less sensitive to this type of mobility.

Experimental

For the reasons given above, a backscattering geometry has proved to be essential for the routine measurement of the ROA spectra of typical biomolecules in aqueous solution, especially since these samples often show a high background in the conventional Raman spectrum from solvent water together with residual fluorescence. In the current Glasgow backscattering ROA spectrometer *(6)*, an incident visible argon ion laser beam is focused into the sample, which is held in a small rectangular cell. The cone of backscattered light is collected by means of a 45° mirror with a small central hole to allow passage of the incident laser beam. The Raman light is thereby reflected into the collection optics of a single grating spectrograph. The CCD camera used as a multichannel detector permits measurement of the full spectral range in a single acquisition. To measure the tiny ROA signals, the spectral acquisition is synchronized with an electro optic modulator used to switch the polarization of the incident laser beam between right- and left-circular at a suitable rate.

Typical sample concentrations are in the range 20-100 mg/ml, but concentrations of virus samples can be as low as 5 mg/ml. The green 514.5 nm line of the argon ion laser is used since this provides a compromise between reduced fluorescence from sample impurities with increasing wavelength and the increased scattering power with decreasing wavelength due to the Rayleigh λ^{-4} law. A typical laser power at the sample is 700 mW. Under these conditions, high quality ROA spectra of small biomolecules such as peptides and carbohydrates may be collected in an hour or two, polypeptides, proteins and nucleic acids in ~10 h and viruses in ~24 h.

In order to study dynamic aspects of biomolecular structure, it is necessary to perform measurements over an appropriate temperature range. This is accomplished by directing dry air downwards over the sample cell from the nozzle of a device used to cool protein crystals in X-ray diffraction experiments, which enables the temperature to be varied in the range 0 to 60 °C.

Polypeptides and Proteins: Misfolding and Disease

ROA is an excellent technique for studies of polypeptide and protein structure in aqueous solution since, as mentioned above, their ROA spectra are often dominated by bands originating in the peptide backbone which directly reflect the solution conformation. Furthermore the special sensitivity to dynamic aspects of structure makes ROA a valuable new source of information on nonnative protein states and on native states containing mobile regions. A few examples of polypeptide and protein ROA spectra are shown and discussed below.

The largest ROA signals from polypeptides and proteins are usually associated with vibrations of the backbone in three main regions of the Raman spectrum (7). These are the backbone skeletal stretch region ~870-1150 cm^{-1} originating mainly in C_α–C, C_α–C_β and C_α–N stretch coordinates; the extended amide III region ~1230-1350 cm^{-1} which involves various combinations of N–H and C_α–H deformations and the C_α–N stretch, and the amide I region ~1630-1700 cm^{-1} which arises mostly from the C=O stretch. The extended amide III region is particularly important for ROA studies because coupling between N–H and C_α–H deformations is very sensitive to backbone geometry and generates a rich and informative ROA band structure. Side chain vibrations also generate many characteristic Raman bands (7): although less prominent in ROA spectra, some side chain vibrations do give rise to useful ROA signals, especially those of tryptophan since the indole ring is usually held in just one conformation relative to the peptide backbone, unlike smaller side chains for which many rotameric conformations are available which tends to wipe out the ROA.

Poly(L-lysine) at alkaline pH and poly(L-glutamic acid) at acid pH have neutral side chains and so are able to support α-helical conformations stabilized by internal hydrogen bonds; whereas poly(L-lysine) at neutral and acid pH and poly(L-glutamic acid) at neutral and alkaline pH have charged side chains which repel each other thereby encouraging a disordered structure. The backscattered Raman and ROA spectra of these samples are shown in Figure 2. There are clearly sufficiently large differences between the ROA spectra of the corresponding α-helical and disordered conformations to enable ROA to distinguish between these two states. It is interesting that the positive amide III ROA band at ~1297 cm^{-1} in α-helical poly(L-lysine) is not observed in α-helical poly(L-glutamic acid); instead the spectrum is dominated by a positive band at ~1345 cm^{-1} similar to one at ~1342 cm^{-1} in α-helical poly(L-lysine). Since the peptide backbone is expected to be more accessible to solvent water in poly(L-glutamic acid), this and other observations (vide infra) suggest that positive ROA bands at ~1340-1345 cm^{-1} originate in a hydrated form of α-helix whereas positive ROA bands at ~1297-1312 cm^{-1} originate in α-helix in a hydrophobic environment (7). In which case α-helical poly(L-lysine) contains both forms but α-helical poly(L-glutamic acid) contains only the hydrated form. New insight into these two α-helix variants may have been provided by recent ESR studies of double spin-labelled alanine rich peptides which identified a new more open form of the α-helix which may be preferred in aqueous solution (10) since it allows main-chain hydrogen bonding with water molecules. Hence the canonical form of α-helix may be responsible for the positive ~1297 cm^{-1} ROA band and the more open form may be responsible for the positive ~1345 cm^{-1} ROA band. The strong positive ROA band at ~1321 cm^{-1} in the disordered states of the two polypeptides has been assigned to polyproline II (PPII) helix since these samples are thought to contain a substantial amount of this conformational element (11).

41

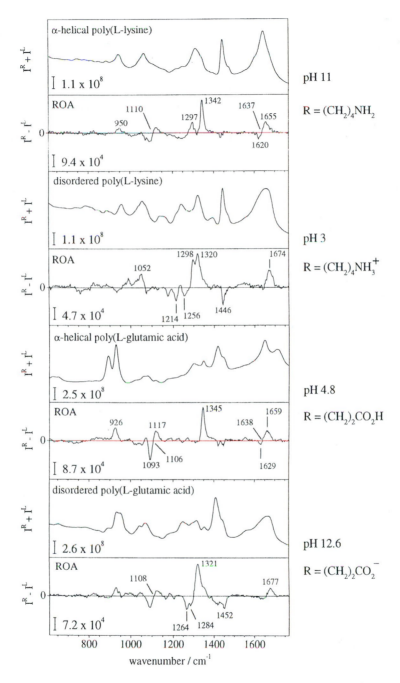

Figure 2. Backscattered Raman and ROA spectra of poly(L-lysine) in α-helical (top pair) and disordered (second pair) conformations, and of poly(L-glutamic acid) in α-helical (third pair) and disordered (bottom pair) conformations in aqueous solution.

Two examples of protein ROA spectra are shown in Figure 3. That of human serum albumin, an α-helical protein, has a similar appearance to that of α-helical poly(L-lysine) in Figure 2. The ROA is dominated by a strong positive band at ~1340 cm^{-1}. The assignment of this band to hydrated α-helix is reinforced by the fact that it disappears rapidly on dissolving the protein in D$_2$O, due to H–D exchange of the amide proton, whereas the other α-helix bands do not *(7)*. Despite the large size of the β-sheet protein human immunoglobulin G (~150 kDa), its ROA spectrum shown in Figure 3 is rich and informative and demonstrates that there is no upper size limit for protein ROA studies. A few specific assignments are the negative band at ~1245 cm^{-1} to β-sheet, the positive band at ~1314 cm^{-1} to PPII helix in the longer loops, and the negative bands at ~1346 and 1366 cm^{-1} to the tight β-turns in hairpin bends. The distinctive ROA band pattern in the extended amide III region is characteristic overall of the immunoglobulin type of Greek key β-barrel fold.

There is considerable current interest in nonnative states of proteins on account of their importance in studies of folding, stability and function. The term 'nonnative' or 'denatured' embraces a plethora of structures ranging from the ideal extended random coil at one extreme to collapsed molten globules at the other. The heterogeneity of such states has made their detailed characterization very difficult, with NMR and UVCD providing most information to date. ROA is proving to be an especially incisive probe of the 'complexity of order' in nonnative protein states *(7)*.

A recent example is a study of a partially denatured form of human lysozyme which has provided new insight into protein misfolding and disease *(12)*. In 'conformational diseases' such as the prion encephalopathies, Alzheimer's, Parkinson's, etc. a common theme is the non-reversible conversion of α-helix or some flexible loop sequence into β-sheet with the resultant build up of fibrils with a cross-β structure. As well as the mutant human lysozymes responsible for fatal amyloidosis, normal human lysozyme has recently been shown to generate amyloid fibrils by incubation at 57 °C at pH 2.0. The fibrils can take up to 24 hours to start to form, which provides sufficient time for ROA measurements to be made on the monomeric amyloidogenic prefibrillar intermediate. The ROA spectra of the native protein and the partially denatured intermediate are shown in Figure 4. Apart from some broadening, there is little change in the negative β-sheet ROA band at ~1241 cm^{-1} indicating that the β-domain remains at least partially structured in the prefibrillar intermediate. However, the positive ROA band at ~1345 cm^{-1} assigned to hydrated α-helix is not present in the prefibrillar intermediate, where a new strong positive band at ~1318 cm^{-1} appears instead that is assigned to PPII helix. The disappearance of a positive ROA band at ~1551 cm^{-1} assigned to vibrations of tryptophan sidechains indicates that major conformational changes have occurred among most of the five tryptophans present in human lysozyme, four of which are

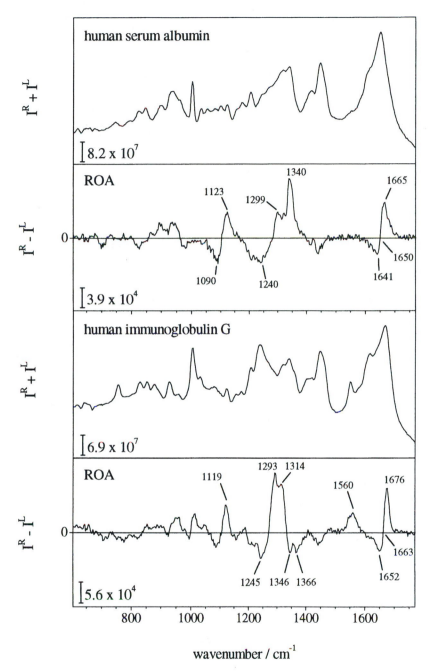

Figure 3. Backscattered Raman and ROA spectra of the α-helical protein human serum albumin (top pair) and the β-sheet protein human immunoglobulin G (bottom pair) in aqueous solution.

44

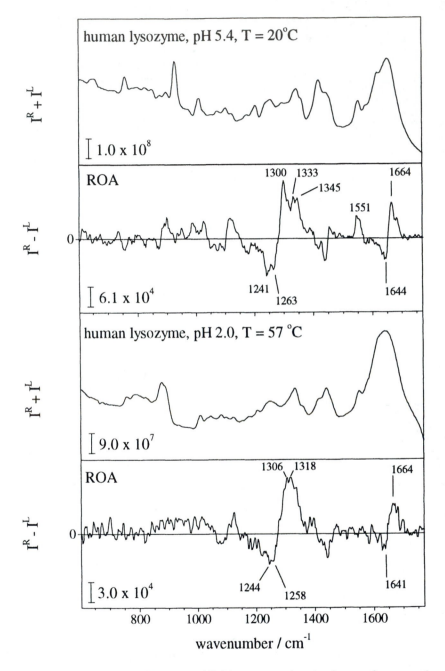

Figure 4. Backscattered Raman and ROA spectra of native human lysozyme (top pair) and of the amyloidogenic prefibrillar intermediate (bottom pair) in aqueous solution.

located in the α-domain. Hence the various ROA data suggest that a major loss of tertiary structure has occurred in the α-domain. The ROA spectrum of hen lysozyme, which does not form amyloid fibrils so readily, remains much more native-like on heating to 57 °C at pH 2.0. The thermal behaviour of the alanine-rich α-helical peptide AK21 in aqueous solution was found to be similar to that of human lysozyme.

Hydrated α-helix therefore appears to readily undergo a conformational change to PPII helix on heating, which may be a key step in the conversion of α-helix into β-sheet in the formation of amyloid fibrils in human lysozyme. Since it is extended, flexible, lacks intrachain hydrogen bonds and is fully hydrated in aqueous solution, PPII helix has the appropriate characteristics to be implicated as the 'killer conformation' in many conformational diseases *(12)*. Relatively static disorder of the PPII type may be a *sine qua non* for the formation of regular fibrils; whereas the more dynamic disorder of the true random coil may lead only to amorphous aggregates.

Carbohydrates and Glycoproteins

Carbohydrates in aqueous solution are highly favourable samples for ROA studies, giving rich and informative band structure over a wide range of the vibrational spectrum from which the central components of carbohydrate stereochemistry can be deduced. These include the dominant anomeric configuration, the ring conformation, the relative orientation of ring substituents, the conformation of the glycosidic link and secondary structure in polysaccharides.

Since both carbohydrates and proteins are excellent samples for ROA studies, it is not surprising that ROA is able to provide new information on intact glycoproteins, which are difficult to study using X-ray crystallography and NMR. This information includes the polypeptide and carbohydrate structure and the mutual influence these two moities exert on each other's conformation and stability. Further details of carbohydrate and glycoprotein ROA studies may be found in *(7)*.

Nucleic Acids

The study of nucleic acid structure and function remains a central element of molecular biology and the ever increasing sophistication of methods used continues to reveal deeper levels of complexity. Although ROA studies of nucleic acids are still at an early stage, the results obtained so far are encouraging. Nucleic acid ROA spectra can be divided approximately into three regions *(13)*. The base stacking region ~1550-1750 cm^{-1}, which contains bands characteristic for each base and which are sensitive to the base-stacking arrangement; the sugar–base region ~1200-1550 cm^{-1}, which contains bands associated with mixing of vibrational coordinates from both the base and the sugar rings and which reflect their mutual orientation; and the sugar–phosphate region ~900-1150 cm^{-1} which contains bands from vibrations localized mainly in the sugar rings and which reflect the sugar ring and phosphate backbone conformations.

ROA has been used to detect a delicate glass-like transition in synthetic polyribonucleotides at ~15-18 °C which may be associated with freezing of motions between conformational substates, perhaps in conjunction with changes in water stucture within and around the polynucleotides *(14)*. In another study, the ROA spectra of phenylalanine-specific transfer RNA and the same tRNA^{Phe} but with the Mg^{2+} ions removed were compared *(15)*. Clear differences were seen in the sugar–phosphate spectral region which originate in the changes in sugar puckers associated with structural changes from the compact L-shaped form of the folded RNA in the presence of Mg^{2+} ions to the open cloverleaf form the RNA adopts in the absence of Mg^{2+} ions.

Viruses

Conventional Raman spectroscopy is a valuable technique for studies of viruses at the molecular level on account of its ability to provide structural information about both protein and nucleic acid constituents of intact virions and viral precursors. We have recently demonstrated that viruses are accessible to ROA measurements.

The first virus ROA spectra were obtained on filamentous bacteriophages *(16)*. These are long slender rods consisting of a loop of single-stranded DNA surrounded by a shell made up of several thousand copies of the same major coat protein, each containing ~50 amino acids residues, which X-ray fibre diffraction has shown to be largely helical. The ROA spectra of two strains, Pf1 and M13 are displayed in Figure 5. The extra incisiveness of ROA compared with conventional Raman is strikingly demonstrated by the fact that, although the parent Raman spectra of the two bacteriophages are very different owing to the different protein side chains and nucleic acid base compositions, the ROA spectra have the same general appearance, being very similar to those of α-helical poly(L-lysine) and human serum albumin shown above, which reflects the fact that the major coat proteins have the same basic extended helical fold. Both Pf1 and M13 have two distinct long stretches of α-helix, one in a hydrophobic environment and the other exposed to solvent water. This was important evidence for assigning the positive ~1300 and 1340 cm^{-1} ROA bands in

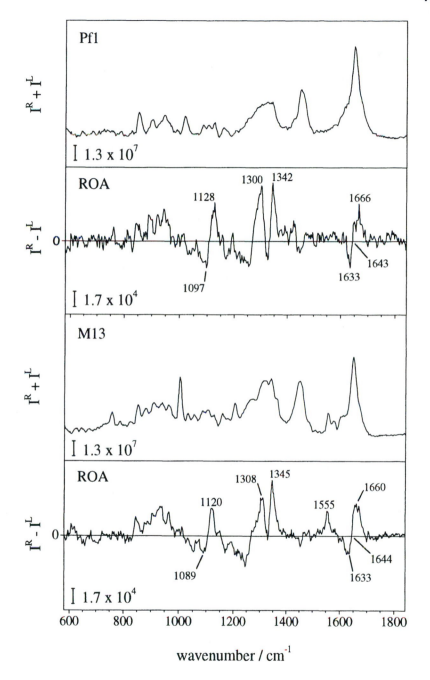

Figure 5. Backscattered Raman and ROA spectra of the filamentous bacteriophages Pf1 (top pair) and M13 (bottom pair) in aqueous solution.

polypeptides and proteins to hydrophobic and hydrated α-helix, respectively. Although the concentrations used are almost 10 times lower than those normally used in ROA studies of proteins and nucleic acids, the signal-to-noise ratio is similar. This may result from restrictions on the mobility of components packed in the virions. The small differences of detail between the ROA spectra of Pf1 and M13 are reproducible and reflect subtle differences in conformation and hydration.

At the time of writing, we have measured the ROA spectra of two other quite different virus samples, namely tobacco mosaic virus (TMV) and the empty protein capsid of MS2. As with the bacteriophages, the ROA spectra reflect the fold type of the major coat proteins, namely a four helix bundle in TMV and up-and-down β-sheet motif in MS2. This suggests that most types of virus should be amenable to ROA studies which will provide valuable new information including the folds of the major coat proteins.

Acknowledgements

We thank the BBSRC for research grants, and the EPSRC for a Senior Fellowship for LDB.

References

1. Barron, L. D. *Molecular Light Scattering and Optical Activity;* Cambridge University Press: Cambridge, 1982.
2. Polavarapu, P. L. *Vibrational Spectra: Principles and Applications with Emphasis on Optical Activity;* Elsevier: Amsterdam, 1998.
3. Nafie, L. A. *Ann. Rev. Phys. Chem.* **1997**, *48*, 357.
4. Lord Kelvin. *Baltimore Lectures;* C. J. Clay and Sons: London, 1904.
5. Barron, L. D.; Bogaard, M. P.; Buckingham, A. D. *J. Am. Chem. Soc.* **1973**, *95*, 603.
6. Hecht, L.; Barron, L. D.; Blanch, E. W.; Bell, A. F.; Day, L. A. *J. Raman Spectrosc.* **1999**, *30*, 815.
7. Barron, L. D.; Hecht, L.; Blanch, E. W.; Bell, A. F. *Prog. Biophys. Mol. Biol.* **2000,**.*73*, 1.
8. Atkins, P. W.; Barron, L. D. *Mol. Phys.* **1969**, *16*, 453.
9. Barron, L. D.; Buckingham, A. D. *Mol. Phys.* **1971**, *20*, 1111.
10. Bolin, K. A.; Millhauser, G. L. *Acc. Chem. Res.* **1999**, *32*, 1027.
11. Woody, R. W. *Adv. Biophys. Chem.* **1992**, *2*, 37.
12. Blanch, E. W.; Morozova-Roche, L. A.; Cochran, D. A. E.; Doig, A. J.; Hecht, L.; Barron, L. D. *J. Mol. Biol.* In press.

13. Bell, A. F.; Hecht, L.; Barron, L. D. *J. Am. Chem. Soc.* **1997,** *119*, 6006.
14. Bell, A. F.; Hecht, L.; Barron, L. D. *J. Raman Spectrosc.* **1999,** *30*, 651.
15. Bell, A. F.; Hecht, L.; Barron, L. D. *J. Am. Chem. Soc.* **1998,** *120*, 5820.
16. Blanch, E. W.; Bell, A. F.; Hecht, L.; Day, L. A.; Barron, L. D. *J. Mol. Biol.* **1999,** *290*, 1.

Chapter 4

Chirality in Peptide Vibrations: Ab Initio Computational Studies of Length, Solvation, Hydrogen Bond, Dipole Coupling, and Isotope Effects on Vibrational CD

Jan Kubelka[1], Petr Bour[2], R. A. Gangani D. Silva[1], Sean M. Decatur[3], and Timothy A. Keiderling[1]

[1]Department of Chemistry, University of Illinois at Chicago, 845 West Taylor Street, Chicago, IL 60607–7061
[2]Institute of Organic Chemistry and Biochemistry, Academy of Sciences of the Czech Republic, Flemingovo nam. 2, 16610 Prague, Czech Republic
[3]Department of Chemistry, Mt. Holyoke College, South Hadley, MA 01075

Conformational chirality of the repeating peptide chain serves as an important tool for analyzing polypeptide and protein structure. Vibrational circular dichroism (VCD) utilizes this chirality with the localized nature of vibrational motion to sample conformation over short peptide segments. VCD, as an electronic ground state property, can be accurately simulated with density functional theory (DFT) electronic structure calculations using the methods of Stephens and coworkers. Simulation results are presented for several model peptide systems increasing in size up to a heptamer, the latter showing very good qualitative agreement with experiment. Effects of hydrogen bonding, simulated solvation and isotopic substitution are explored. Spectra for larger oligopeptides are simulated by transferring DFT calculated parameters from a related short fragment. Effects of long range interactions which cannot be easily obtained from first principles, are here approximated by transition dipole coupling. Simulations of the VCD of ^{13}C isotopically labeled 17-mer peptides, are compared to experimental spectra. Both calculations and experiment demonstrate site-specific stereochemical sensitivity due to the isotopic substitution, as well as signal enhancement from the ^{13}C labeled part.

Introduction

Most biologically important molecules are chiral and consequently chirality based measurement techniques have long been important for their characterization. However, configuration is rarely an issue, for example, in proteins the absolute configuration derives from L amino acids. The repeating sequence of similarly configured residues creates stereo-specific interactions, which give the achiral parts of the chain, such as the planar peptide linkages in proteins (or bases in nucleic acids), a chiral nature. This conformational chirality is the basis for most circular dichroism (CD) studies of biopolymers. Historically, CD is measured in the UV for electronic transitions (1), but also (for 20 years (2)) in the IR for vibrational transitions (3, 4).

Electronic CD (ECD) of peptides and proteins (5) focuses on n-π^* and π-π^* transitions of the amide group (190-220 nm) and yields the most intense signals for long α-helices. Other conformations have different but overlapping bandshapes without any site-specific resolution of the structural elements. Furthermore, theoretical modeling of ECD is severely challenged by the need to accurately compute the electronic excited states.

Vibrational circular dichroism (VCD), due to more localized nuclear motion and lower dipole strengths, exhibits narrower line widths than ECD. This allows detection of different conformations in globular proteins through both bandshape and frequency properties (4, 6). Most studies of VCD of proteins and peptides have focused on the amide I at ~1650 cm^{-1} (mainly amide C=O stretch) (4, 7-9), but the lower energy amide II and III transitions (mixes of NH deformation and C-N stretch) also give conformationally sensitive VCD bandshapes (10, 11). However, this natural resolution is only by conformational type; specific sites still cannot be distinguished.

Site-specific conformational resolution in peptide VCD can be accomplished by isotopic labeling (12), where the amide C=O is substituted with ^{13}C leading to a shift down of ~40 cm^{-1} for the amide I' (N-H deuterated) band. Labeling of less than 10% can easily be detected as a side-band in the IR spectra (13, 14). Frequency alone is not sufficient for determining the conformation (6), for example, solvated α-helices are a problem (15-17), but the shape of the associated VCD can define the secondary structure of the labeled site.

Since VCD is an electronic ground state property, accurate *ab initio* calculations of the VCD (18) can be used to examine the detailed nature of chirality in small oligopeptides. We have calculated VCD for α-helical, 3$_{10}$-helical and ProII-helical (left-handed poly-L-proline II-like) alanine-based peptides of different sizes (up to a heptamer, fully *ab initio*). For polymeric computations we use a property transfer method (19, 20), wherein the *ab initio* parameters computed for a short peptide are transferred onto a longer one to simulate its VCD. Simulations of the VCD of the Ala-based 17-mer are discussed, inspired by our studies of site-selective isotope-labeled peptides. It is not yet possible to properly account for solvent and any interactions beyond the size of the short fragment. However, model calculations on a small peptide with explicit water and on longer peptides with simple long-range interaction corrections provide a qualitative picture of these effects.

Materials and Methods

Materials. Blocked 17-mer peptides of the general form Ac-Y(AAKAA)$_3$H-NH$_2$ were prepared by solid state synthesis (*14*). An isotopic variant contains ^{13}C-substituted Ala (on the C=O) at the N-terminal AA sequence. The peptides contain a TFA impurity (trifluoroacetic acid) which partially interferes with the amide I' IR, but not the VCD (since TFA is achiral), but whose spectra could be subtracted as a baseline. All the peptides were dissolved in D$_2$O in the same concentration, 25 mg/ml (pH ~ 3), and allowed to fully exchange overnight at room temperature.

Spectral measurements. VCD and IR absorption spectra were measured on the UIC dispersive VCD instrument (*21*). The samples were placed in a homemade demountable cell with CaF$_2$ windows separated by a 50 µm Teflon spacer and held in a variable temperature mount. Spectra were recorded with 10 cm^{-1} spectral resolution and 10 s time constant as an average of 6 to 8 scans. Mid-IR measurements were repeated with 4 cm^{-1} resolution using a BIORAD FTS-60A FTIR spectrometer.

Computations. Geometry optimizations and analytical harmonic force fields (FF), atomic polar (APT) and axial tensors (AAT) were calculated at the DFT level using Gaussian 98 (*22*). Our own set of programs (*19, 20*) was used to perform the isotopic substitutions and simulation of the IR absorption and VCD spectra from the *ab initio* parameters. All spectra were simulated with Lorentzian bandshapes with a uniform line-width (FWHM) of 20 cm^{-1}.

L-alanyl-L-alanine (Ala-Ala), containing a single amide bond, was fully optimized from a previously reported minimum energy conformation (*23*). FF, APT and AAT parameters were calculated at the BPW91/6-31G** level.

Model "pseudo-dipeptides", Ac-Ala-NH-CH$_3$ were constructed with ϕ, ψ torsional angle values characteristic of the α-helix (-57°, -47°), 3$_{10}$-helix (-60°, -30°) and ProII-helix (-78°, 149°). Peptide bonds were set to be planar and trans (ω=180°). These structures were optimized with ϕ, ψ and ω held fixed and their FF, APT and AAT calculated at the BPW91/6-31G** level.

An α-helical (ϕ =-57°, ψ =-47°, ω=180°) "pseudo-tripeptide" (Ac-(Ala)$_2$-NH-CH$_3$) VCD was calculated *in vacuo*, and with five water molecules: three H-bonded to the carbonyl =O and two to the amide N-H. A sixth water molecule, H-bonded to the remaining N-H group, did not sterically fit in the model. Both the *in vacuo* and "solvated" structures were partially optimized with just the peptide ϕ, ψ and ω fixed, and the FF, APT and AAT computed at the BPW91/6-31G** level.

A "heptapeptide" (Ac-(Ala)$_6$-NH-CH$_3$) constrained to be α-helical and two "pentapeptides" (Ac-(Ala)$_4$-NH-CH$_3$) constrained to be 3_{10}- and ProII-helical were used as minimal fragments. In the case of α- and 3_{10}- helices, the middle peptide group has both its N and O hydrogen bonded to the terminal groups. The geometries were taken from *ab initio* minimized alanine decamers (*24*) and optimized with all torsional angles frozen at the respective decamer values. FF, APT and AAT were calculated at the BPW91/6-31G* level.

Geometries of the α-helical and ProII-helical 17-mer peptides, Ac-Ala$_{17}$-NH$_2$, were also based on the decamer calculations (*24*). The FF, APT and AAT values were obtained by transfer of the *ab initio* calculated parameters from the heptamer (α-helix) and pentamer (ProII-helix) described above. Parameters for hydrogen bonded middle residues were transferred from the fully H-bonded residue in the center of the fragment. Force constants with respect to atom pairs not encompassed in the fragment are normally set to zero. To test for long range coupling effects on the FF we used the transition dipole coupling (TDC) approximation, based on *ab initio* APT values:

$$F_{A\lambda,B\xi}^{TDC} = \frac{\partial U^{TDC}}{\partial q_\lambda^A \partial q_\xi^B} = \frac{1}{\varepsilon R^3}\left[\sum_{i=x,y,z}\frac{\partial \mu_i^A}{\partial q_\lambda^A}\frac{\partial \mu_i^B}{\partial q_\xi^B} - 3\sum_{i,j=x,y,z}\left(\frac{\partial \mu_i^A}{\partial q_\lambda^A}e_i\right)\left(\frac{\partial \mu_j^B}{\partial q_\xi^B}e_j\right)\right]$$

for all pairs of atoms A, B whose force constants are absent in the transferred FF. Here q_λ^A, $\lambda=\{x,y,z\}$ are cartesian displacements of the atom A, ε was chosen as 1, R is the distance between atoms A and B, e_j are the components of a unit vector from atom A to B, and $\partial\mu_i/\partial q_\lambda$ are the components of the APT.

Computational Results

The computed IR and VCD spectra for Ala-Ala, deuterated in various positions, are shown in Figure 1. Ala-Ala has only a single amide group but is chiral by virtue of its two C_α asymmetric centers and internal H-bonded ring. Due to an H-bond between the amide C=O and the carboxyl terminal O-H, the amide I mode computed at ~1660 cm^{-1} (for isolated amides it is normally >1700 cm^{-1} with this FF type (*20, 25*)). In the all deuterated case (d), the amide I' (~1654 cm^{-1}) mode yields a weak positive VCD, whereas if the amide group is protonated (N-H), it is weak negative (case b, c). In the totally protonated calculations (a), the amide I VCD is very weak positive but obscured by negative VCD from the terminal NH$_2$ scissor deformation mode (~1626 cm^{-1}). On the other hand, the amide II mode is quite strong and positive, even when deuterated.

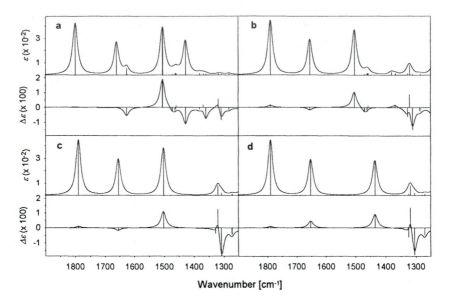

Figure 1. Ab initio Simulated IR and VCD spectra for Ala-Ala: a) fully protonated, b) OD, ND_2, c) ND_2, OD, CD_3, d) OD, CD_3, ND, ND_2.

Typical amide I VCD, which experimentally has a couplet shape arising from peptide-peptide interaction, is most simply modeled with a "pseudo-dipeptide", such as Ac-Ala-NH-CH_3. In this model, there is only one C_α group but two coupled peptide bonds which we have constrained to α-, 3_{10}- or ProII-helical conformations. In Figure 2 are presented the *ab initio* simulated VCD for the pseudo-dipeptide in these 3 conformations in the N-protonated form but with all methyl groups converted to $-CD_3$ to reduce interference. For all three

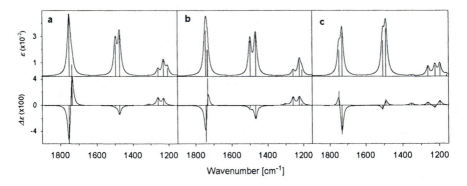

Figure 2. Ab initio simulated IR and VCD spectra for alanine "pseudo dipeptide" in a) α-helical, b) 3_{10}-helical, and c) ProII-helical conformation.

structures, the amide I band is computed at ~1760-1740 cm^{-1}, while the amide II is below 1500 cm^{-1}, except for the ProII. The α-helical geometry leads to a strong positive couplet amide I VCD and weak negative and positive amide II and III VCD respectively, qualitatively agreeing with experiment. The 3_{10} result is very similar except the amide I is weaker and more negatively biased and the amide II is stronger. This only partially reflects experimentally observed 3_{10}-helical bandshapes (*26, 27*). Finally the ProII "pseudo-dipeptide" VCD has a negative couplet amide I VCD, weaker in intensity than for the α-helix, qualitatively reflecting experimental results. The predicted positive couplet amide II VCD has not been seen experimentally but, of course, an amide II does not occur in Pro oligomers (*28*). We note that qualitatively the same results were found earlier for this model with HF/4-31G level simulations (29).

Effects of the solvent on the α-helical tripeptide VCD simulation are shown in Figure 3. Both N-protonated (a) and N-deuterated (b) states *in vacuo* and with

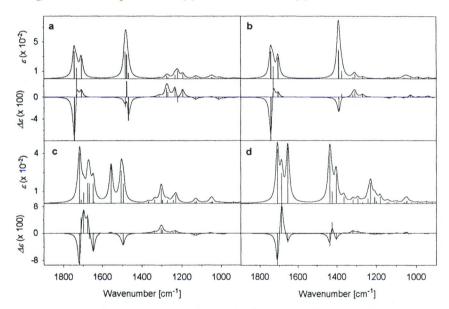

Figure 3. Ab initio simulated IR and VCD spectra for alanine "pseudo-tripeptide" a) N-protonated, b) N-deuterated, c) N-protonated with 5 H$_2$O, d) N-deuterated with 5 D$_2$O.

5 explicit H$_2$O (c) and D$_2$O (d) molecules are compared. Methyl groups of all species are deuterated for simplification (*vide supra*). For the protonated peptide-water complex (Figure 3c) APT and AAT values for atoms in the water molecules were set to zero, so that localized H$_2$O vibrations do not obscure the spectra. This simulates the effect of subtracting the baseline. Without the solvent, no significant changes in the VCD bandshapes (other than shift of the

amide II' frequency) occur upon deuteration. The deuterated tripeptide with explicit D_2O reproduces the experimentally observed amide I' bandshape, but a split "W"-type VCD for the amide I of the protonated complex is also observed. Hydrogen bonding in the "supermolecule" also improves the computed amide I and II frequencies (amide I/I' shifting lower and amide II/II' higher as compared to the vacuum calculation) as well as absorption intensities (amide II/II' being too intense in vacuum).

To account for intramoleular H-bonds, the IR and VCD spectra of a "pseudo-heptapeptide", Ac-Ala$_6$-NH-CH$_3$, constrained to an α-helical geometry and a "pseudo-pentapeptide", Ac-Ala$_4$-NH-CH$_3$, constrained to the 3_{10} and ProII conformations, were simulated. The α- and 3_{10}-helical peptides contain one peptide group H-bonded at both its C=O and N-H groups and several others at either the C=O or N-H. All three peptides are fully chiral (L-Ala). To our knowledge, these represent VCD and IR simulations for the largest peptide oligomers attempted *ab initio* thus far. For consistency, all methyl groups were -CD$_3$ substituted. The spectra (Figure 4) are in even better agreement with experiment than previously reported for di- and tripeptide results (*19, 20, 29*), especially for the amide I of the α-helix and ProII helix, which both yield computed oppositely signed, negatively biased couplets. The 3_{10} simulation has its amide II VCD stronger than the amide I, as seen experimentally, but has a distorted amide I. The details of the amide II frequency distribution are less satisfactory which may reflect coupling to other modes. In the fully protonated simulation (not shown) all the amide II modes experience CH$_3$ interference. The frequency of the amide I remains quite high ($\sim$1730 cm^{-1}) despite the internal hydrogen bonds.

In order to compare these calculations directly to experimental data we have transferred the FF, APT and AAT parameters from the Ac-Ala$_6$-NH-CH$_3$ and Ac-Ala$_4$-NH-CH$_3$ results onto an oligomer of 17 alanines (A), Ac-A$_{17}$-NH$_2$, which can be compared to experimental data obtained for Ac-Y-(AAKAA)$_3$-H-NH$_2$ (*14*). In Figure 5 (a and c) are compared the amide I' VCD and IR spectra predicted for the unlabeled peptide in the α-helical and ProII-helical conformations which can be compared with the experimental VCD and IR for the α-helical and unordered (coil-like) 17-mer (Figure 6, dashed lines). The relation of ProII and coil-like VCD spectral patterns has been established (*30*).

Since even a dipeptide generates a characteristic VCD pattern, it is reasonable that the band associated with sequential ^{13}C labels will yield an interpretable result. This is indeed true as can be seen for the amide I' VCD of a 17-mer having two ^{13}C labeled (on C=O) A residues near the N-terminal. The α-helical and ProII-helical peptide simulated VCD (Figure 5b and d) can be compared with the corresponding high and low temperature experimental results, respectively (Figure 6, solid lines). To compare these, it is necessary to judge the frequencies of the VCD component bands with respect to the absorbance maximum. The IR frequency of the ^{12}C amide I peak shifts up in going from helix ($\sim$1637 cm^{-1}, for fully solvated helix instead of the traditional 1650 cm^{-1} (*15*)) to unordered ($\sim$1645 cm^{-1}). At the same time the VCD negative

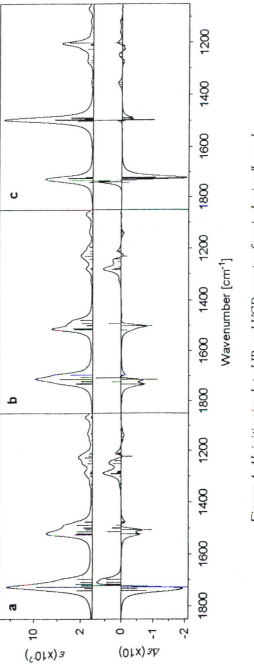

Figure 4. Ab initio simulated IR and VCD spectra for a) alanine "pseudo-heptapeptide" in an α-helical conformation b), c) alanine "pseudo-pentapeptide" in a 3₁₀- and a ProII-helical conformation, respectively.

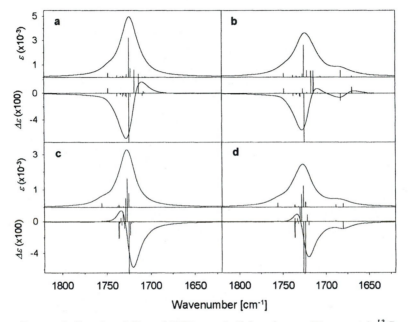

Figure 5. Simulated IR and VCD amide I' for alanine 17-mer with ^{13}C isotopic labels: a) α-helical, unlabeled, b) α-helical, N-teminally labeled, c) ProII-helical, unlabeled, d) ProII-helical, N-teminally labeled.

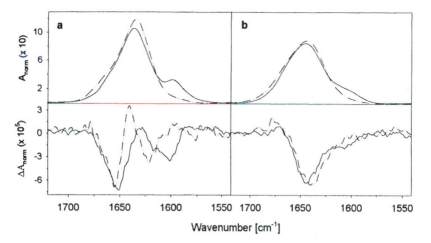

Figure 6. Experimental amide I' spectra for an unlabeled (dash-dot-dot) and N-terminally ^{13}C labeled (solid) alanine 17-mer a) IR 5°C (α-helix), b) IR 50°C (unordered), c) VCD 5°C (α-helix), d) VCD 50°C (unordered).

lobe shifts <u>down</u> in frequency, indicating that the VCD couplet (relative to the absorption) <u>changes sign</u> in going from helix to coil. The same behavior occurs for the ^{13}C component, but is harder to discern due to its weaker intensity. The simulations (N-deuterated, Figure 5) for this particular isotopically labeled peptide follow this pattern. While these 17-mer calculations also provide predictions for other modes, just as was shown above for shorter peptides, there are no experimental data for comparison at this time.

The transferred FF for the 17-mer peptide is limited to pairs of atoms that lie in a band along the diagonal within the range spanned by the fully computed, small (hepta- and penta-) peptide, and the remaining terms are set to zero. We examined effects of these long-range interactions, by approximating these 'zeros' with dipole-dipole coupling (TDC) interactions computed from the APT. The results for the amide I' of the unlabeled and N-terminally ^{13}C labeled α-helical 17-mer peptide are shown in Figure 7. To emphasize the character of change, the TDC magnitude was doubled. The TDC interaction causes the amide I' to shift slightly down in frequency and develop a more intense IR absorption and less intense VCD. In the isotopically labeled species, the TDC causes the ^{13}C modes to become more intense relative to the ^{12}C modes. This enhancement is more pronounced in VCD than in IR absorption, which fits the experimentally observed pattern: in the low temperature spectra (α-helix) the ^{13}C intensity is higher than would be expected from the fraction (~ 11%) of labeled residues. Thus long-range couplings, approximated here by TDC, evidently play a non-negligible role in the α-helical peptide FF. By contrast, no effect was observed upon including the TDC in the transferred FF for the 17-mer constrained to the ProII conformation (not shown), and all other spectral bands, aside from amide I/I', remained unaffected.

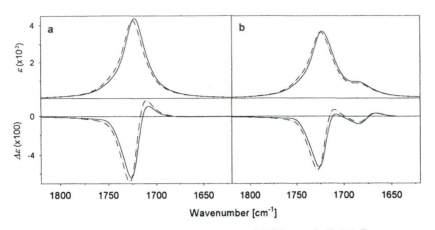

Figure 7. Comparison of IR absorption and VCD amide I', TDC correction (solid) and without TDC correction (dash-dot-dot) for a) unlabeled and b) N-terminally ^{13}C labeled α-helical 17-mer.

Discussion

The above examples demonstrate the conformational basis of vibrational optical activity in polypeptides. It is the chirality of the oligomer conformation, rather than the chirality of an amino acid unit, and the coupling of vibrations of repeating amide groups (especially in the amide I mode), that gives VCD its sensitivity to the peptide backbone conformation. As the Ala-Ala example shows, the amide I VCD for a single peptide is very weak, almost an order of magnitude smaller than that of a dipeptide. It may perhaps be surprising that a non-zero VCD intensity is computed at all. As an analysis of the normal modes shows, the amide modes are not localized, but virtually all atoms in the molecule, including the chiral centers, are involved to a small degree. While various isotopic substitutions seem to cause drastic changes in the simulated amide I VCD, on an absolute scale these are very small. For example, the difference between the spectra in Figure 1c and d, where the amide I VCD flips sign, arises from small variations in amide N-H and N-D motions leading to slight differences in transition moment vectors. Furthermore, the computed VCD is a result of one particular (minimum energy) geometry, whereas the VCD measured for a real sample would be a conformational average of several close-lying, local-minimum-energy structures (23). By contrast, amide II is more intense and consistent, indicating chirality arising from local environment.

Simulated spectra for a system as small as our model "pseudo-dipeptide" contain the primary conformational VCD features observed experimentally for larger oligopeptides. For such calculations, the conformation of the dipeptide fragment must be constrained, since such short helices are unstable, and therefore the molecule is not in the true energy minimum. As shown previously (19, 20) this has a negligible effect on the stretching and bending modes, such as amide I and II. In each case the two amide I modes are delocalized, the higher frequency one in-phase and lower out-of-phase. Thus, for two peptide bonds, large amide I VCD intensities result as compared to the amide II mode, which is much less coupled and therefore more localized, and has about the same intensity per amide as the monomer. The higher frequency amide II mode is mostly localized on the N-terminal, and the lower on the C-terminal peptide group. Consistent with this, in the hepta- and pentapeptide calculations, the most intense VCD components correspond to delocalized modes in the amide I, while the weaker amide II was more localized.

These *ab initio* simulations calculate the amide I frequencies too high, overestimating the amide I-II separation ($\square$ 200 cm^{-1}, while experimentally it is ~100 cm^{-1}). This flaw is common to all density functional methods we have tested. In particular, the hybrid functionals, such as B3LYP and B3PW91, while having the same amide I-II separation, give even higher frequencies for the amide I (~1800 cm^{-1} and 1600 cm^{-1} for the amide I and II, respectively). Moreover, the hybrid functionals scale much worse computationally with the size of the system than do "pure" density functionals (without HF exchange) such as BPW91 (31).

As can be seen in Figure 1 and Figure 3, hydrogen bonding to the C=O shifts the amide I and amide II frequencies toward more reasonable values, but larger peptide models, with explicit intra-molecular hydrogen bonds, do not show much amide I frequency improvement. However, the geometries used for these longer peptides are based on calculations with implicit solvent (COSMO model (32)) and were torsionally constrained for the VCD calculations done in vacuum. Our recent results (25) for smaller peptides show that a DFT/COSMO derived FF also improves the calculated amide frequencies. Unfortunately, such calculations are still impractical for large molecules. Moreover the AAT calculations with implicit solvent have not yet been implemented in commercial program packages. Similar improvement in the amide I frequency is obtained by use of larger basis sets, namely those with diffuse functions (24). Additionally, including TDC in the force field of larger peptides also shifts the amide I and II frequencies slightly closer together, but only for the α-helices, not for ProII.

Another challenge to our *ab initio* simulations is to reproduce the qualitative bandshape change seen between the amide I and amide I' (N-D) for α-helices. While that amide I is a negatively biased positive couplet, the corresponding amide I' exhibits an asymmetric "W" shape (negative-positive-negative), with weaker lower frequency negative lobe. This deuteration effect has not been observed in other structures (33). As seen in Figure 3, the same couplet bandshape is calculated for amide I and I' VCD of the α-helical tripeptide *in vacuo.* Adding H-bonded water molecules results in the "W"-shape both for both the amide I and I'. However, the strongly split amide I VCD pattern in the protonated case is a result of a single, rigid "supermolecule" whose amide I mode is coupled with O-H bending in H_2O. (Even with no water contribution to the spectral intensity, through FF interactions, the 3 amide I modes devolve into 7 overlapped transitions.) In reality, due to thermal motion, and exchange of water from the first hydration sphere with the bulk, the effect of this coupling in VCD would probably average out. On the other hand, in the deuterated case this coupling is essentially removed, and the changes in the observed spectral pattern result entirely from the peptide (though in the presence of hydrogen bonds to the solvent).

The *ab initio* modeling of peptide VCD has proven very useful for comparison with experiment in studying the ^{13}C isotopically labeled oligopeptides. These molecules are too large to be calculated fully *ab initio*, but less expensive semiempirical methods are not capable of accurately simulating VCD. Using our property transfer method, we can simulate the VCD of these large oligopeptides with reasonable accuracy. The bandshape patterns associated with different positions of the isotope label (*12*) are in excellent agreement with experiment for both the low temperature α-helical, and high temperature disordered (coil) conformations, the latter modeled with the ProII structure (*30*). In these 17-mer simulations the spectra fundamentally change sign in going from α-helical to ProII-helical (Figure 5), just as seen for the shorter oligomer spectra simulated *ab initio* (Figure 4). The same patterns persist even in the dipeptide (Figure 2), showing that the dominant VCD interaction is essentially local coupling of peptides. This predicted bandshape change between helix and coil is just what is experimentally seen in temperature-induced denaturation of the 17-mer peptide, as shown in Figure 6, and in longer peptides (*12*).

The VCD intensity enhancement of the ^{13}C modes for the isotopically labeled molecules can be qualitatively (but not quantitatively) reproduced by adding TDC to the larger peptide FF, which increases coupling between the otherwise separated ^{13}C and ^{12}C modes. The effect of long-range TDC underlines the importance of weakly- and non-bonded interactions for oligopeptide vibrational optical activity, especially in the case of α–helix. Moreover, this phenomenon is more intense in the VCD than in the IR, which demonstrates the greater sensitivity of VCD to small perturbations.

Conclusion

The conformational origin of the chirality of the amide vibrations provides VCD with unique sensitivity to the local oligopeptide chain stereochemistry. The local nature of the VCD is evident from the "pseudo-dipeptide" simulations, but experimentally it is not accessible in such a simple form, since the dipeptide does not form any well defined secondary structure. However, such dipeptides isolated in a longer peptide via ^{13}C labeling are accessible, as we have shown. The weakly bonded and non-bonded long range interactions significantly contribute to both frequency and bandshape in the VCD spectra as do the interactions with solvent. Isotopic substitution has a more pronounced effect in the VCD than it has in the IR absorption, in that aside from the frequency shift, the labeled signal is significantly enhanced. This intensity enhancement, especially in α-helical peptides, is consistent with long-range coupling of amide I vibrations. Theoretical VCD simulations, when compared to experiment, demonstrate their validity and usefulness for analysis of the experimental results. Our transfer of property tensors model has no intrinsic limits in terms of extension to longer peptides or to heterogeneous structures, provided the appropriate basis FF, APT and AAT parameters are determined.

Acknowledgement: This work was supported by a UIC internal grant (to TAK), by the Petroleum Research Fund sponsored by the American Chemical Society (35443-AC4 to TAK), by NIH (GM/OD 55897 to SMD) and by the Grant Agency of the Czech Republic (grant no. 203/97/P002 to PB).

References

1. Fasman, G. D. *Circular Dichroism and the Conformational Analysis of Biomolecules*; Plenum: New York, 1996.
2. Singh, R. D.; Keiderling, T. A. *Biopolymers* **1981,** *20*, 237-240.
3. Keiderling, T. A. In *Circular Dichroism and the Conformational Analysis of Biomolecules*; Fasman, G. D., Ed.; Plenum: New York, 1996, pp 555-598.
4. Keiderling, T. A. In *Circular dichroism: principles and applications*; Nakanishi, K., Berova, N., Woody, R. A., Eds.; Wiley: New York, 2000, pp 621-666.
5. Woody, R. W. In *Circular Dichroism and the Conformational Analysis of Biomolecules*; Fasman, G. D., Ed.; Plenum Press: New York, 1996, pp 25-67.
6. Pancoska, P.; Wang, L.; Keiderling, T. A. *Protein Sci.* **1993,** *2*, 411-419.
7. Paterlini, M. G.; Freedman, T. B.; Nafie, L. A. *Biopolymers* **1986,** *25*, 1751-1765.
8. Freedman, T. B.; Nafie, L. A.; Keiderling, T. A. *Biopolymers* **1995,** *37*, 265-279.
9. Yasui, S. C.; Keiderling, T. A. *J. Am. Chem. Soc.* **1986,** *108*, 5576-5581.
10. Gupta, V. P.; Keiderling, T. A. *Biopolymers* **1992,** *32*, 239-248.
11. Baello, B. I.; Pancoska, P.; Keiderling, T. A. *Anal. Biochem.* **1997,** *250*, 212-221.

12. Silva, R. A. G. D.; Kubelka, J.; Decatur, S. M.; Bour, P.; Keiderling, T. A. *Proc. Natl. Acad. Sci. U. S. A.* **2000**, *127*, 8318-8323.
13. Fabian, H.; Chapman, D.; Mantsch, H. H. In *Infrared Spectroscopy of Biomolecules*; Mantsch, H. H., Chapman, D., Eds.; Wiley-Liss: Chichester, 1996, pp 341-352.
14. Decatur, S. M.; Antonic, J. *J. Am. Chem. Soc.* **1999**, *121*, 11914-11915.
15. Martinez, G.; Millhauser, G. *J. Struct. Biol.* **1995**, *114*, 23-27.
16. Williams, S.; Causgrove, T. P.; Gilmanshin, R.; Fang, K. S.; Callender, R. H.; Woodruff, W. H.; Dyer, R. B. *Biochemistry* **1996**, *35*, 691-697.
17. Yoder, G.; Pancoska, P.; Keiderling, T. A. *Biochemistry* **1997**, *36*, 15123-15133.
18. Stephens, P. J.; Devlin, F. J.; Ashvar, C. S.; Chabalowski, C. F.; Frisch, M. J. *Faraday Discuss.* **1994**, *99*, 103-119.
19. Bour, P.; Sopkova, J.; Bednarova, L.; Malon, P.; Keiderling, T. A. *J. Comput.Chem.* **1997**, *18*, 646-659.
20. Bour, P.; Kubelka, J.; Keiderling, T. A. *Biopolymers* **2000**, *53*, 380-395.
21. Keiderling, T. A. In *Practical Fourier Transform Infrared Spectroscopy*; Krishnan, K., Ferraro, J. R., Eds.; Academic Press: San Diego, 1990, pp 203-284.
22. Frisch, M. J.; Trucks, G. W.; Schlegel, H. B.; Scuseria, G. E.; Robb, M. A.; Cheeseman, J. R.; Zakrzewski, V. G.; Montgomery, J., J. A.; Stratmann, R. E.; Burant, J. C., et al. *Gaussian 98*, Revision A.6 ed.; Gaussian, Inc.: Pittsburgh PA, 1998.
23. Knapp-Mohammady, M.; Jalkanen, K. J.; Nardi, F.; Wade, R. C.; Suhai, S. *Chem. Phys.* **1999**, *240*, 63-77.
24. Bour, P.; Kubelka, J.; Keiderling, T. A. *unpublished*.
25. Kubelka, J.; Keiderling, T. A. *unpublished*.
26. Yoder, G.; Polese, A.; Silva, R. A. G. D.; Formaggio, F.; Crisma, M.; Broxterman, Q. B.; Kamphuis, J.; Toniolo, C.; Keiderling, T. A. *J. Am. Chem. Soc.* **1997**, *119*, 10278-10285.
27. Yasui, S. C.; Keiderling, T. A.; Bonora, G. M.; Toniolo, C. *Biopolymers* **1986**, *25*, 79-89.
28. Dukor, R. K.; Keiderling, T. A.; Gut, V. *International Journal of Peptide and Protein Research* **1991**, *38*, 198-203.
29. Bour, P.; Keiderling, T. A. *J. Am. Chem. Soc.* **1993**, *115*, 9602-9607.
30. Dukor, R. K.; Keiderling, T. A. *Biopolymers* **1991**, *31*, 1747-1761.
31. Stratmann, R. E.; Burant, J. C.; Scuseria, G. E.; Frisch, M. J. *Journal of Chemical Physics* **1997**, *106*, 10175-10183.
32. Klamt, A.; Schuurmann, G. *J. Chem. Soc. Perkin Trans.* **1993**, *2*, 799-805.
33. Baumruk, V.; Keiderling, T. A. *J. Am. Chem. Soc.* **1993**, *115*, 6939-6942.

Chapter 5

Vibrational Transition Current Density: Visualizing the Origin of Vibrational Circular Dichroism and Infrared Intensities

Teresa B. Freedman, Eunah Lee, and Taiping Zhao

Department of Chemistry, Syracuse University, Syracuse, NY 13244–4100

Vibrational transition current density (TCD) maps and charge-weighted nuclear displacement vectors can be used to visualize the electronic and nuclear contributions to infrared absorption (IR) and vibrational circular dichroism (VCD) intensities. Vibrational TCD, a vector-field plot related to the integrand of the electronic contribution to the velocity-form electric dipole transition moment, shows the flow of electron density produced by nuclear motion. Examples are presented showing a variety of current patterns for modes of $(2S,3S)$-oxirane-d_2, (S)-methyl lactate, L-alanine, and (S)-methyl chloropropionate, which provide insight into the origin of the observed VCD and IR intensities.

Introduction

Vibrational circular dichroism (VCD) is the differential absorbance of left and right circularly polarized infrared radiation by a chiral molecule during vibrational excitation (1). Infrared absorption intensity arises from vibrationally-generated linear oscillation of charge, producing an oscillating electric dipole moment (1). VCD intensity arises from vibrationally-generated angular oscillation of charge occurring about the direction of linear charge oscillation. These linear and angular charge oscillations result in oscillating electric and magnetic dipole moments, respectively, and both are required for non-zero VCD intensity. Present computational chemistry methodology yields calculated IR and VCD intensities that agree well with experiment (1,2). However, these scalar intensity values arise from integration over electronic and nuclear coordinates, and do not provide insight into the nature and origin of the charge oscillations responsible for the intensities. Instead, a way to directly calculate and view nuclear and electronic currents produced by the vibrational motion is required. Numerous molecular modeling programs incorporate routines to animate the nuclear vibrational motion, which provides some understanding of the nuclear contribution to VCD and IR intensities when the nuclear displacements are weighted by the nuclear charge. We have recently developed a method for calculating and viewing vibrationally-generated electron density motion, termed vibrational transition current density (TCD) (3-6). We provide here a theoretical development for VCD and IR intensity that identifies the charge flow contributions, and demonstrate the relationship of vibrational TCD to IR and VCD intensities. Finally, we provide examples of charge-weighted nuclear displacement and TCD vector field maps for a variety of vibrational modes, and demonstrate how these maps provide an understanding of the origin of IR and VCD intensities.

Theoretical Background

Infrared absorption intensity is proportional to the dipole strength, the absolute square of the electric dipole transition moment. As seen from eq 1, the position-form dipole strength, D_r^a, for a harmonic fundamental transition for normal mode a in the electronic ground-state g, at angular frequency ω_a, depends on the derivative of the position-form electric dipole moment, $\vec{\mu}_r$, with respect to normal mode Q_a. Non-zero dipole strength thus requires linear oscillation of charge during the vibration (1).

$$D_r^a(g0 \rightarrow g1) = (\hat{\mu}_r)_{01} \bullet (\hat{\mu}_r)_{10} = \frac{\hbar}{2\omega_a} \left| \left(\frac{\partial \vec{\mu}_r}{\partial Q_a} \right)_0 \right|^2 \qquad (1)$$

Vibrational circular dichroism intensity is proportional to the rotational or rotatory strength, the scalar product of the electric and magnetic dipole transition moments. The expression shown in eq 2 for the position-form rotational strength, R_r^a, involves the derivative of the magnetic dipole moment, $\vec{m}$, with respect to the conjugate vibrational momentum P_a. VCD intensity thus requires vibrationally-generated circular or angular oscillation of charge about the direction of linear charge oscillation (1).

$$R_r^a(g0 \to g1) = \text{Im}(\hat{\vec{\mu}}_r)_{01} \bullet (\hat{m})_{10} = \frac{\hbar}{2}\left(\frac{\partial \vec{\mu}_r}{\partial Q_a}\right)_0 \bullet \left(\frac{\partial \vec{m}}{\partial P_a}\right)_0 \qquad (2)$$

The dipole and rotational strengths can also be expressed in terms of the velocity-form electric dipole moment, $\vec{\mu}_v$, in which case the electric dipole moment derivative is with respect to P_a. The velocity-form rotational strength is given by

$$R_v^a(g0 \to g1) = \omega_a^{-1}\,\text{Re}(\hat{\vec{\mu}}_v)_{01} \bullet (\hat{m})_{10} = \frac{\hbar}{2}\left(\frac{\partial \vec{\mu}_v}{\partial P_a}\right)_0 \bullet \left(\frac{\partial \vec{m}}{\partial P_a}\right)_0 \qquad (3)$$

The Cartesian components of the electric and magnetic dipole transition moments are further expressed in terms of the atomic polar tensor (APT), $P_{\alpha\beta}^A$ [position-form implied] or $P_{v,\alpha\beta}^A$ [velocity-form], or the atomic axial tensor (AAT), $M_{\alpha\beta}^A$, for atom A, shown in eqs 4-6. The electronic and nuclear contributions to the atomic polar tensor and atomic axial tensor elements are compiled in Table 1. In these expressions, summation over Cartesian directions for repeated Greek subscripts is implied, $R_{A\alpha}$ is nuclear position, $\dot{R}_{A\alpha}$ is nuclear velocity, $Z_A e$ is nuclear charge, $\delta_{\alpha\beta}$ is the Kronecker delta, and $\varepsilon_{\alpha\beta\gamma}$ is the rank 3 antisymmetric tensor ($\varepsilon_{\alpha\beta\gamma} = 0$ if any of $\alpha\beta\gamma$ are the same and $\varepsilon_{\alpha\beta\gamma} = 1$ or -1 if $\alpha\beta\gamma$ is an even or odd permutation of the order xyz, respectively), which is used in expressing vector cross product elements. A tilde denotes a complex quantity; the velocity-form electric dipole transition moment and the magnetic dipole transition moment are pure imaginary quantities, whereas the APT and AAT elements are real. The s-vector component, $s_{A\alpha,a} = (\partial R_{A\alpha}/\partial Q_a)_0 = (\partial \dot{R}_{A\alpha}/\partial P_a)_0$, is expressed either in terms of nuclear displacement or velocity.

$$\left(\hat{\mu}_{r,\beta}\right)_{01} \equiv \left\langle \Psi_{g0}^a \left| \hat{\mu}_{r,\beta} \right| \Psi_{g1}^a \right\rangle = \left(\frac{\hbar}{2\omega_a}\right)^{1/2} \sum_A P_{\alpha\beta}^A \, s_{A\alpha,a} = \left(\frac{\hbar}{2\omega_a}\right)^{1/2} \left(\frac{\partial \mu_{r,\beta}}{\partial Q_a}\right)_0 \qquad (4)$$

In our expressions for the atomic polar and atomic axial tensors (Table 1), we use a parallel definition for both tensors in terms of a dipole moment

$$\left(\hat{\mu}_{v,\beta}\right)_{01} \equiv \left\langle \tilde{\Psi}_{g0}^{a} \left| \hat{\mu}_{v,\beta} \right| \tilde{\Psi}_{g1}^{a} \right\rangle$$

$$= -i\left(\frac{\hbar\omega_a}{2}\right)^{1/2} \sum_A P_{v,\alpha\beta}^A \, s_{A\alpha,a} = -i\left(\frac{\hbar\omega_a}{2}\right)^{1/2} \left(\frac{\partial\mu_{v,\beta}}{\partial P_a}\right)_0 \tag{5}$$

$$\left(\hat{m}_{\beta}\right)_{10} \equiv \left\langle \tilde{\Psi}_{g1}^{a} \left| \hat{m}_{\beta} \right| \tilde{\Psi}_{g0}^{a} \right\rangle = i\left(\frac{\hbar\omega_a}{2}\right)^{1/2} \sum_A M_{\alpha\beta}^A \, s_{A\alpha,a} = i\left(\frac{\hbar\omega_a}{2}\right)^{1/2} \left(\frac{\partial m_{\beta}}{\partial P_a}\right)_0 \tag{6}$$

Table 1. Electronic and Nuclear Contributions to the APT and AAT

Tensor	Nuclear Contribution	Electronic Contribution
Atomic Polar Tensor (APT) Position Form $P_{\alpha\beta}^A = \left(\dfrac{\partial\mu_{r,\beta}}{\partial R_{A\alpha}}\right)_0$ $= N_{\alpha\beta}^A + E_{r,\alpha\beta}^A$	$N_{\alpha\beta}^A = Z_A e\,\delta_{\alpha\beta}$	$E_{r,\alpha\beta}^A =$ $\left(\dfrac{\partial}{\partial R_{A\alpha}}\left\langle \psi_g \left| \hat{\mu}_{r,\beta}^E \right| \psi_g \right\rangle\right)_{R=0}$
Atomic Polar Tensor (APT) Velocity Form $P_{v,\alpha\beta}^A = \left(\dfrac{\partial\mu_{v,\beta}}{\partial\dot{R}_{A\alpha}}\right)_0$ $= N_{\alpha\beta}^A + E_{v,\alpha\beta}^A$	$N_{\alpha\beta}^A = Z_A e\,\delta_{\alpha\beta}$	$E_{v,\alpha\beta}^A =$ $\left(\dfrac{\partial}{\partial\dot{R}_{A\alpha}}\left\langle \tilde{\psi}_g \left| \hat{\mu}_{v,\beta}^E \right| \tilde{\psi}_g \right\rangle\right)_{R,\dot{R}=0}$
Atomic Axial Tensor (AAT) $M_{\alpha\beta}^A = \left(\dfrac{\partial m_{\beta}}{\partial\dot{R}_{A\alpha}}\right)_0$ $= J_{\alpha\beta}^A + I_{\alpha\beta}^A$	$J_{\alpha\beta}^A = \dfrac{Z_A e}{2c}\varepsilon_{\alpha\beta\gamma}R_{A\gamma}^0$	$I_{\alpha\beta}^A =$ $\left(\dfrac{\partial}{\partial\dot{R}_{A\alpha}}\left\langle \tilde{\psi}_g \left| \hat{m}_{\beta}^E \right| \tilde{\psi}_g \right\rangle\right)_{R,\dot{R}=0}$

derivative with respect to nuclear position or velocity. Our definition of the AAT differs slightly from that of Stephens (2,7).

The molecular dipole moments can be viewed as arising from perturbations of the molecular Hamiltonian by an electric field (position-form electric dipole moment), vector potential (velocity-form electric dipole moment), or magnetic field (magnetic dipole moment) (1,8). The electronic contributions to the APT and AAT are then derived from second-order perturbation theory as second derivatives of the electronic energy, as shown in Table 2. The position-form APT electronic contribution is calculated with standard coupled-perturbed-Hartree-Fock (CPHF) methods. The electronic contributions to the velocity-form APT and the AAT require derivatives with respect to nuclear velocity, and can be calculated either with the field perturbation a priori formalism of Stephens (MFP) (7,9) or by the formally equivalent vibronic coupling theory (VCT) sum-over-states a priori formalism of Nafie and Freedman (10,11). These expressions (Table 2) lie beyond the Born-Oppenheimer approximation, since they require correlation of electronic and nuclear velocities (12). In this case, the wavefunctions remain adiabatic and separable into a product of electronic and vibrational wavefunctions, but the electronic wavefunction depends on the nuclear velocity as well as the nuclear position, and includes imaginary terms (12).

From eqs 4-6 and Table 1, we see that the nuclear contribution to the electric dipole transition moment in either the position or velocity form involves nuclear charge-weighted nuclear displacement vectors (s-vectors), and that the nuclear contribution to the magnetic dipole transition moment involves the cross-products of nuclear position with the charge-weighted displacement vectors. These nuclear contributions are equivalently expressed with charge-weighted nuclear velocity vectors, and thus represent nuclear linear or angular current, respectively.

The vibrationally generated electron density current for normal mode a is described by vibrational transition current density (3), $J_{g0,g1}^{a}(r)$, defined as

$$J_{g0,g1}^{a}(r) = (2\hbar\omega_a)^{1/2} \sum_{e \neq g} \left[\frac{\left\langle \psi_e \left| \left(\frac{\partial \psi_g}{\partial Q_a} \right)_0 \right\rangle J_{ge}(r) \right.}{\omega_{eg}} \right] , \qquad (7)$$

where the transition current density, $J_{ge}(r)$, for the g→e pure electronic transition is $J_{ge}(r) = (\hbar/2m)[\psi_g \nabla \psi_e - \psi_e \nabla \psi_g]$ and $\omega_{eg} = (E_e - E_g)/\hbar$.

From eqs 5 and 6 and Table 2, we see that vibrational TCD is related to the integrand of the velocity-form electric dipole transition moment and vibrational angular transition current density is related to the integrand of the magnetic

Table 2. Perturbation Theory Expressions for Electronic Contributions to Atomic Polar Tensors and Atomic Axial Tensor

Perturbation expression	*A Priori Formalism Expressions*
$$E^A_{r,\alpha\beta} = -\left(\frac{\partial^2 E_{el}}{\partial \mathbf{E}_\beta \partial R_{A\alpha}}\right)_0$$ ($\mathbf{E}$ = electric field)	$$E^A_{r,\alpha\beta} =$$ $$-2e\left(\left\langle \psi_g \left\| \sum_j r_{j\beta} \left\| \frac{\partial \psi_g}{\partial R_{A\alpha}} \right\rangle\right)_{R=0}\right.$$ (r_j denotes position of electron j)
$$E^A_{v,\alpha\beta} = -c\left(\frac{\partial^2 E_{el}}{\partial \mathbf{A}_\beta \partial \dot{R}_{A\alpha}}\right)_0$$ ($\mathbf{A}$ = vector potential)	$$E^A_{v,\alpha\beta} =$$ $$\frac{-2e\hbar^2}{m}\sum_{e\neq g}\left\{\left\langle \psi_g \left\| \sum_j \frac{\partial \psi_e}{\partial r_{j\beta}} \right\rangle\left\langle \psi_e \left\| \frac{\partial \psi_g}{\partial R_{A\alpha}} \right\rangle E^{-1}_{eg}\right\}_{R=0}\right.$$ (sum-over-states; $E_{eg} = E_e - E_g$) $$= -2i\hbar c\left(\left\langle \frac{\partial \tilde{\psi}_g}{\partial R_{A\alpha}} \left\| \frac{\partial \tilde{\psi}_g}{\partial \mathbf{A}_\beta} \right\rangle\right)_{R,\mathbf{A}=0}\right.$$ (field perturbation)
$$I^A_{\alpha\beta} = -\left(\frac{\partial^2 E_{el}}{\partial \mathbf{H}_\beta \partial \dot{R}_{A\alpha}}\right)_0$$ ($\mathbf{H}$ = magnetic field)	$$I^A_{\alpha\beta} =$$ $$\frac{-e\hbar^2}{mc}\sum_{e\neq g}\varepsilon_{\beta\gamma\delta}\left\{\left\langle \psi_g \left\| \sum_j r_{j\gamma} \left\| \frac{\partial \psi_e}{\partial r_{j\delta}} \right\rangle\left\langle \psi_e \left\| \frac{\partial \psi_g}{\partial R_{A\alpha}} \right\rangle E^{-1}_{eg}\right\}_{R=0}\right.\right.$$ (sum-over-states) $$= -2i\hbar\left(\left\langle \frac{\partial \tilde{\psi}_g}{\partial R_{A\alpha}} \left\| \frac{\partial \tilde{\psi}_g}{\partial \mathbf{H}_\beta} \right\rangle\right)_{R,\mathbf{H}=0}\right.$$ (field perturbation)

dipole transition moment, expressed in the VCT sum-over-states formalism, as shown in eqs 10 and 11,

$$\left\langle \tilde{\Psi}_{g1}^a \left| \hat{\mu}_v \right| \tilde{\Psi}_{g0}^a \right\rangle^{electronic} = -ie \int J_{g0,g1}^a(r) dr \tag{10}$$

$$\left\langle \tilde{\Psi}_{g1}^a \left| \hat{m} \right| \tilde{\Psi}_{g0}^a \right\rangle^{electronic} = \frac{-ie}{2c} \int r \times J_{g0,g1}^a(r) dr \tag{11}$$

Vibrational TCD is a vector field showing electron density current at each point in space due to the vibrational transition. In contrast, the vector field arising from the integrand of the position-form electric dipole transition moment, eq 4 and Tables 1 and 2, involves vectors all radiating from the origin, and does not provide information on charge flow. The integrated values of the rotational strength derived from the field perturbation and sum-over-states formalisms show similar agreement with experiment, but the integrands in the field perturbation formulation do not lead to vector fields describing vibrationally-generated electronic current.

Methods

To implement the theoretical formalism, vibrational TCD has been calculated from eq 7 over a grid of points encompassing the molecule, utilizing code that links to the output of a Gaussian 98 (Gaussian, Inc., Pittsburgh, PA) frequency calculation. Geometry optimizations and force field calculations were carried out with Gaussian 98. VCD intensities from eq 2 were calculated with the MFP formalism utilizing Gaussian 98 (9) and with the VCT formalism utilizing code previously described (4-6), which links to Gaussian output. The TCD was plotted using networks developed with AVS 5 software (Advanced Visual Systems, Burlington, MA), with the TCD vectors summed over grid points perpendicular to the plane of the figure and projected into the plane, along with ball-and-stick representations of the molecule and charge-weighted nuclear displacement vectors at the atomic positions. For all modes shown here, the molecule has been rotated such that the total electric dipole transition moment lies along the positive z-axis, with a center of mass origin, and with grids 16 x 21 x 20 (oxirane), 38 x 39 x 21 (methyl lactate), 36 x 37 x 27 (alanine) and 27 x 29 x 31 (methyl chloropropionate). From these maps, patterns of linear and angular charge flow are readily identified, which can be compared with the calculated dipole and rotational strengths and transition moment components to understand the origin of the IR and VCD intensities.

Results

To illustrate the relationship between TCD maps and IR and VCD intensities, we present calculations for two modes of (2S,3S)-oxirane-d_2 (6) and the methine stretches of (S)-methyl-d_3 lactate-Cd_3 (5), L-alanine-Nd_3-Cd_3 and (S)-methyl-d_3 chloropropionate-Cd_3, shown schematically in Figure 1. The modes of (2S,3S)-oxirane-d_2 provide examples of hydrogen-deformation, ring stretching and ring deformation motion, and the methine stretches allow comparison of TCD patterns for modes with similar nuclear displacements, but widely differing VCD intensities.

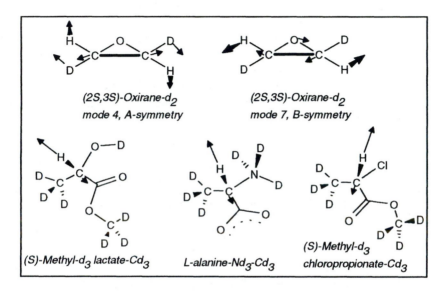

Figure 1. Structures of molecules studied, showing approximate mass-weighted nuclear displacements

(2S,3S)-Oxirane-d_2 is chiral by virtue of isotopic substitution. This C_2-symmetry molecule has modes of A and B symmetry. We have carried out TCD calculations for two basis sets, 6-311(2+,2+)G** (13), an augmentation of 6-311G with a diffuse s-shell on hydrogen and two diffuse sp-shells on oxygen and carbon (110 basis functions for oxirane) and a modified 6-31G(d) basis set (53 basis functions). The geometry optimization, force field and APTs were calculated at the DFT/B3LYP level. Both the magnetic field perturbation and vibronic coupling theory methods were used to calculate the AATs. Overall agreement with experiment was similar for the two formalisms, with slightly better agreement for the larger basis set. We present in Table 3 the results for vibrations 4 and 7 for the VCT formalism calculation with the 6-311(2+,2+)G** basis set (6) and the gas-phase experiment (14).

Table 3. Observed and Calculated Intensities and Calculated Dipole Transition Moment z-Component Contributions for Vibrations 4 and 7 of (2S,3S)-Oxirane-d₂.

Property	Vibration 4	Vibration 7
Frequency, obs (cm^{-1})	885	1102
Frequency, calc (cm^{-1})	891	1118
D, calculated	144.9	7.7
D, observed		8.6
R, calculated	+4.5	+5.6
R, observed	~5	+11.1
Total μ_z^T	0.182	0.047
Electronic μ_z^E	−0.134	−0.105
Nuclear μ_z^N	0.316	0.153
Total m_z^T	0.038	0.183
Electronic m_z^E	0.172	−1.548
Nuclear m_z^N	−0.133	1.731

NOTES: D, dipole strength (10^{-40} esu^2 cm^2); R, rotational strength (10^{-44} esu^2 cm^2); E, electronic, N, nuclear, T, total; μ, m, electric and magnetic dipole transition moments, respectively; dipole moment components in atomic units. Data from refs. 6 and 14.

Vibration 4 is an A-symmetry mode that is symmetric to rotation about the C_2-axis. This mode consists of CCH and CHD deformation, C-C elongation and C-O contraction. The charge-weighted nuclear displacements and TCD map in Figure 2a illustrate the z-allowed electric-dipole selection rule and the relationship between nuclear and electronic motion. Since electron current is plotted, opposite in direction to electron displacement, we see that the electronic motion roughly follows the nuclear displacement in this plane. The origin of the magnetic dipole character of the vibration is seen in Figure 2b. The nuclear contribution is dominated by the clockwise (m_z negative) circulation of nuclear charge at the carbon atoms. The electronic contribution is a sum of clockwise circulation for electronic motion that follows the deuterium displacement and counterclockwise local electron circulation about the oxygen atom. The net electric and magnetic dipole transition moments are parallel, yielding positive rotational strength (Table 3). The origin of the

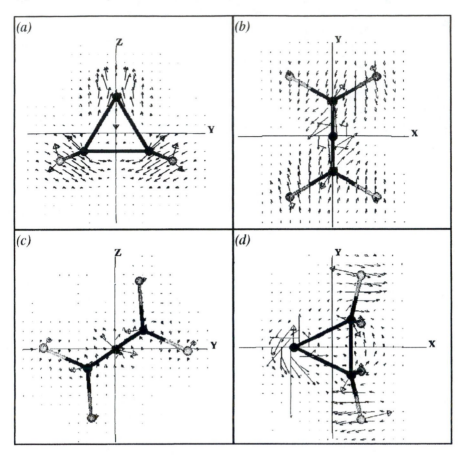

Figure 2. Charge-weighted nuclear displacements and TCD maps for (2S,3S)-oxirane-d₂. (a) mode 4, summed over x-direction; (b) mode 4, summed over z-direction; (c) mode 7, summed over x-direction; (d) mode 7, summed over z-direction. Electric dipole transition moment is along +z-axis for each mode.

much smaller IR intensity for the B-symmetry mode 7 (primarily hydrogen motion out of the CHD plane) is clear from the smaller electronic and nuclear contributions seen in Figure 2c. The large electronic and nuclear charge circulations apparent in Figure 2d give rise to the large z-component electronic and nuclear contributions to the magnetic dipole transition moment (Table 3). For this mode, electronic motion that follows the hydrogen displacement is accompanied by local electronic charge circulation about the oxygen atom.

The methine-stretching VCD in amino acids and α-hydroxy acids serves as a marker for absolute configuration (15,16). The calculated anisotropy ratios

$(g = \Delta A/A = \Delta\varepsilon/\varepsilon = 4R/D)$ for the methine stretch in both L-alanine-Nd_3-Cd_3 and (S)-methyl-d_3 lactate-Cd_3 (~1 x 10^{-4}) (16) compare favorably with the large observed values (~2 x 10^{-4}) (16,17). Both the observed and calculated dipole and rotational strengths for the methine stretching modes are smaller for L-alanine compared to (S)-methyl lactate. The large positive VCD intensity at 2882 cm^{-1} for (S)-methyl-d_3 lactate arising from the methine stretch dominates the VCD spectrum, whereas in (S)-methyl chloropropionate no such feature is observed near the expected frequency, ~2990 cm^{-1} (Figure 3).

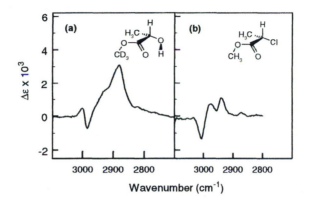

Figure 3. CH-stretching VCD spectra of (S)-methyl-d_3 lactate and (S)-methyl chloropropionate, 0.01 M in CCl$_4$ solution.

The origins of the differences in VCD intensity for the methine stretches for these three molecules are clear from analysis of the TCD maps in Figure 4 and the values of the transition moment components in Table 4. For the calculations on all three molecules, deuterium has been substituted for all hydrogens except the methine C*H, and as a result, the nuclear motion is restricted to the methine bond. However, as seen from the TCD plots in Figure 4, this localized nuclear motion results in electron density current throughout each molecule, including circulation about oxygen and carbon centers. Linear electron charge flow off the C*H axis (Figure 4b) results in electronic and nuclear contributions to the electric dipole transition moment that are not anti-parallel. As a result, there are components of the charge-weighted nuclear displacement vectors in the xy-plane, when the molecule is oriented with the +z-axis in the direction of the net electric dipole transition moment (Figure 4a,c,d). For all three molecules, these vectors have a clockwise sense of circulation about the center of mass, giving rise to a nuclear contribution to the magnetic dipole transition moment along +z, into the plane of Figure 4a, c and d. The

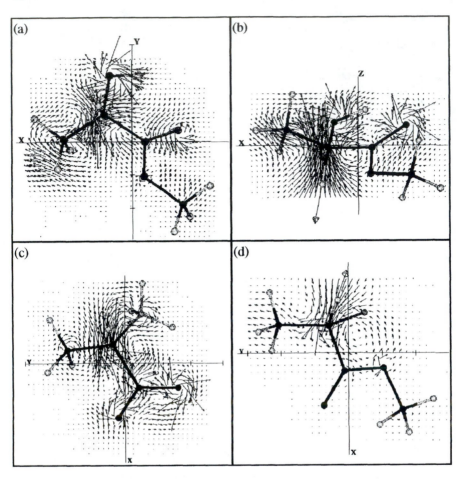

Figure 4. TCD maps for methine stretching modes. (a)(S)-methyl-d₃ lactate-Cd₃, summed over z-direction; (b) (S)-methyl-d₃ lactate-Cd₃, summed over y-direction; (c) L-alanine-Nd₃-Cd₃, summed over z-direction; (d) (S)-methyl-d₃ chloropropionate-Cd₃, summed over z-direction. HF/6-31G$^{(0.3)}$ calculations.

xy-projection of the charge-weighted hydrogen displacement is considerably larger for (S)-methyl-d_3 chloropropionate-Cd_3, and the m_z^N contribution is over seven times larger than for the other two molecules. For (S)-methyl-d_3 lactate-Cd_3, clockwise TCD circulation about the methyl group and hydroxyl oxygen is balanced by counterclockwise circulation about the carbonyl oxygen and between the methine and ester groups. The resulting net electronic contribution m_z^E is quite small. The total m_z^T is dominated by the nuclear contribution

Table 4. Calculated Intensities and Dipole Transition Moment z-Component Contributions for Methine Stretches

	(S)-Methyl-d_3 lactate-Cd$_3$	L-alanine-Nd$_3$-Cd$_3$	(S)-methyl-d_3 chloroproprionate-Cd$_3$
μ_z^N	−0.447	−0.394	−0.306
μ_z^E	+0.612	+0.488	+0.371
μ_z^T	+0.165	+0.094	+0.065
m_z^N	+0.088	+0.076	+0.581
m_z^E	−0.003	−0.024	−0.595
m_z^T	+0.085	+0.052	−0.015
D	33.7	10.7	4.98
R	+9.1	+3.2	−0.61

NOTE: Units and notation same as Table 3. Methyl lactate data from reference 5.

(Table 4). The TCD patterns for (S)-methyl-d_3 lactate-Cd$_3$ and L-alanine-Nd$_3$-Cd$_3$, are similar, but the sense of circulation about the carboxylate oxygens is clockwise (Figure 4c), in contrast to the counterclockwise circulation about the carbonyl oxygen (Figure 4a). The larger net m_z^E produces some cancellation between the nuclear and electronic contributions to the magnetic dipole transition moment for L-alanine-Nd$_3$-Cd$_3$. For (S)-methyl-d_3 chloropropionate-Cd$_3$, no charge circulation is apparent about the chlorine center (Figure 4d). The extended counterclockwise charge circulation between the methine bond and ether oxygen produces large negative m_z^E that now is larger than m_z^N. The calculated rotational strength is small and negative. For these three molecules, the differences in methine-stretching rotational strengths are related to the balance between electronic and nuclear contributions to the magnetic dipole transition moment. The TCD maps allow clear identification of the differences in electronic charge flow that affect the magnitude of the electronic contributions.

Conclusions

These examples have demonstrated how TCD maps and charge-weighted nuclear displacements can be used to understand vibrational selection rules and to identify the linear oscillation of charge that produces the electric dipole transition moment and the IR intensity, and the angular and circular oscillation of charge that produces the magnetic dipole transition moment in the direction of the electric dipole transition moment, leading to VCD intensity.

78

Acknowledgments

The authors acknowledge support of this work from the National Institutes of Health (GM-23567).

References

1. Nafie, L. A.; Freedman, T. B.   In *Circular Dichroism: Principles and Applications, Second Edition*; Berova, N., Nakanishi, K. and Woody, R. W., Ed.; John Wiley and Sons: New York, 2000, pp 97-131.
2. Stephens, P. J.; Devlin, F. J.; Ashvar, C. S.; Chabalowski, C. F.; Frisch, M. J. *Faraday Discuss.* **1994**, 103-119.
3. Nafie, L. A. *J. Phys. Chem. A* **1997**, *101*, 7826-7833.
4. Freedman, T. B.; Shih, M.-L.; Lee, E.; Nafie, L. A. *J. Am. Chem. Soc.* **1997**, *119*, 10620-10626.
5. Freedman, T. B.; Lee, E.; Nafie, L. A. *J. Phys. Chem.* . **2000**, *104*, 3944-3951.
6. Freedman, T. B.; Lee, E.; Nafie, L. A. *J. Mol. Struct.* **2000**, *(in press)*.
7. Stephens, P. J. *J. Phys. Chem.* **1985**, *89*, 748-752.
8. Freedman, T. B.; Nafie, L. A.   In *Modern Nonlinear Optics, Part 3*; Evans, M. and Kielich, S., Ed.; John Wiley & Sons: New York, 1994; Vol. 85, pp 207-263.
9. Cheeseman, J. R.; Frisch, M. J.; Devlin, F. J.; Stephens, P. J. *Chem. Phys. Lett.* **1996**, *252*, 211-220.
10. Nafie, L. A.; Freedman, T. B. *J. Chem. Phys.* **1983**, *78*, 7108-7116.
11. Yang, D.; Rauk, A.   In *Reviews in Computational Chemistry*; Lipkowitz, K. B. and Boyd, D. B., Ed.; VCH Publishers: New York, 1996; Vol. 7, pp 261-301.
12. Nafie, L. A. *J. Chem. Phys.* **1983**, *79*, 4950-4957.
13. Forseman, J. B.; Head-Gordon, M.; Pople, J. A. *J. Phys. Chem.* **1992**, *96*, 135-149.
14. Freedman, T. B.; Spencer, K. M.; Ragunathan, N.; Nafie, L. A.; Moore, J. A.; Schwab, J. M. *Can. J. Chem.* **1991**, *69*, 1619-1629.
15. Freedman, T. B.; Nafie, L. A.   In *Topics in Sterochemistry*; Eliel, E. L. and Wilen, S. H., Ed.; John Wiley & Sons: New York, 1987; Vol. 17, pp 113-206.
16. Gigante, D. M. P.; Long, F.; Bodack, L.; Evans, J. M.; Kallmerten, J.; Nafie, L. A.; Freedman, T. B. *J. Phys. Chem. A* **1999**, *103*, 1523-1537.
17. Zuk, W. M.; Freedman, T. B.; Nafie, L. A. *J. Phys. Chem.* **1989**, *93*, 1771-1779.

Chapter 6

The Use of Dual Polarization Modulation in Vibrational Circular Dichroism Spectroscopy

Laurence A. Nafie[1,2] and Rina K. Dukor[2]

[1]Department of Chemistry, Syracuse University, Syracuse, NY 13244–4100
[2]BioTools, Inc., 657 South Fairfield, Elmhurst, IL 60126

Typically, vibrational circular dichroism (VCD) spectrometers operate with one source of polarization modulation. One of the earliest VCD spectrometers employed two polarization modulators, one for the VCD signal and the other to scramble asynchronously the polarization of the infrared radiation to reduce baseline artifacts. In this paper, we report the use of two polarization modulators to eliminate most VCD baseline artifacts simultaneously across the entire spectrum through the real-time synchronous measurement of both polarization modulation signals. Examples are provided to demonstrate the effectiveness of one particular dual polarization modulation setup using a mid-infrared Fourier transform VCD spectrometer and no second polarizer.

Introduction

Vibrational circular dichroism (VCD) is one of two principal forms of vibrational optical activity (VOA), the other being vibrational Raman optical activity (ROA) (*1,2*). Both VCD and ROA are highly sensitive to stereochemical features of molecular structure and combine the stereo-sensitivity of optical activity with the structural richness of vibrational spectroscopy. Although both VCD and ROA were first measured in the early 1970s (*3-6*), their widespread application has been hindered by the instrumental challenge of measuring reliably spectra that are three to six orders of magnitude smaller than the parent infrared or Raman spectra.

Overall sensitivity, however, has not been the principal obstacle to obtaining accurate experimental VOA results. Rather it has been spectral artifacts arising from optical imperfections or misalignments in the instrument that have been the main problem. Artifacts result both in the displacement of the zero-VOA baseline from true zero and in the distortion of spectral bandshapes that prevents the discernment of the true VOA bands from overall observed features.

While much progress has been achieved over the years in artifact reduction in both VCD (*7-13*) and ROA (*14-18*), the remainder of this paper will focus on a new approach to baseline correction for VCD spectra. The most effective way to remove artifacts from a VCD spectrum is to subtract the VCD spectrum of one enantiomer from that of its opposite enantiomer and dividing the resulting spectrum by two. However, the opposite enantiomer is not always available, thus preventing this as a routine solution to the problem of artifacts. In addition, measurement of the opposite enantiomer can be problematic if instrumental drift is present in time resulting in a changing baseline between the two spectral measurements. Also, one needs to be careful in changing samples between enantiomers to ensure that precisely the same sampling conditions are achieved, i.e. pathlength, concentration and sample cell position. If the opposite enantiomer is not available, the racemic mixture can serve to define the baseline but without the advantage of providing any VCD information during the measurement.

Another, less accurate, approach to baseline correction is to use the empty beam, solvent spectrum or some other racemic mixture in the solvent as the VCD baseline and then subtract this baseline from the measured VCD. In the case of more severe baseline artifacts, the absorption characteristics of the sample may have an effect on the baseline, thus making the baseline for each sample unique and not accurately correctable.

In this paper, we describe the use of a method we call dual polarization modulation (DPM) to reduce significantly baseline deviations and artifacts in VCD spectroscopy. A more comprehensive account of the theory of DPM circular dichroism spectroscopy can be found elsewhere (*19*). The key element of the DPM approach is the simultaneous measurement of the CD signals of two photoelastic modulators (PEMs), one placed before the sample and one placed after the sample. By subtracting these two signals in real time, baseline corrected VCD spectra can be measured with greatly reduced need for a second measurement with a different sample at a later point in time.

Single Polarization Modulation VCD Measurement

The traditional optical setup for the measurement of VCD is linear polarizer, PEM, sample and detector. The signal from the detector can be written in terms of a DC term associated with the ordinary transmission spectrum of the sample, and an AC term representing the raw VCD transmission spectrum. In an FT-IR spectrometer, the DC term is obtained from the Fourier transform of the DC interferogram signal of the detector. The DC term consists of an empty beam transmission spectrum multiplied by a factor representing the decrease in the transmission intensity by the IR absorbance $A(\bar{\nu})$. The AC term depends linearly on four factors. These are the DC transmission term, 2 times the first-order Bessel function dependence of the PEM, the numerical factor (1.5153), and the VCD intensity, $\Delta A(\bar{\nu})$. The resulting detector signal is given by (20)

$$I_D(\bar{\nu}) = I_{DC}(\bar{\nu}) + I_{AC}(\bar{\nu}) = I_0(\bar{\nu})10^{-A(\bar{\nu})}\left[1 + 2J_1[\delta_M^0(\bar{\nu})](1.1513)\Delta A(\bar{\nu})\right] \qquad (1)$$

where the argument of the Bessel function, $\delta_M^0(\bar{\nu})$, is the angle of maximum retardation of the PEM at the wavenumber frequency $\bar{\nu}$. This is the ideal expression for the detector signal with a perfect VCD baseline. In practice, small levels of birefringence are present in the optical path in the form of stress in optical elements such as the PEM, cell windows, the detector window, and focusing lenses. The birefringence at the detector is cumulative through the entire optical path, and for small levels of birefringence, the location of different sources of birefringence relative to one another does not alter the final effect. Birefringence enters the expression for the detector signal as birefringence angle. There are two distinct contributions to the birefringence for an optical setup employing a PEM. One is the projection of the birefringence angle along the direction of the stress direction of the PEM, designated $\delta_B(\bar{\nu})$, and the other is the projection 45° with respect to the PEM stress direction, designated $\delta_B'(\bar{\nu})$. In order for birefringence to lead to an artifact interfering with the VCD spectrum, there needs to be linear polarization sensitivity at the detector. The sensitivity is typically on the order of 1% of the efficiency of a linear polarizer. Its magnitude and directional dependence transverse to the IR beam are given by $\varepsilon(\theta)$, where θ is an angle relative to the direction of the linear polarizer just prior to the PEM. If linear birefringence and linear polarization sensitivity of the detector are included in the traditional VCD optical set up, the intensity at the detector is now given by the following equation.

$$I_D(\overline{\nu}) = I_{DC}(\overline{\nu}) + I_{AC}(\overline{\nu}) = I_0(\overline{\nu})10^{-A(\overline{\nu})}\Big[1 + \varepsilon(\theta)J_0[\delta^0_{M1}(\overline{\nu})]\cos 2\theta$$

$$+2J_1[\delta^0_{M1}(\overline{\nu})]\big\{(1.1513)\Delta A(\overline{\nu}) - \varepsilon(\theta)\big[\delta_B(\overline{\nu})\cos 2\theta - \delta'_B(\overline{\nu})\sin 2\theta\big]\big\}\Big] \qquad (2)$$

There is now a new small contribution to the DC term arising just from the linear polarization sensitivity of the detector. For small birefringence typical of a VCD optical setup, this is a negligible contribution compared to the magnitude of the ordinary sample transmission spectrum. For the AC term, there is the same VCD dependent term as before, but in addition, there are two birefringent terms. The first is sensitive to linear polarization sensitivity of the detector parallel to the linear polarizer placed just before the PEM in the optical train and a second term due to linear polarization sensitivity at the detector at 45° relative to the same polarizer. In general, these two additional terms are responsible for deviations of the VCD baseline from zero as well as the frequency dependence of this baseline deviation across the spectrum.

Dual Polarization Modulation VCD Measurement

To the setup above, we now add a second PEM after the sample. The optical configuration is now linear polarizer, PEM1, sample, PEM2, detector, as shown in Fig.1 below.

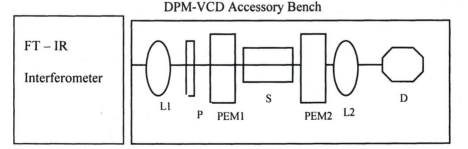

Fig. 1 Optical layout for the DPM-VCD measurements showing the FT-IR bench and the VCD accessory bench including the first lens (L$_1$), a vertical polarizer (P), the first PEM (PEM1), the sample (S), the second PEM (PEM2), the focusing lens (L$_2$), and the detector (D).

We assume the same expressions for the presence of linear birefringence and linear polarization sensitivity of the detector. In contrast to a more general description of DPM VCD (*19*), we have not included the option of placing a second polarizer after the second PEM. For the setup shown in Fig. 1, the detector signal now consists of three terms, one DC term and two AC terms, with one AC term for each PEM. The intensity at the detector is given by the following equation.

$$I_D(\bar{\nu}) = I_{DC}(\bar{\nu}) + I_{AC1}(\bar{\nu}) + I_{AC2}(\bar{\nu})$$

$$= I_0(\bar{\nu})10^{-A(\bar{\nu})}[1 + \varepsilon(\theta)J_0[\delta^0_{M1}(\bar{\nu})]J_0[\delta^0_{M2}(\bar{\nu})]\cos 2\theta$$

$$+2J_1[\delta^0_{M1}(\bar{\nu})]\left\{(1.1513)\Delta A(\bar{\nu}) - \varepsilon(\theta)\left[\delta_B(\bar{\nu})J_0[\delta^0_{M2}(\bar{\nu})]\cos 2\theta + \delta'_B(\bar{\nu})\sin 2\theta\right]\right\}$$

$$-2J_1[\delta^0_{M2}(\bar{\nu})]\varepsilon(\theta)\{(1.1513)\Delta A(\bar{\nu}) + \delta_B(\bar{\nu})J_0[\delta^0_{M1}(\bar{\nu})]\}\cos 2\theta] \qquad (3)$$

For this equation, the DC term now has a linear polarization correction term that contains the product of two zeroth-order Bessel functions, one for each PEM. These Bessel functions are always less than one in magnitude except when the PEM is turned off, in which case the value is one exactly. Hence, the DC correction term is even less important with two PEMs than it was with one. Each of the two AC terms depends on the first-order Bessel function of the PEM with which it is associated. The VCD term of PEM1 is the same as the corresponding VCD term for only one PEM. The VCD term for PEM2 is diminished by the detector polarization factor $\varepsilon(\theta)$ and hence is typically two orders of magnitude smaller than the VCD term from PEM1 and therefore is not of much significance. The next terms in each of the two AC expressions are equivalent and numerically equal if the two PEMs are set to the same retardation value. The third term in the expression for the AC term from PEM1 has no counterpart in the terms for PEM2. However, this additional term is zero if the polarization sensitivity of the detector is either parallel or perpendicular to the polarization direction of the first linear polarizer in the optical train. If the two AC terms are subtracted under the condition that the two PEMs have the same values of the zeroth- and first-order Bessel function, the two equivalent linear birefringence terms from the two PEMs cancel giving the expression for the two AC terms in the expression above, as the following equation.

$$I_{AC}(\overline{\nu}) = I_{AC1}(\overline{\nu}) + I_{AC2}(\overline{\nu}) = I_0(\overline{\nu})10^{-A(\overline{\nu})}$$

$$\times 2J_1[\delta_M^0(\overline{\nu})]\{(1.1513)[1 + \varepsilon(\theta)\cos 2\theta]\Delta A(\overline{\nu}) - \varepsilon(\theta)\delta_B'(\overline{\nu})\sin 2\theta\} \qquad (4)$$

This expression for the AC terms from both PEMs is primarily a pure VCD contribution. There is a single birefringence term remaining of the three that were originally present. It is thought not to be large because it requires polarization sensitivity of the detector at 45° relative to the linear polarizer before the PEMs. It is worth noting that of the two birefringence terms that cancel, their magnitudes can be greatly reduced prior to subtraction by setting the first-order Bessel functions somewhat past their maximum values where the zero-order Bessel functions go through a value of zero on their way to small negative values. This increases the efficiency of the cancellation of the two near-identical terms (within the accuracy of the setting each of the two PEMs to the same effective values). The cancellation of these two terms is effective across the entire spectrum simultaneously. Since the two PEM AC-detector signals can be subtracted from one another in real time, this baseline correction technique is instantaneous and can be effected with only one measurement of one sample. Electronically, it is equivalent to performing two separate measurements with two separate samples (enantiomer and racemic mixture, when available) at two separate periods of time.

Examples of DPM-VCD spectra

In Fig. 2 we present two VCD spectra of (-)-α-pinene as a neat liquid using a cell with a pathlength of 75 microns in region between 1350 and 800 cm^{-1}. The spectra were collected using a modified version of the Chiral*ir* from Bomem-BioTools in Quebec, Canada and Elmhurst, Illinois, respectively. The VCD sample bench was configured as shown in Fig. 1 with two ZnSe lenses, a ZeSe wire grid polarizer, two ZnSe PEMs from Hinds Instruments in Hillsboro, Oregon and a liquid-nitrogen-cooled MCT detector with a 2 mm square detector element. The spectral resolution was 4 cm^{-1} and otherwise the experimental conditions were identical for the two spectra except for the upper VCD spectrum the second PEM was turned off, while for the lower VCD spectrum, the retardation was matched to the first PEM such that the linear birefringence terms from each PEM cancel. The upper spectrum represents a VCD spectrum on a baseline that is distorted from true zero by linear birefringence in the instrument. The sharp decrease in the baseline starting near 850 cm^{-1} is due to response cut-off of the MCT detector and has caused an absorption-like artifact in the VCD spectrum. Use of the second PEM in the DPM not only brings the main part of the baseline to zero but also corrects the absorption artifact tail at the low-frequency end of the spectrum.

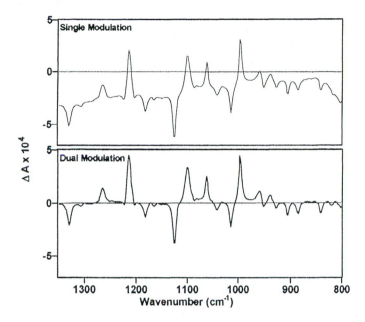

Fig. 2 The VCD spectrum of (–)-α-pinene as a neat liquid using the DPM-VCD setup under the conditions of single modulation where the second PEM was turned off and dual modulation where both PEMs are operating and the signals of each PEM are subtracted to remove baselines offsets and artifacts.

In Fig. 3, we present a second example. In this case we show the VCD spectrum of R-(–)-ibuprofen dissolved in $CDCl_3$ at a pathlength of 75 microns between 1700 and 950 cm^{-1} collected in three different ways. The VCD spectrum of ibuprofen in this region is approximately four times smaller than the VCD spectrum of (–)-α-pinene in this same region, and it therefore represents a more stringent challenge for baseline correction. The uppermost spectrum is the result of subtracting the spectrum of the S-(+)-ibuprofen from that of the R-(–)-ibuprofen and dividing by two. This eliminates all common artifacts between the VCD measurements of the two enantiomers. The lowermost VCD spectrum, labeled single modulation is for only the R-enantiomer of ibuprofen and was collected using the same DPM setup as described for Fig. 2 and shown in Fig.1, but with the second PEM turned off. The baseline for this VCD measurement is displaced significantly below zero with a positive slope toward lower frequencies. The higher frequency end of the spectrum contains absorption artifacts due to water vapor in the atmosphere since the VCD optical bench was open to the laboratory and not purged with nitrogen. Throughout this VCD

86

spectrum there is evidence of distortion of the true (enantiomerically subtracted) VCD spectrum. In particular there is a negative VCD absorption artifact band 1000 cm^{-1}. The middle VCD spectrum labeled dual modulation is the result of a DPM-VCD measurement of only R-(−)-ibuprofen with both PEMs on and optimized for the cancellation of the VCD artifacts. Due to the wide spectral range shown, it was not possible, at this level of sensitivity, to bring the baseline completely to zero. Nevertheless, nearly all the baseline offset of the single modulation VCD has been corrected. In addition, all spurious VCD features due to absorption, including the water vapor artifact features, have been removed from the dual modulation VCD spectrum. A slight two-point linear slope correction for the high-frequency end of the spectrum would bring the dual modulation VCD spectrum into complete agreement of the enantiomer subtracted VCD spectrum to within the noise limit of the spectrum.

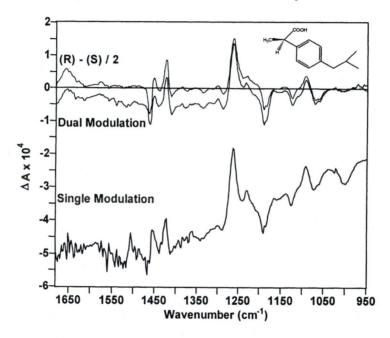

Fig. 3. The VCD spectrum of R-(−)-ibuprofen determined by three different methods. The uppermost spectrum is half the difference of the VCD spectrum of the R-enantiomer minus that of the S-enantiomer. The lowermost spectrum, single modulation, was obtained with the DPM-VCD setup and second PEM turned off showing baseline offset and artifacts. The third spectrum, dual modulation, was obtained with the DPM-VCD setup with both PEMs operating and all artifacts and nearly all baseline offsets removed.

Conclusions

In this paper, we have presented theory of DPM-VCD for the case of no second polarizer. This is the most practical embodiment of the DPM-VCD method to date, and the only case for which VCD spectra have been measured. We conclude that the use of a second PEM in the optical train after the sample can significantly reduce the magnitude and seriousness of artifacts encountered in VCD spectral measurements. For the case of no second polarizer featured here, the first PEM carries the VCD spectrum and also two artifact terms. The second PEM carries only this artifact plus a greatly reduced contribution to the VCD spectrum. By subtracting the VCD spectra of the two PEM Ac-signals, one of the two original artifact terms is eliminated. In experiments performed to date, the remaining artifact has been small and has not posed as a serious deterrent to obtaining VCD spectra of single enantiomers that rival, in accuracy and suppression of all VCD artifacts, the VCD spectra obtained by subtracting the uncorrected VCD of the two opposing enantiomers of the target chiral molecule.

Acknowledgements

The authors would like to thank Drs. Allan Rilling, Jean-Rene Roy and Henry Buijs of Bomem, Inc. for advice and access to instrumentation in performing early feasibility studies of the DPM-VCD method. We also wish to thank Dr. Baoliang (Bob) Wang of Hinds Instruments for arranging for the loan of a PEM to perform the measurements reported in this paper.

References

1. Nafie, L. A. *Ann. Rev. Phys. Chem.* **1997**, *48*, 357-386.
2. Nafie, L. A. Raman Optical Activity: Theory. In *Encyclopedia of Spectroscopy and Spectrometery*; Lindon, J., Tranter, G., Holmes, J., Eds.; Academic Press, Ltd: London, 2000; pp 1976-1985.
3. Holzwarth, G.; Hsu, E. C.; Mosher, H. S.; Faulkner, T. R.; Moscowitz, A. *J. Am. Chem. Soc.* **1974**, *96*, 251-252.
4. Nafie, L. A.; Cheng, J. C.; Stephens, P. J. *J. Am. Chem. Soc.* **1975**, *97*, 3842.

5. Barron, L. D.; Bogaard, M. P.; Buckingham, A. D. *J. Am. Chem. Soc.* **1973**, *95*, 603-605.

6. Hug, W.; Kint, S.; Bailey, G. F.; Scherer, J. R. *J. Am. Chem. Soc.* **1975**, *97*, 5589-5590.

7. Cheng, J. C.; Nafie, L. A.; Stephens, P. J. *J. Opt. Soc. Amer.* **1975**, *65*, 1031-1035.

8. Lipp, E. D.; Zimba, C. G.; Nafie, L. A.; Vidrine, D. W. *Appl. Spectrosc.* **1982**, *36*, 496-498.

9. Malon, P.; Keiderling, T. A. *Applied Spectroscopy* **1988**, *42*, 32-38.

10. Polavarapu, P. L. *Appl. Spectrosc.* **1989**, *43*, 1295.

11. Xie, P.; Diem, M. *Appl Spectrosc* **1996**, *50*, 675-680.

12. Malon, P.; Keiderling, T. A. *Appl Spectrosc* **1996**, *50*, 669-674.

13. Malon, P.; Keiderling, T. A. *Appl. Opt.* **1997**, *36*, 6141-6148.

14. Hug, W. *Applied Spec.* **1981**, *35(1)*, 115-124.

15. Hecht, L.; Jordanov, B.; Schrader, B. *Appl. Spectrosc.* **1987**, *41*, 295-307.

16. Che, D.; Nafie, L. A. *Appl. Spectrosc.* **1993**, *47*, 544-555.

17. Hecht, L.; Barron, L. D. *J. Raman Spectrosc.* **1994**, *25*, 443-451.

18. Hug, W. Raman Optical Activity: Spectrometers. In *Encylcopedia of Spectroscopy and Spectrometry*; Lindon, J., Tranter, G., Holmes, J., Eds.; Academic Press, Ltd.: London, 1999; pp 1966-1976.

19. Nafie, L. A. *Appl. Spectrosc.* **2000**, *(submitted for publication)*.

20. Nafie, L. A.; Keiderling, T. A.; Stephens, P. J. *J. Am. Chem. Soc.* **1976**, *98*, 2715-2723.

Chapter 7

Conformational Study of Gramicidin D in Organic Solvents in the Presence of Cations Using Vibrational Circular Dichroism

Chunxia Zhao and Prasad L. Polavarapu

Department of Chemistry, Vanderbilt University, Nashville, TN 37235

Vibrational circular dichroism (VCD) study of gramicidin D is carried out in the presence of monovalent and divalent cations in methanol-d_4, 1-propanol and $CHCl_3$-methanol solutions. The relation between VCD spectra and different conformations of gramicidin is established. The influence of solvents and cations on gramicidin conformations is discussed.

Introduction

Gramicidin is a hydrophobic linear polypeptide that is produced by *Bacillus brevis*.[1] It is made up of 15 amino acids with alternating D- and L-configurations, and the amino acid sequence[2] of gramicidin is: HCO-L-Val-Gly-L-Ala-D-Leu-L-Ala-D-Val-L-Val-D-Val-L-Trp-D-Leu-L-XXX-D-Leu-L-Trp-D-Leu-L-Trp-NHCH$_2$CH$_2$OH. It is designated as gramicidin A, when XXX=Trp, as gramicidin B when XXX=Phe and as gramicidin C when XXX=Tyr. The mixture of gramicidin A, B and C in the ratio of 80:5:15 is designated as gramicidin D.[3] Gramicidin has attracted a lot of interest because of its complex conformational behavior, making it a good model for the study of polypeptide folding. When incorporated into phospholipid membranes gramicidin serves as an ion channel, specifically for the transfer of monovalent cations,[4-8] and this attribute makes it an ideal candidate for modeling of ion transport through biomembranes.

The structure of gramicidin is very flexible and sensitive to the surroundings.[9] However, due to its alternating D- and L- configurations, gramicidin has shown a unique conformational character compared to the all L-amino acid polypeptides, where the most commonly found secondary structures are α- helix and β-sheet. For gramicidin in organic solvents, a four-species family of double helical structure has been proposed[10] and well accepted. These species differ from each other in the handedness of their helices, the relative orientation of strands and the overlap stagger between the monomers. Species 1 and 2 were proposed to have left handed parallel double helical structures, ↑↑ππ 5.6(left), differing in the stagger between their chains, Species 3 to have left handed anti parallel double helical structure, ↓↑ππ 5.6(left), and Species 4 to have right handed parallel double helical structure, ↑↑ππ 5.6 (right), all with 5.6 residues per turn.[11] Results obtained from NMR,[11] electronic circular dichroism (ECD)[10] and vibrational circular dichroism (VCD)[12] show that all four species are present in alcoholic solutions with different proportions.[13]

The conformations of ion-free gramicidin, and their proportions, in solutions are known to change with solvents. When cations are present in methanol solution the structure of gramicidin has been reported to shift from the equilibrium mixture to a single double-helix component. For example, both NMR and ECD results support the conclusion that in the presence of cesium cations the predominant structure of gramicidin is a right handed anti parallel double helix with 7.2 residues per turn,[14,15] and in the presence of Ca^{2+} ions gramicidin adopts a left handed parallel double helix with 5.6 residues per turn[16,17]. Then these ion dependent studies provide an approach to obtain the characteristic VCD spectra for individual double helical conformations.

The ion dependent experiments in the literature were carried out only in methanol solution. Besides, there is a discrepancy between NMR and ECD results upon addition of Li$^+$ ions. ECD spectra[15] suggest that Li$^+$ ions have a similar influence as Cs$^+$ ions do on gramicidin, whereas NMR results[18] showed that the presence of Li$^+$ ions destroyed the double helical structure and converted it into a random coil structure.

To understand the conformational behavior of gramicidin it is necessary to investigate in greater detail the following three factors: (a). change in structure as a function of solvent, (b). ion dependent structural changes in a given solvent, and (c). ion dependent structural changes as a function of solvent. In this paper, VCD is used to investigate the gramicidin structures. As has been demonstrated,[19-26] VCD has distinct advantage over ECD in studying polypeptides such as gramicidin in that the electronic transitions from tryptophan residues obscure the peptide transitions, whereas such interference is not present for the amide group vibrations. Numerous VCD studies have been carried out on peptides and proteins to establish the relation between VCD signs and secondary structures.[22] However, almost all of these studies dealt with most commonly found secondary structures such as α-helix and β-sheet, and not with the dimeric structures made up of single strands. The well-known double-helical structures of deoxyribonucleic acids (DNA) and ribonucleic acids (RNA) are very different from polypeptide double-helical structures and show VCD features[27-30] that are different from those for gramicidin (vide infra), so they can not be used as references for interpreting the VCD spectra of gramicidin.

We measured the VCD spectra of ion-free gramicidin in organic solvents before.[12] Also a preliminary study on ion dependent studies in methanol-d$_4$ solution was reported.[31] Here we report a detailed investigation and analysis of ion dependent structural changes of gramicidin in methanol-d$_4$, 1-propanol and CHCl$_3$-methanol (v/v 4:1) mixture. By comparison with the results obtained from model calculations, the relation between VCD signals and a particular conformation of gramicidin has been established. Gramicidin in 1-propanol and CHCl$_3$ solutions showed similar absorption and VCD spectral features that are different from those obtained[12] in methanol-d$_4$. Since cations cannot dissolve in CHCl$_3$, methanol was added to it to increase the solubility of cations. The salts used were CaCl$_2$, BaCl$_2$, MgCl$_2$, CsCl, and LiCl, depending on their solubility in different solvents. Finally, spectral simulations were carried out using coupled oscillator theory.[32] This theory has been successful[29,32] in qualitatively predicting the sign pattern of the dominant VCD spectral features for duplex forms of DNA and RNA and is expected to be helpful in the interpretation of the double helix conformations of gramicidin.

Experimental

Gramicidin D was purchased from ICN Biochemicals Inc. Methanol-d$_4$ was purchased from Cambridge Isotope Labs. Chloroform and 1-propanol were obtained from Fisher Scientific company. CsCl was obtained from Aldrich

Chemical Company. $CaCl_2$, $BaCl_2$, $MgCl_2$ and LiCl were obtained from Sigma Chemical Company. FTIR and VCD spectra were recorded using a commercial Chiralir spectrometer (Bomem-Biotools, Canada). All gramicidin concentrations were 4mg/ml and the temperature was 20 ^{0}C. All spectra were collected at a resolution of 8 cm^{-1}, and the contribution of solvents to the spectra has been subtracted. The data collection time is one hour. All samples were held in a variable path length cell with BaF_2 windows.

Results and Discussion

Methanol-d$_4$:

The vibrational absorption and VCD spectra of gramicidin in methanol-d$_4$ in the presence of $CaCl_2$ are shown in Fig. 1a. All measurements were made under the same experimental conditions. The contribution of solvent to the spectra was subtracted for all the spectra. In the ion-free methanol-d$_4$ solution[12], an amide I absorption band was observed at 1635 cm^{-1} with a strong shoulder at 1640-1660 cm^{-1}, and an amide II absorption band at 1454 cm^{-1}. The VCD spectrum showed a positive couplet in the amide I region with the positive VCD component at 1628 cm^{-1} and the negative component at 1651 cm^{-1}, and a weak negative couplet in the amide II region with the positive component at 1452 cm^{-1} and the negative component at 1431 cm^{-1}, respectively. Significant changes are observed in the amide I region in both absorption and VCD spectra when $CaCl_2$ is added to the solution. With the increase of $CaCl_2$ concentration, the intensity of the absorption band at 1628 cm^{-1} increases and the shoulder band at 1640-1660 cm^{-1} decreases and a band at 1655 cm^{-1} is resolved. A new positive VCD band at 1666 cm^{-1} and a weak negative VCD band at 1685 cm^{-1} become significant with increase in the $CaCl_2$ concentrations. The intensities of VCD peaks at 1685, 1666, 1651, 1628, 1454, 1431 cm^{-1} increase. Not only do the absorption and VCD band intensities change, the absolute values of the dissymmetry factors also increase with increasing $CaCl_2$ concentrations. The maximum change in dissymmetry factor occurs when ~5 mM or higher concentration $CaCl_2$ is added to the solution (Table 1). The presence of $BaCl_2$ in the solution gives similar absorption and VCD spectra (not shown) to those in the presence of $CaCl_2$ but the magnitudes of changes are smaller. The changes seen in the presence of $MgCl_2$ (not shown) are very small. Thus the influence of divalent cations on the structure of gramicidin decreases in the order $Ca^{2+} > Ba^{2+} > Mg^{2+}$.

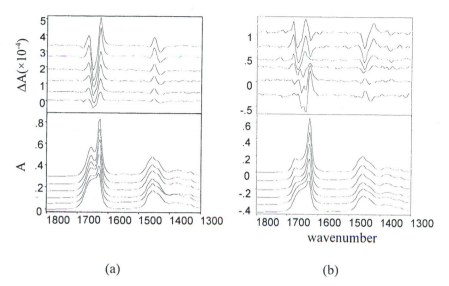

(a)   (b)

Figure 1*. Infrared absorption (bottom six traces) and VCD (top six traces) spectra of gramicidin D in methanol-d₄ solution in the presence of increasing concentration (from bottom to top) of (a). CaCl₂; (b). CsCl. The spectra are shifted upwards from each other for clarity.*

When CsCl is added to the solution (Fig. 1b, Table 2), the intensity of absorption band at 1628 cm^{-1} increases and the shoulder absorption band at 1640-1660 cm^{-1} decreases with increase in the concentration of CsCl. The amide I absorption bands get resolved into two peaks at 1628 cm^{-1} and 1674 cm^{-1} as the concentration of CsCl is increased, and significant changes occur in the associated VCD. No significant changes in the amide II absorption bands are apparent but the associated VCD changes significantly. When the concentration of CsCl is higher than 40 mM, the original positive VCD couplet at 1628 and 1651 cm^{-1} converts to a negative couplet and a new negative couplet is observed with the positive component at 1682 cm^{-1} and the negative component at 1666 cm^{-1}. Furthermore, the negative VCD couplet in the amide II region converts to a positive VCD couplet with enhanced intensities.

Computed absorption and VCD spectra are shown in Fig. 2a. Calculations were carried out for a left-handed anti parallel double helix with 5.6 residues per turn, for which the coordinates were obtained from the crystal structure data in Protein Data Bank (PDB). The calculated absorption spectra show a large

Table 1. Dissymmetry factors ($\Delta A/A$ x 10^4) for gramicidin D in the presence of $CaCl_2$ in methanol-d_4

Concentration (mM)	1685 cm^{-1}	1666 cm^{-1}	1651 cm^{-1}	1628 cm^{-1}	1454 cm^{-1}	1431 cm^{-1}
0	-----	-----	-1.60	1.68	0.70	-1.09
0.3	-----	1.12	-2.62	2.39	1.21	-1.29
0.6	-----	1.50	-3.30	2.50	1.71	-1.79
1.2	-0.34	2.75	-4.02	2.62	2.35	-2.00
3.0	-3.91	3.17	-5.70	3.04	2.61	-2.64
4.8	-4.47	3.07	-6.29	3.16	2.71	-3.25
12	-2.87	3.17	-5.50	3.00	2.84	-3.21

Table 2. Dissymmetry factors ($\Delta A/A$ x 10^4) for gramicidin D in the presence of CsCl in methanol-d_4

Concentration (mM)	1682 cm^{-1}	1666 cm^{-1}	1651 cm^{-1}	1643 cm^{-1}	1639 cm^{-1}	1628 cm^{-1}	1454 cm^{-1}	1431 cm^{-1}
0	-----	-----	-1.60	-----	-----	1.68	0.70	-1.09
1	-----	-----	-0.97	-----	-1.11	1.40	0.60	-0.59
2	-----	-----	-0.94	-----	-0.61	0.91	0.37	-----
4	0.82	-0.99	-1.06	-----	-----	0.20	-0.66	0.55
10	1.15	-1.43	-----	0.96	-----	0	-1.08	0.90
40	2.57	-2.52	-----	1.22	-----	-0.40	-1.85	1.76
56	1.77	-2.18	-----	1.16	-----	-0.31	-1.88	1.89

absorption band at 1632 cm^{-1}, a strong shoulder at ~1655 cm^{-1} and a weak absorption band at ~1700 cm^{-1}. These features match the experimental spectrum for anti parallel double helix structure of ion-free gramicidin[10]. VCD spectra show a negative (weak)-positive-negative-positive pattern, with the negative VCD bands at ~1720 cm^{-1} and ~1655 cm^{-1} and the positive VCD bands at ~1700 cm^{-1} and 1628 cm^{-1}, respectively.

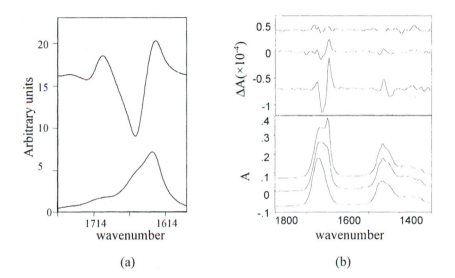

(a)   (b)

Figure 2. *(a). Calculated vibrational absorption (bottom trace) and VCD (top trace) spectra of gramicidin. (b). Infrared absorption (bottom three traces) and VCD (top three traces) spectra of gramicidin D in methanol-d₄ solution in the presnece of increasing (from top to bottom) of LiCl.*

As mentioned before, the VCD sign pattern relates directly to the handedness of the helical structure. Comparing the spectra obtained in the presence of CaCl$_2$ (Fig. 1a) and CsCl (Fig. 1b), we find that the final VCD spectra have the opposite signs for all the bands, which indicates that the predominant components in these two cases have different sense, one is right-handed and another is left-handed. In Fig. 1a the VCD sign pattern is the same as that in Fig. 2a, indicating that predominant conformation of gramicidin in the presence of excessive Ca^{2+} ions is left-handed double helix structure. And in contrast, in the presence of excessive Cs^{1} ions gramicidin adopts right-handed double helix structure predominantly. The absorption spectrum in Fig. 1a is similar to that in Fig. 2a. But the weak absorption band at ~1700 cm^{-1} observed in Fig. 2a, which is characteristic for anti parallel structure[10], is not seen in Fig.

1a. This indicates that the predominant structure in Fig. 1a is parallel structure. In contrast, the absorption spectrum in Fig. 1b has a band at 1670 cm^{-1} (corresponding to 1700 cm^{-1} band in Fig. 2a), indicating that gramicidin structure in this case is anti parallel structure.

Based on above observations, it can be concluded that in methanol-d$_4$ solution gramicidin structure shifts from equilibrium mixture to a left-handed parallel double helix in the presence of Ca^{2+} and to a right-handed anti parallel double helix in the presence of Cs$^+$ ions. This conclusion is consistent with the conclusion obtained from ECD [15,17] and NMR [14,16] studies. The investigation using ECD also showed that the divalent alkaline earth metal cations had similar effect on the conformational change of gramicidin, with the sizes of the cations affecting the magnitude of spectral change. The addition of excessive Ca^{2+} ions changed ECD spectra most, and Ba^{2+} ions changed less[17]. Smaller cations such as Mg^{2+} did not affect ECD spectrum significantly. These are consistent with our VCD observations.

The spectrum obtained in the presence of LiCl is shown is Fig. 2b. Compared to the ion-free gramicidin spectra[12], the amide I absorption band at 1628 and 1651 cm^{-1} disappear gradually with increasing concentration of LiCl and finally only one absorption band at 1663 cm^{-1} was observed. No significant changes are apparent in the amide II absorption bands. The VCD signals in both amide I and amide II regions disappear when concentration of LiCl reaches 101 mg/mL (2.38 mM) or higher. These observations suggest that the conformation of gramicidin in the presence of Li$^+$ ions changes to unordered structure, possibly a random coil.

1-Propanol:

Due to the low solubility of other cations in 1-propanol, only Ca^{2+} and Li$^+$ were used for the investigation. An experimental procedure similar to that in methanol-d$_4$ solutions was adopted but only the results obtained at higher concentrations of cations are presented here. The vibrational absorption and VCD spectra of gramicidin in ion-free 1-propanol and in the presence of CaCl$_2$ are shown in Fig. 3a. Since 1-propanol has large absorption in the amide II region which interferes with gramicidin absorption, only the amide I region is shown. The VCD spectrum in the presence of Ca^{2+} ions resembles that in Fig. 1a; a negative-positive-negative-positive VCD band shape is observed with the negative components at 1693 cm^{-1} and 1655 cm^{-1} and the positive components at 1670 cm^{-1} and 1636 cm^{-1}. This indicates that the predominant structure of gramicidin is left-handed conformation. Compared to the absorption spectrum of ion-free gramicidin in 1-propanol, a band at 1655 cm^{-1} becomes significant and well resolved in the presence of ions, as was also observed in the case of methanol-d4 solutions. However, the intensity ratio of the band at 1655 cm^{-1} to the band at 1636 cm^{-1} is higher than the corresponding ratio in the methanol-d4 solution. Besides, a strong shoulder band is present at 1675-1700 cm^{-1}, which is characteristic of anti parallel structure. According to Veatch et al[10], both the ion-free parallel and anti parallel conformations (with 5.6 residues per turn) have the absorption band at 1655 cm^{-1}. If there is only anti parallel structure in the presence of Ca^{2+} in 1-propanol solution, the absorption band should show a well-resolved band around 1680 cm^{-1}, as observed in Fig. 1b and in the literature[10]. In contrast, the band in this region is weak in Fig. 3a and appears as a shoulder. So it is more likely that both parallel and anti parallel structures exist in 1-propanol solution in the presence of Ca^{2+}.

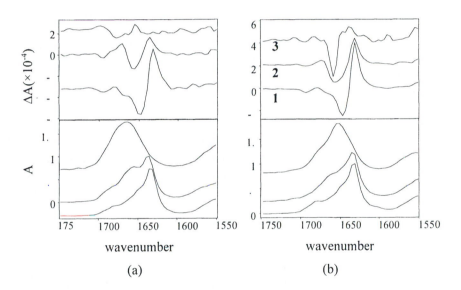

Figure 3. Vibrational absorption (bottom three traces) and VCD (top three traces) spectra of gramicidin D in (a) 1-propanol. Bottom to top: ion-free solution; in the presence of 40 mM Ca^{2+} ions; in the presence of 1.52 M Li^{+} ions. (b) CHCl$_3$/methanol solution (v/v 4:1). From bottom to top: ion-free CHCl$_3$ and trace amount of ethanol solution; in ion-free CHCl$_3$/methanol mixture solution; in mixture solution of CHCl$_3$ and methanol in the presence of 30 mM Ca^{2+} ions. The spectra are shifted upwards from each other for clarity.

Comparison of Fig. 1a and Fig. 3a leads to the conclusion that the presence of Ca^{2+} affects the sense of double helical structures of gramicidin and switches them to the left-handed double helix structure. The alignment of the two strands, however, depends on the solvents. In methanol-d_4 solution environment only the left-handed parallel double helix appears stable whereas in 1-propanol solution environment both parallel and anti parallel left-handed double helices appear stable. These strand alignments do not change upon the presence or absence of Ca^{2+} ions.

Fig. 3a also shows the vibrational absorption and VCD spectra of gramicidin in 1-propanol in the presence of 1.5 M Li^+ ions. A broad amide I absorption band is observed at 1663 cm^{-1} and no significant VCD band appears. These features are similar to those obtained in methanol-d_4.

Mixture of CHCl₃-methanol:

The solvent effect is extraordinary when a mixture of $CHCl_3$-methanol is used as solvent. The commercial $CHCl_3$ solvent contains a small amount of ethanol to protect $CHCl_3$ from photodecomposition. Due to the presence of ethanol, gramicidin shows similar conformational behavior in $CHCl_3$ as that observed in 1-propanol[12]. In the absence of ethanol, aggregation of gramicidin occurs in $CHCl_3$ as suggested by the unusual[12] VCD spectrum. Obviously the presence of ethanol in $CHCl_3$ is responsible for the formation of double helices (which may be due to the formation of inter-strand hydrogen bonds). However, it is not clear how $CHCl_3$ influences the gramicidin structure in the presence of a co-solvent. The conformational change in $CHCl_3$-methanol mixture upon the addition of cations is expected to shed light on this solvent effect. If $CHCl_3$ does not have any effect on the conformational behavior of gramicidin, gramicidin would be expected to have similar conformational behavior upon addition of cations in $CHCl_3$-methanol mixture as in alcoholic solutions.

To increase the solubility of cations in $CHCl_3$ a small amount of methanol (20%) was mixed with it. After the equilibrium is reached, the absorption and VCD spectra of gramicidin in $CHCl_3$-methanol mixture are found to be different from those in 1-propanol and $CHCl_3$. The absorption and VCD spectra of gramicidin in $CHCl_3$ containing trace amounts of ethanol, in $CHCl_3$-methanol mixture solvent (v/v 4:1), and in the presence of 30 mM Ca^{2+} at equilibrium are shown in Fig. 3b. Only the amide I region is shown here. In the absence of $CaCl_2$, the absorption spectrum in the solvent mixture shows a strong absorption at 1636 cm^{-1} and a shoulder band at 1650-1700 cm^{-1}. The VCD spectrum shows a positive-negative-positive shape with the positive components at 1674 cm^{-1} and 1632 cm^{-1} and negative component at 1659 cm^{-1}, respectively. A small negative VCD band is also observed at 1693 cm^{-1}. In the presence of $CaCl_2$ changes in both absorption and VCD spectra occur. A single large and broad absorption centered at 1651 cm^{-1} is observed. VCD spectrum shows a positive VCD couplet with a sharp intense negative component at 1659 cm^{-1} and a weak positive component at 1639 cm^{-1}. The VCD magnitudes in the presence of Ca^{2+} cations in CHCl$_3$-methanol mixture (trace 3 in Fig. 3b) are an order of magnitude smaller (VCD spectrum was amplified by a factor of 8 for the sake of clarity) than those without ions in CHCl$_3$ (trace 1, Fig. 3b), CHCl$_3$-methanol mixture (trace 2, Fig. 3b) or in methanol-d$_4$ in the presence of Ca^{2+} ions (Fig. 1a).

We believe that the predominant composition of gramicidin in this system is α-helix with a small amount of β-sheet structure. This conclusion is based on the following analysis. First, in Fig. 3b in the presence of Ca^{2+}, the characteristic absorption band of double helix structure at 1632 cm^{-1} totally disappears. And the new broad absorption band at 1651 cm^{-1} suggests this to be a monomer structure. This monomer must be ordered structure because a VCD couplet is observed in this region. Second, the broad absorption band centered at 1651 cm^{-1} was reported to be the characteristic absorption band of α-helix or β-sheet structures. Third, it has been reported that the protein with a high right-handed α-helix content gives rise to a positive couplet amide I VCD with the negative component more intense than the positive component.[34-36] This is consistent with Fig. 3b except that in Fig. 3b the negative VCD component is sharp and symmetric whereas the counter part in α-helix is somewhat broader and extends to the higher frequency region. The β-sheet structure was reported to give rise to a dominant low-energy negative VCD band in the amide I region and a small positive VCD band on the high frequency side of the absorption maximum. The negative VCD component of β-sheet extends to the lower frequency region. It has been shown that the high α + low β structure gives rise to a positive VCD couplet in the amide I region with an intense symmetric negative component[36]. And this is just what we observed in Fig. 3b.

Now come back to see the spectra in the absence of cations. The absorption spectrum of gramicidin in ion-free CHCl$_3$-methanol solution resembles the summation spectrum absorption spectrum of gramicidin in CHCl$_3$ and in CHCl$_3$/methanol in the presence of Ca^{2+}. Most important, the VCD spectrum in ion-free CHCl$_3$-methanol solution resembles well the summation VCD spectra of the other two cases. The broad negative VCD band at 1659 cm^{-1} can be attributed to the overlapping of the negative VCD bands at 1647 cm^{-1} in spectrum 1 and 1659 cm^{-1} in spectrum 3 in Fig. 3b. Above observations indicate that in ion-free CHCl$_3$-methanol solution the ordered monomer structures (α+β) and double helix structures coexist.

In this case, when both monomer and dimer structures exist in the solution, monomer becomes the stable one upon addition of Ca^{2+} ions and the equilibrium mixture shifts to the monomer form. This is due to the existence of CHCl$_3$ solvent, indicating that solvent also plays an important role in the conformational behavior of gramicidin in the presence of ions.

Conclusion

By carrying out VCD study of gramicidin D in methanol-d$_4$, 1-propanol and CHCl$_3$-methanol solutions, in the presence of Ca^{2+}, Ba^{2+}, Mg^{2+}, Cs$^+$ and Li$^+$ ions, the conformational behavior of gramicidin D in organic solvents was investigated. And the relation between VCD spectra and different conformations of gramicidin is established. The study also shows that both solvents and cations play important roles in the formation of secondary structure of gramicidin D. When Ca^{2+} ions are added to methanol-d$_4$, 1-propanol and CHCl$_3$-methanol solutions, respectively, the predominant conformations of gramicidin are parallel left-handed double helix, mixture of parallel and anti parallel left-handed double helices, and high α+ low β structures. In the presence of Cs$^+$ ions, gramicidin D in methanol-d$_4$ solution adopts a right-handed anti parallel double helix. And in the presence of Li$^+$, gramicidin D in methanol-d$_4$ and 1-propanol solutions adopts unordered monomer form.

Acknowledgements: Grants from NSF (CHE9707773) and Vanderbilt University are gratefully acknowledged. We thank Dr. Max Diem, Hunter College for providing us a copy of his coupled oscillator program.

References

1. Hotchkiss, R.D.; Dubos, R. J. *J. Biol. Chem.* **1940**, *132*, 791-792.
2. Sarges, R.; Witkop, B. *J. Am. Chem. Soc.* **1965**, *87*, 2011-2020.
3. Weinstein, S.; Wallace, B. A.; Morrow, J.; Veatch, W. R.; *J. Mol. Biol.* **1980**, *143*, 1-19.
4. Hladky, S. B.; Haydon, D. A. *Nature* (London) **1970**, *225*, 451-453.
5. Krasne, S.; Eisenman, G.; Szabo, G. *Science* **1971**, *174*, 412-425.
6. Hladky, S. B.; Haydon, D. A. *Biochim. Biophys. Acta.* **1972**, *274*, 294-312.
7. Myers, V. B.; Haydon, D. A. *Biochim. Biophys. Acta.* **1972**, *274*, 313-322.
8. Anderson, O. S. *Annu. Rev. Physiol.*, **1984**, *46*, 531-548.
9. Wallace, B. A. *Annu. Rev. Biophys. Chem.* **1990**, *19*, 127-157.
10. Veatch, W. R.; Fossel, E. T.; Blout, E. R. *Biochemistry* **1974**, *13*, 5249-5256.
11. Bystrov, V. F.; Arseniev, A. S. *Tetrahedron* **1988**, *44*, 925-940.
12. Zhao, C.; Polavarapu, P. L. *Biospectroscopy* **1999**, *5*, 276-283.
13. Veatch, W. R.; Fossel, E. T.; Blout, E. R. *Biochemistry.* **1974,** *13*, 5257-5264.
14. Arseniev, A. S.; Barsukov, I. L.; Bystrov, V. F. *FEBS. Lett.* **1985**, *180*, 33-39.

15. Chen, Y.; Wallace, B. A. *Biophysical Journal* **1996**, *71*, 163-170.
16. Chen, Y.; Tucker, A.; Wallace, B. A. *J. Mol. Biol.* **1996**, *265*, 757-769.
17. Doyle, D. A.; Wallace, B. A. *Biophysical J.* **1998**, *75*, 635-640.
18. See ref. 21 in Ref. 11.
19. Ashvar, C. S.; Devlin, F. J.; Stephens, P. J.; Bak, K. L.; Eggiman, T.; Weiser, H. *J. Phys. Chem.* **1998**, *102*, 6842-6857.
20. Barron, L. D. *Molecular Light Scattering and Optical Activity*; Cambridge University Press: London , 1982.
21. Diem, M. *Introduction to Modern Vibrational Spectroscopy*; John Wiley & Sons: New York 1993.
22. Keiderling, T. A. In *Circular dichroism and the conformational analysis of biomolecules*; Fasman, G. D., Ed.; Plenum press: New York, 1996; pp.555-597.
23. McCann, J.; Rauk, A.; Shustov, G. V.; Wieser, H.; Yang D. *Appl. Spectrosc.* **1996**, *50*, 630-641.
24. Nafie, L. A. *Ann. Rev. Phys. Chem.* **1997**, *48*, 357-386.
25. Polavarapu, P. L. *Vibrational Spectra: Principles and Applications with Emphasis on Optical Activity*; Elsevier Publications: New York, 1998.
26. Yasui, S. C.; Keiderling, T. A.; Bonora, G. M.; Toniolo, C. *Biopolymers.* **1986**, *25*, 79-89.
27. Birke, S. S.; Moses, M.; kagalovsky, B.; Jano, D.; Gulotta, M.; Diem, M. *Biophysical Journal* **1993**, *65*, 1262-1271.
28. Wang, L.; Keiderling, T. A. *Biochemistry* **1992**, *31*, 10265-10271.
29. Xiang, T.; Goss, D. J.; Diem, M. *Biophysical Journal* **1993**, *65*, 1255-1261.
30. Zhong, W.; Gulotta, M.; Goss, D. J.; Diem, M.; *Biochemistry* **1990**, *29*, 7485-7491.
31. Zhao, C.; Polavarapu, P. L. *J. Am. Chem. Soc.* **1999**, *121*, 11259-11260.
32. Gulotta, M.; Goss, D. J.; Diem, M. *Biopolymers* **1989**, *28*, 2047-2058.
33. Roux, B.; Bruschweiler, R.; Ernst, R. R. *Eur. J. Biochem.* **1990**, *194*, 57-60.
34. Sen, A. C.; Keiderling, T. A. *Biopolymers* **1984**, *23*, 1519-1532.
35. Lal, B. B.; Nafie, L. A. *Biopolymers* **1982**, *21*, 2161-2183.
36. Baumruk, V.; Keiderling, T. A. *J. Am. Chem. Soc.* **1993**, *115*, 6939-6942.
37. Wallace, B. A.; Ravikumar, K. *Science.* **1988**, *241*, 182-187.

Advances in Chiral Spectroscopy

Chapter 8

Optical Activity: From Structure–Function to Structure Prediction

David N. Beratan[1], Rama K. Kondru[2,3], and Peter Wipf[2]

[1]Department of Chemistry, Box 90346, Duke University,
Durham, NC 27708–0346
[2]Department of Chemistry, University of Pittsburgh, Pittsburgh, PA 15260
[3]Current address: UCB-Research, Inc., 840 Memorial Drive,
Cambridge, MA 02139

Optical rotation is easily measured and provides a comprehensive probe of molecular dissymmetry. Reliable calculations of optical rotation angles are now accessible for organic molecules. These calculations have allowed us to establish new computational approaches to assign the absolute stereochemistry of complex natural products. These methods also allow us to pinpoint chemical group contributions to optical rotation, resulting in assignment of how particular dissymmetric structural elements influence the sign and the magnitude of optical rotations. This chapter reviews our recent developments in computing optical rotation angles for natural products, in using this data to assign absolute stereochemistry, and in establishing structure-property relations for rotation angles.

I.   Introduction

A promising approach in the search for new antiviral, antibiotic, and anticancer drugs relies upon the collection, isolation, and structure determination of novel natural products [1,2].   One of the most challenging aspects of molecular structure determination is the assignment of relative and absolute stereochemistry [3].   Molecules containing N stereogenic centers present 2^N possible diastereomers.   In the case of complex structures, stereochemical assignment is tedious, involving a battery of optical spectroscopy, synthetic modification or degradation, NMR spectroscopy, X-ray crystallography, and ultimately total synthesis [3,4].   Structures containing a large number of stereocenters - in particular stereocenters in flexible molecules possessing novel bonding that lack simple chromophores - present a particular challenge in the absence of X-ray structural data. This paper reviews our recent progress in using optical rotation angle calculations to assign absolute stereochemistry and to determine structure-property relations for optical rotation angles.

Electronic and vibrational spectroscopy methods are used widely to assist in absolute stereochemical assignment [5]. Circular dichroism (CD) and exciton-coupled CD methods [4] have focused mainly upon rigid structures with stereocenters in close proximity to chromophores. Although well established, these methods are not entirely reliable in complex structures of common interest. Problems are known to arise in flexible molecules and in structures with degenerate electronic transitions. The spectroscopic methods are of little use in structures lacking well accessible chromophores. Optical rotation has the additional advantage of probing comprehensively all aspects of molecular dissymmetry, rather than being limited to dissymmetry surrounding a single chromophore. In the electronically nonresonant regime, qualitative connections between optical rotatory dispersion (ORD) spectra and stereochemistry have been established *via* empirical rules (octant [4], Kirkwood [6], Brewster [7], and Applequist [8] rules are well known). Established "rules," too, are reliable only within families of closely related rigid structures [3].

Until a few years ago, explicit quantum chemical calculations played a limited role in absolute stereochemistry assignment. While formal expressions for computing CD and ORD spectra have been known for decades [9,10,11,12], their actual computation is extremely challenging [13]. The calculation of reliable excited state wave functions (and from them individual rotational strengths) – essential for CD analysis - remains formidable for even modest sized structures. And the calculation of optical rotation, which can be expressed as arising from a linear combination of rotational strengths from "all" excited states, seems even more formidable.

Several years ago, we turned our combined theoretical and synthetic attention to the optical rotation problem. In spite of the greater role played by CD spectroscopy in assessing stereochemistry, we believed that access to quantitatively reliable specific rotation computations provided great practical and fundamental promise, in particular for flexible molecules lacking choromophores with well localized excited states. Practically speaking, optical rotation measurements are simple to make and are performed routinely (huge volumes of specific rotation data are available in the CRC Handbook, Aldrich catalogue, chemical online databases, etc.). Moreover, optical rotations (at frequencies removed from electronic resonances) are a molecular composite property, derived from all molecular dissymmetry. From a conceptual as well as theoretical perspective, we thought (and still believe) that reliable general structure-function relations governing both the sign and the magnitude of optical rotation angles remain to be discovered. Optical rotation is therefore important and interesting from fundamental and practical perspectives. We restrict the discussion in this chapter to recent advances in optical rotation angle computation. Descriptions of other techniques appear elsewhere in this book.

II. Optical Rotation

One of the most familiar manifestations of molecular dissymmetry is optical rotation: rotation of the polarization plane of linearly polarized light on passing through an enantiomerically enriched solution.

Optical rotation caused by dissymmetric structures is readily interpreted as arising from differences in the indices for left and right circularly polarized light, which comprise the linearly polarized light. That is, the left and right-circularly polarized components of the light develop a phase lag Δt as they propagate [14].

$$\Delta t = \left(n_R - n_L \right) \ell / c \tag{1}$$

This phase lag gives rise to an accompanying rotation angle $\Delta \theta$:

$$\Delta \theta = (n_R - n_L) 2\pi \ell / \lambda \tag{2}$$

Parts per million differences in the two indices give rise to readily detectable rotation angles on the order of radians.

The quantum mechanical description of optical activity begins with consideration of the matter-radiation interaction hamiltonian [6]. The usual series expansion of this interaction term gives rise to electric dipole, magnetic dipole, quadrupole, and higher-order interaction terms. Optical activity is intrinsically linked with the spatial variation of the electric field on the length scale of the molecule [12]. That is, a description of isotropic solution polarization - in the electric-dipole approximation - does not give rise to optical rotation. The next-order term in the radiation-matter interaction must be included, the magnetic-dipole term. Building a time-dependent perturbation theory of molecular polarization in the presence of the harmonically varying electric dipole and magnetic dipole interactions leads to the well-known formulation of linear chiro-optical phenomena [11,12]. The time varying polarization induced by the light is [12]

$$< \vec{P} > \propto \alpha \vec{E} + \beta \frac{\partial \vec{B}}{\partial t} \tag{3}$$

where

$$\alpha(\omega) \propto \sum_{i=x,y,z} \sum_{ex} \frac{< \Psi_g^{(0)} | \vec{r} | \Psi_{ex}^{(0)} >_i < \Psi_{ex}^{(0)} | \vec{r} | \Psi_g^{(0)} >_i}{\omega_{ex,g}^2 - \omega^2} \tag{4}$$

$$\beta(\omega) \propto Im \sum_{i=x,y,z} \sum_{ex} \frac{< \Psi_g^{(0)} | \vec{r} | \Psi_{ex}^{(0)} >_i < \Psi_{ex}^{(0)} | (\vec{r} \times \vec{p}) | \Psi_g^{(0)} >_i}{\omega_{ex,g}^2 - \omega^2} \tag{5}$$

The specific rotation angle is proportional to β.

$$[\alpha]_\omega \propto \omega^2 \beta(\omega) \tag{6}$$

There are numerous ways to interpret eq 5, in which the physical foundation of optical activity, and its structure dependence, are buried. One can view the electric-dipole matrix element as inducing linear polarization of charge in the molecule, while the angular momentum matrix element drives circulation of charge. Combined, the polarization rotation is nonzero if the light field induces helical motion of charge [11,15]. An alternative perturbative view of electronic response can be developed in the regime far from electronic resonances, $\omega << \omega_{ex,g}$ [16]. In this limit, eq 5 simplifies to the inner product of two first-order wave function corrections.

$$\beta(\omega) \propto \mathrm{Im} \sum_{i=x,y,z} \left\langle \frac{\partial \Psi}{\partial E_i} \middle| \frac{\partial \Psi}{\partial B_i} \right\rangle \qquad (7)$$

The computational challenge, in this regime at least, is to compute the *first-order* changes in the *ground state* induced by combined electric and magnetic dipole perturbations. Coupled Hartree-Fock (CHF) methods compute these changes analytically (rather than by performing multiple calculations at many different field strengths). The CHF calculation appears complicated compared to textbook first-order perturbation theory because the effective-potential that the electrons "feel" changes to first-order itself, which in turn leads directly to first-order corrections to the ground state [17]. Determining the first-order wave function corrections (the derivatives in eq 7) amounts to solving a set of linear equations following the initial Hartree-Fock (HF) analysis in zero field [17]. One strategy for developing structure-function relations in optical activity is to map the wave function derivatives of eq 7 for occupied molecular orbitals. This CHF strategy is a type of "analytical derivative" method, and is very closely related to Hartree-Fock finite-field strategies. In finite-field calculations, wave function perturbations are determined by repeatedly solving the Schrödinger equation (in the HF approximation) for varied applied-field strengths. The CHF calculation has the advantage of generating the response property from a single HF ground state calculation and some linear algebra.

The wave function derivative of eq 7 can be used to identify the contributions of specific molecular orbitals to optical rotation. For example, Figure 1 maps the E- and B-field derivatives of the π-symmetry HOMO of ethylene. Note that the overlap of these two wave function derivatives is exactly zero, as ethylene is achiral. Summing these overlaps over all occupied molecular orbitals produces zero in the case of achiral structures.

A frequency-dependent response-theory formulation of β, somewhat more complex than the CHF analysis can be carried out as well. This theory, often called coupled-perturbed HF (CPHF), leads to explicitly frequency-dependent response properties [18,19,20]. Such calculations account for specific electronic resonances. As such, one makes additional approximations in the analysis that amount to building increasingly complex descriptions of the ground and excited states. The most common approach is to base the description on a single configurational reference state with a double configuration interaction (CI) ground state and single CI excited states. This level of CPHF computation, known as the random-phase approximation (RPA), includes the influence of some ground state correlation in the calculation [21]. Simpler CPHF strategies, such as the Tamm-Dancoff approximation, have no ground state dynamical correlation built in. CPHF computations are formulated to satisfy the hypervirial theorem. A significant consequence is that the oscillator

Fig. 1. *Wave function derivative maps for ethylene. Top: the HOMO π orbital of ethylene. Middle: the electric-field derivative of the HOMO is the orthogonal LUMO π* molecular orbital. Bottom: the magnetic-field derivative of the HOMO is also an orthogonal π* orbital (LUMO+1). As such, the overlap of derivatives in eq 7 for the HOMO is exactly zero, and makes no contribution to optical rotation.*

strength and the rotational strength sum rules are valid [12,22]. These are important constraints, leading to molecular response properties that one expects to be more "balanced" than might result from unconstrained variational calculations in a finite basis with an approximate description of correlation. The sum-on-states expression for β suggests the importance of maintaining balance among a very large set of transition element properties in order to make reliable property predictions.

Alternative density functional approaches to computing response properties are becoming available for rotation angle computation. One method performs a Kramers-Kronig transformation of a simulated CD spectrum [23]. Linear-response DFT methods analogous to CHF and CPHF appear promising as well [24].

III. Computation of optical rotations: Small Molecules

Before implementing computational methods to assign the absolute stereochemistry of natural products, it is important to point out that the CHF and CPHF methods are satisfactory in describing the optical rotations of small molecules. For a summary of these calculations the reader is referred to refs 16 and 25-35. These rotations were computed using the standard CADPAC software [36] for off-resonance results and DALTON [37] for explicit frequency dependent analysis. The DALTON package has the added advantage of being able to employ gauge independent (London) atomic orbitals. We have found that the results for smaller molecules resulting from placing the gauge origin at the molecular center of mass produce satisfactory results in such systems [33]. However, this is probably not the case for larger molecules. We have also found that for small molecules (and for composite descriptions of larger structures) a 6-31G* basis set is adequate. Cheeseman and coworkers are performing important investigations of the limitations of modest basis set computation, as well as gauge origin dependence in small rigid structures [24].

IV. Computation of optical rotation: Natural Products

Our computational approach to determining the optical rotation of natural products containing multiple stereocenters is: divide the structure of interest into fragments, each with stereogenic centers several bonds removed from one another; use Monte Carlo sampling (based on molecular mechanics force fields) to build a family of thermally accessible structures; compute the rotation angle for this family of structures; Boltzmann weight the resulting angles to predict a composite-molecule rotation angle. We have demonstrated the viability of this strategy for the marine natural product hennoxazole A, making a stereochemical prediction consistent with experiment. In this case, theory had to choose one of eight possible absolute configurations. As described in ref. 3, we computed theoretical molar rotation angles for three molecular fragments (shown in Table 1). This allowed us to add together contributions to the rotation angle and to predict the optical rotation of all 8 possible enantiomers and diasteriomers. This strategy is a kind of "van't Hoff summation" [3] of contributions from the distinct fragments. Summations of this kind are known to succeed when the motion of the individual units is essentially independent. In structures where the assumption of unconstrained relative motion is not adequate, larger fragments are required in the calculations (helicenes present an extreme example where the strategy of additivity is problematic). Table 2 shows the predictions of fragment computations of molar rotations based on CPHF RPA gauge independent atomic orbital (DALTON) analysis of hennoxazole A based on the fragments below [33]. Analysis of the predicted rotation angle for the 8 composite hennoxazole A stereoisomers appears in ref. 33.

A similar analysis of pitiamide A – performed in advance of chemical synthesis - produced excellent agreement between computed and measured optical rotations for the synthesized diastereomers. Interestingly, the NMR analysis of the natural product and the synthetic diastereomers leads to an assignment of $(7R,10R)$ or $(7S,10S)$, and a stereochemistry that is inconsistent with the reported specific rotation of -10.3. It is possible that the natural product contained degradation products, a mixture of enantiomers or diasteromers, or chiral impurities. The fragmentation strategy is shown below [29].

Table 1. Calculated molar rotations for the three fragments (I-III) in the specific enantiomeric configurations (6-31G basis), averaged from four calculations. From examining the 8 possible linear combinations of these three numbers we were able to assign the hennoxazole stereochemistry correctly as (2R,4R,6R,8R,22S) [33]. Fragment I was known to be 2R,4R,6R or 2S,4S,6S from NMR experiments [33].

Fragment	$[M]_D \pm 2\sigma$ (Configuration, Sign)
I	162 ± 11 $(2R,4R,6R, \text{-ve}; \ 2S,4S,6S, \text{+ve})$
II	105 ± 22 $(8R, \text{+ve}; \ 8S, \text{-ve})$
III	146 ± 17 $(22R, \text{+ve} ; \ 22S, \text{-ve})$

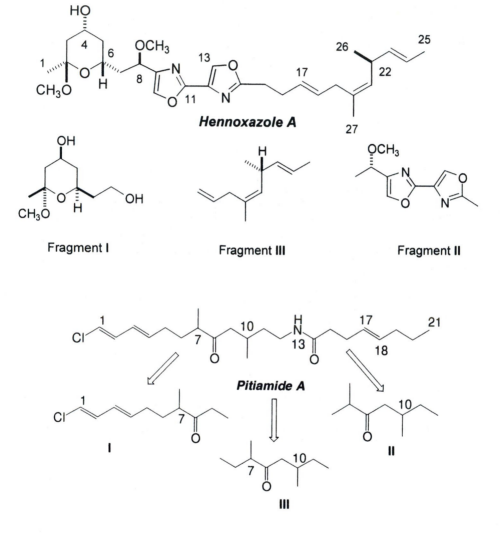

Hennoxazole A

Fragment I

Fragment III

Fragment II

Pitiamide A

I

III

II

113

Table 2. Computed pitiamide fragment molar rotation values. Results in the final row represent the calculation ½(Fragments I + II + III).

	[M]$_D$ for indicated stereoisomer	
Fragment **I**	+111 (7*S*)	
Fragment **II**	+66 (10*S*)	
Fragment **III**	+123 (7*S*,10*S*)	+17 (7*S*,10*R*)
Pitiamide A	**+150** (7*S*,10*S*)	**+31** (7*S*,10*R*)

V. Optical rotation computation as a probe of reaction mechanism

With confidence in our ability to compute the sign and magnitude of rotation angles in organic molecules, we turned to unanswered questions of reaction mechanism that can be approached from optical rotation. In Curran's group at Pittsburgh, new methods are under study to use radical cyclization to generate chiral products. The reaction of achrial atropisomer to form an enantioenriched indolinone was examined (Fig. 2), with the hypothesis that the P- stereochemistry of the

starting material would be preserved (assuming rotation around the CN bond is slow compared to the rate of ring closure) in the product. Calculations of rotation angles for the product specific rotation at the sodium D line of −17 were fully consistent with optical rotation of −16 measured for the dominant enantiomer [30]. Experimental attempts to test this prediction *via* crystallization or derivatization have, so far, not met with success.

VI. Toward structure-function relationships for optical rotation

The non-resonant rotation angles in eq 5 can be written in terms of individual occupied molecular orbital contributions:

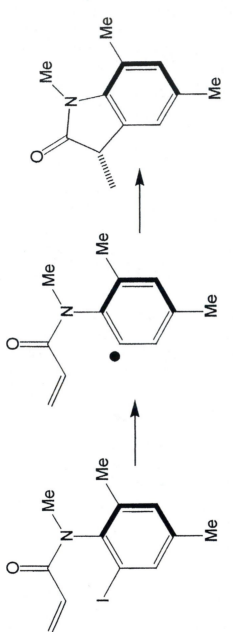

Fig. 2. The optical rotation of the indolinone formed by cyclization of the atropisomer indicates that the stereochemistry is retained in the course of this reaction.

$$\beta(\omega) \propto \text{Im} \sum_{n_{occ}} \sum_{i=x,y,z} \left\langle \frac{\partial \Psi_n}{\partial E_i} \middle| \frac{\partial \Psi_n}{\partial B_i} \right\rangle \propto \sum_{n_{occ}} \sum_{p,q} C_{pq}^n S_{pq} \qquad (8)$$

where C_{pq}^n is a product of atomic orbital coefficients associated with the perturbation of each polarized occupied molecular orbital and S is the overlap between atomic orbitals [32]. This is a Mulliken-like analysis of optical rotation. We have made atomic maps of these contributions, combining the diagonal (p=q) contributions with equally divided off-diagonal terms (p≠q) [31,32]. This mapping reveals, for example, the relative contributions to optical rotation that arise from stereogenic centers as opposed to twisted chains that may be remote from these centers [32].

Atomic contribution map analysis for an indoline [31] indicate that - for a specific conformer – the contribution of a side-chain twist to the rotation angle is comparable to the contribution from the stereogenic center itself. We have used this analysis to probe the decay of optical rotation contributions as a function of distance of groups from a stereocenter [32] and to understand the origin of sign differences in the rotation angles for (+)-calyculin A vs. (-)-calyculin B, which differ only in the Z vs. E configuration of a CN substituent that is nine bonds removed from the nearest stereogenic center [28]. Atomic contribution maps for achiral oxirane and chiral chiral chloro-oxirane appear in Fig. 3.

VII. Prospects

Having access to reliable computational tools for determining optical rotation angles has established new strategies for absolute stereochemical assignment. Moreover, the new tools are leading toward structure-function relations for a rather poorly understood macroscopic property. We have shown that molecular fragmentation, geometry sampling, linear-response computation, and Boltzmann weighting of contributions leads to reliable rotation-angle predictions. Future directions in our research program will include extension to molecules of increasing complexity. For example, reducing the number of plausible stereoisomers in a natural product with 10 stereocenters from 2^{10} (1024) to a few dozen will place the laboratory synthesis of the structure within the realm of possibility. Complementary studies of the "chiral imprint" imposed on solvent by a dissolved chiral solute must be understood. Also, the competing influence of tetrahedral stereogenic centers vs. twisted chain contributions to optical rotation angles, predicted by theory, must now be tested in carefully

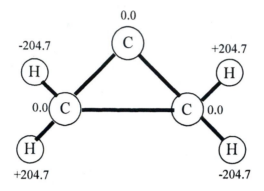

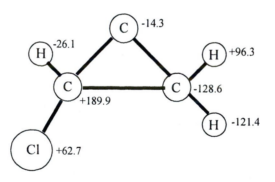

Fig 3. Maps of atomic contributions to [α]$_D$ for achiral oxirane (top) and chiral chloro-oxirane (bottom). Note the canceling contributions to the angle in oxirane and the non-zero sum in the chiral species. A 6-31G** basis was used in CADPAC calculations of optical rotation with 6-31G** based geometry optimization.

designed experiments. In the near future, increasingly reliable electronic structure methods will undoubtedly emerge that will prove valuable for making predictions of increasing reliability on structures of ever increasing size and complexity. The methods discussed in this chapter complement other chiro-optical methods [38,39] – and were developed specifically to address stereochemical assignments in complex natural products containing multiple stereogenic centers, conformational flexibility, and novel chemical bonding.

Acknowledgments

We thank ACS-PRF (33532-AC), NSF (CHE-9727657,CHE-9453461) and NIH (GM-55433) for support of this research project. DNB also thanks All Souls College, the Burroughs-Wellcome Foundation, and the J.S. Guggenheim Foundation for support while some of this work was conducted. We thank D.P. Curran, D.W. Pratt, A.B. Smith III, and G.K Friestad for their important contributions to our research.

References

1. Cragg, G. M.; Newman, D. J.; Snader, K. M., *J. Nat. Prod.* **1997**, *60*, 52-60.
2. Shu, Y.-Z., *J. Nat. Prod.* **1998**, 61, 1053-1071.
3. Eliel, E.L.; Wilen, S.H.; Mander, L.N. *Stereochemistry of Organic Compounds*; Wiley; New York, 1994.
4. Nakanishi, K.; Berova, N.; Woody, R. W. *Circular Dichroism. Principles and Applications;* VCH Publishers Inc.; New York, 1994.
5. Crews, P.; Rodriguez, J.; Jaspars, M. *Organic structure analysis*, Oxford University Press; New York, 1998.
6. Caldwell, D.J; Eyring, H. *The Theory of Optical Activity*, Wiley-Interscience, New York, 1971.
7. Brewster, J.H. *J. Am. Chem. Soc.* **1959**, *81*, 5475-.
8. Applequest, J. *J. Chem. Phys.* **1973**, *10*, 4251-.
9. Rosenfeld, L.Z. *Physik* **1928**, *52*, 161-.
10. *Selected Papers on Natural Optical Activity, SPIE Milestone Series, Vol. MS 15,* Lakhtakia, A. Ed., SPIE Press, Bellingham, WA 1990.
11. Kauzmann, W. *Quantum Chemistry*, Academic Press, New York, 1957.
12. Atkins, P. *Molecular Quantum Mechanics, 2nd ed.*, Oxford, 1983.
13. Besley, N.A.; Hirst, J.D. *J. Am. Chem. Soc.* **1999**, *121*, 9636-9644.
14. Atkins, P. *Physical Chemistry, 6th ed.*, Freeman, New York, 1998.

118

15. Cantor, C.R. and Schimmel, P.R. *Biophysical Chemistry, Part II*, Chapter 8, WH Freeman and Co, San Francisco, 1980.
16. Polavarapu, P.L. *Molec. Phys.* **1997**, *91*, 551-554.
17. Stevens, R.M.; Pitzer, R.M.; Lipscomb, W.N. *J. Chem. Phys.* **1963**, *38*, 550.
18. Jensen, F. *Introduction to Computational Chemistry*, Wiley; New York, Chapter 10, 1999.
19. McWeeny, R. *Methods of Molecular Quantum Mechanics, 2nd ed.*, Academic Press, New York, 1992.
20. Cook, D.B. *Handbook of Quantum Chemistry*, Oxford Science, Oxford, 1998.
21. Hansen, A.E.; Bouman, T.D., *Adv. Chem. Phys.* **1980**, *44*, 545-644.
22. E. Merzbacher, *Quantum Mechanics, 3rd ed.,* Wiley, New York, 1998.
23. De Meijere, A.; Khlebnikov, A.F.; Kostikov, R.R.; Kozhushkov, S.I.; Schreiner, P.R.; Wittkopp, A.; Yufit, S.S. Angew. Chem. **1999**, *38*, 3473-.
24. Cheeseman, J.R.; Frisch, M.J.; Devlin, F.J.; Stephens, P.J. *J. Phys. Chem. A* **2000**, *104*, 1039-1046.
25. Polavarapu, P.L. *Tetrahedron Asymmetry* **1997**, *8*, 3397-3401.
26. Polavarapu, P.L.; Zhao, C.X. *Chem. Phys. Lett.* **1998**, *296*,105-110.
27. Polavarapu, P.L.; Chakraborty, D. K. *J. Am. Chem. Soc.* **1998**, *120*, 6160-6164.
28. Kondru, R.K.; Beratan, D.N.; Friestad, G.K.; Smith III, A.B.; Wipf, P. *Org. Lett.*, **2000**, 2, 1509-1512.
29. Ribe, S.; Kondru, R.K.; Beratan, D.N.; Wipf, P. *J. Am. Chem. Soc.* **2000**, 122, 4608-4617.
30. Kondru, R.K.; Chen, C.H.-T.; Curran, D.P.; Wipf, P.; Beratan, D.N. *Tetrahedron: Asymmetry* **1999**, *10*, 4143-4150.
31. Kondru, R.K.; Wipf, P.; Beratan, D.N. *J. Phys. Chem. A* **1999**, *103*, 6603-6611.
32. Kondru, R.K.; Wipf, P.; Beratan, D.N. *Science* **1998**, *282*, 2247-2250.
33. Kondru, R.K.; Wipf, P.; Beratan, D.N. *J. Am. Chem. Soc.* **1998**, *120*, 2204-2205.
34. Kondru, R.K.; Lim, S.; Wipf, P.; Beratan, D.N. *Chirality* **1997**, *9*, 469-477.
35. Kondru, R.K.; Wipf, P.; Beratan, D.N. *in preparation*.
36. Amos, R.D.; Rice, J.E. *The Cambridge Analytic Derivative Package*, issue 4.0, 1987.
37. Helgaker, T. *et al.*, *DALTON, an ab initio electronic structure program*, Release 1.0, 1997.
38. F. Furche, R. Ahlrichs, C. Wachsmann, E. Weber, A. Sobanski, F.Vögtle, and S. Grimme, *J. Am. Chem. Soc.* **2000**, *122*, 1717-1724.
39. F.J. Devlin and P.J. Stephens, *J. Am. Chem. Soc.* **1999**, 121, 7413-7414.

Chapter 9

Optical Response of a Chiral Liquid

A. David Buckingham and Peer Fischer

Department of Chemistry, University of Cambridge, Lensfield Road, Cambridge CB2 1EW, United Kingdom

The optical properties of a molecule can be analyzed in terms of the oscillations in the charge and current density induced by the electromagnetic field associated with the optical wave. The induced electric and magnetic moments are related to the optical field through molecular property tensors. We discuss the symmetry of these response tensors and show that the first hyperpolarizability, like the optical rotation tensor, has an isotropic component and is chirally sensitive. The first hyperpolarizability may give rise to second-harmonic generation, sum- and difference-frequency generation, and the Pockels effect. We show that second-harmonic generation and the Pockels effect vanish for any liquid. Sum- and difference-frequency generation are symmetry allowed in chiral liquids, and may give rise to a novel linear electro-optic effect. This effect arises from the interference of the first hyperpolarizability $\beta_{\alpha\beta\gamma}$ with an electric field-induced hyperpolarizability $\gamma_{\alpha\beta\gamma\delta} F_\delta$, where F_δ is a component of an electrostatic field.

Optical response tensors

Optical fields that are incident upon a molecule induce time-varying molecular moments which themselves radiate. The interaction of the electromagnetic field and an uncharged molecule is formally described by the multipolar interaction Hamiltonian, H_I:

$$\hat{H}_I(t) = -\hat{\mu}_\alpha E_\alpha(t) - \frac{1}{3}\hat{\Theta}_{\alpha\beta}E_{\alpha\beta}(t) - \hat{m}_\alpha B_\alpha + \ldots \tag{1}$$

where $E_{\alpha\beta} = \nabla_\alpha E_\beta$.

The molecular dipole moment, electric quadrupole moment and magnetic dipole moment operators are respectively defined by

$$\hat{\mu}_\alpha = \sum_i e_i r_{i_\alpha} \tag{2}$$

$$\hat{\Theta}_{\alpha\beta} = \frac{1}{2}\sum_i e_i(3r_{i_\alpha}r_{i_\beta} - r_i^2\,\delta_{\alpha\beta}) \tag{3}$$

$$\hat{m}_\alpha = \sum_i \frac{e_i}{2\,m_i}\,(l_{i_\alpha} + g_i\,s_{i_\alpha}) \quad . \tag{4}$$

Note that the electric quadrupole moment is symmetric in α and β and traceless.

Most linear optical phenomena such as refraction, absorption of light, Rayleigh and Raman scattering can be interpreted through the oscillating induced dipole moment linear in the electric field [1] :

$$\Delta\mu_\alpha = \alpha_{\alpha\beta}\,E_\beta + \alpha'_{\alpha\beta}\,\dot{E}_\beta/\omega \tag{5}$$

where $E_\beta = E_\beta^{(0)}\cos{(\omega t - k_\gamma\,r_\gamma)}$. $\alpha_{\alpha\beta}$ is the symmetric polarizability tensor and $\alpha'_{\alpha\beta}$ is the antisymmetric polarizability tensor. Summation over repeated tensor suffices is implied.

In order to describe the optical properties of chiral molecules it becomes necessary to consider response tensors other than the electric-dipole polarizability.

More generally the molecular moments linear in the periodic electromagnetic field are [1, 2, 3]:

$$\Delta\mu_\alpha = \alpha_{\alpha\beta} E_\beta + \frac{1}{\omega} \alpha'_{\alpha\beta} \dot{E}_\beta + G_{\alpha\beta} B_\beta + \frac{1}{\omega} G'_{\alpha\beta} \dot{B}_\beta$$

$$+ \frac{1}{3} A_{\alpha,\beta\gamma} E_{\beta\gamma} + \frac{1}{3\,\omega} A'_{\alpha,\beta\gamma} \dot{E}_{\beta\gamma} + \ldots \qquad (6a)$$

$$\Delta m_\alpha = \chi_{\alpha\beta} B_\beta + \frac{1}{\omega} \chi'_{\alpha\beta} \dot{B}_\beta + G_{\beta\alpha} E_\beta - \frac{1}{\omega} G'_{\beta\alpha} \dot{E}_\beta$$

$$+ \frac{1}{3} D_{\alpha,\beta\gamma} E_{\beta\gamma} + \frac{1}{3\,\omega} D'_{\alpha,\beta\gamma} \dot{E}_{\beta\gamma} + \ldots \qquad (6b)$$

$$\Delta\Theta_{\alpha\beta} = A_{\gamma,\alpha\beta} E_\gamma - \frac{1}{\omega} A'_{\gamma,\alpha\beta} \dot{E}_\gamma + C_{\alpha\beta,\gamma\delta} E_{\gamma\delta}$$

$$+ \frac{1}{\omega} C'_{\alpha\beta,\gamma\delta} \dot{E}_{\gamma\delta} + D_{\gamma,\alpha\beta} B_\gamma - \frac{1}{\omega} D'_{\gamma,\alpha\beta} \dot{B}_\gamma + \ldots \qquad (6c)$$

Higher powers of the fields or the simultaneous interaction of several electromagnetic fields give rise to a nonlinear optical response. The dipole oscillating at the sum-frequency $\omega_3 = \omega_1 + \omega_2$ induced by linearly polarized beams $E_{1_\beta} = E_{1_\beta}^{(0)} \cos \omega_1 t$ and $E_{2_\gamma} = E_{2_\gamma}^{(0)} \cos \omega_2 t$ is

$$\Delta\mu_\alpha(-\omega_3; \omega_1, \omega_2) = \frac{1}{2} \beta_{\alpha\beta\gamma} E_{1_\beta}^{(0)} E_{2_\gamma}^{(0)} \cos(\omega_1 + \omega_2)t \qquad (7)$$

where the incident optical fields are $E_{1_\beta}(\omega_1)$ and $E_{2_\gamma}(\omega_2)$, and where we have limited our analysis to electric dipolar interactions. Only the time-even first hyperpolarizability is considered.

In order to describe the optical properties of a liquid, we now need to average the response tensors in Eqns (6) and (7) over the orientational distribution of the ensemble of molecules.

Isotropic tensors

A vector transforms as D_1 and therefore has no isotropic component. A second rank tensor (e.g. the polarizability) transforms as $D_1 \otimes D_1 = D_0 + D_1 + D_2$, where the tensor's isotropic part transforms as D_0, its antisymmetric part as D_1 and its traceless symmetric part as D_2. Similarly, a third and fourth rank tensor transform as $D_1 \otimes D_1 \otimes D_1 = D_0 + 3D_1 + 2D_2 + D_3$ and $D_1 \otimes D_1 \otimes D_1 \otimes D_1 = 3D_0 + 6D_1 + 6D_2 + 3D_3 + D_4$, respectively, so there is one isotropic tensor of the third rank and three independent isotropic tensors of the fourth rank. In an isotropic medium, such as a fluid or a gas, a Cartesian property tensor $< \mathbf{X} >$ may thus be replaced by [4]

$$< X_{\alpha\beta} > = \overline{X}_a \, \delta_{\alpha\beta} \tag{8a}$$

$$< X_{\alpha\beta\gamma} > = \overline{X}_b \, \epsilon_{\alpha\beta\gamma} \tag{8b}$$

$$< X_{\alpha\beta\gamma\delta} > = \overline{X}_{c1} \, \delta_{\alpha\beta}\delta_{\gamma\delta} + \overline{X}_{c2} \, \delta_{\alpha\gamma}\delta_{\beta\delta} + \overline{X}_{c3} \, \delta_{\alpha\delta}\delta_{\beta\gamma} \tag{8c}$$

with

$$\overline{X}_a = \frac{1}{3} X_{\alpha\beta} \, \delta_{\alpha\beta} = \frac{1}{3} X_{\alpha\alpha} \tag{9a}$$

$$\overline{X}_b = \frac{1}{6} X_{\alpha\beta\gamma} \, \epsilon_{\alpha\beta\gamma} \tag{9b}$$

$$\overline{X}_{c1} = \frac{1}{30} [4X_{\alpha\alpha\beta\beta} - X_{\alpha\beta\alpha\beta} - X_{\alpha\beta\beta\alpha}] \tag{9c}$$

$$\overline{X}_{c2} = \frac{1}{30} [-X_{\alpha\alpha\beta\beta} + 4X_{\alpha\beta\alpha\beta} - X_{\alpha\beta\beta\alpha}] \tag{9d}$$

$$\overline{X}_{c3} = \frac{1}{30} [-X_{\alpha\alpha\beta\beta} - X_{\alpha\beta\alpha\beta} + 4X_{\alpha\beta\beta\alpha}] \tag{9e}$$

where $\delta_{\alpha\beta}$ is the Kronecker delta and $\epsilon_{\alpha\beta\gamma}$ is the unit skew-symmetric (Levi-Civita) tensor.

The polarizability and the first hyperpolarizability are second and third rank tensors respectively and have the following isotropic components:

$$< \alpha_{\alpha\beta} > = \overline{\alpha} \, \delta_{\alpha\beta}$$

$$\overline{\alpha} = \frac{1}{3} \alpha_{\alpha\alpha} = \frac{1}{3} (\alpha_{xx} + \alpha_{yy} + \alpha_{zz})$$

$$< \beta_{\alpha\beta\gamma} > = \overline{\beta} \, \epsilon_{\alpha\beta\gamma}$$

$$\overline{\beta} = \frac{1}{6} \epsilon_{\alpha\beta\gamma} \, \beta_{\alpha\beta\gamma}$$

$$= \frac{1}{6} (\beta_{xyz} - \beta_{xzy} + \beta_{yzx} - \beta_{yxz} + \beta_{zxy} - \beta_{zyx}) \tag{10}$$

Note that the isotropic component of a third rank tensor is completely anti-symmetric. As the dipole-quadrupole polarizability $A_{\alpha,\beta\gamma}$ is symmetric in its last two indices, it follows that it does not have an isotropic part (see Table 2).

Symmetry

We consider the symmetry operations of time reversal and parity in order to identify those optical property tensors that may give rise to a chirally sensitive response in a liquid. Parity inverts all coordinates and hence the electric field and the electric dipole are odd under parity whereas the magnetic field is even. Time-reversal inverts the direction of momenta and spins but leaves charges invariant. It follows that an electric field is symmetric under time-reversal, whereas a magnetic field and a magnetic dipole are time-antisymmetric. Table 1 shows the effect of $\hat{P}$ and $\hat{T}$ on various properties and fields.

Table 1: Symmetry of various molecular property tensors and external fields under parity ($\hat{P}$) and time reversal ($\hat{T}$).

external field or property		$\hat{P}$	$\hat{T}$
$\mathbf{E}$	electric field	−	+
$\dot{\mathbf{E}}$	time derivative of electric field	−	−
$\boldsymbol{\mu}$	electric dipole moment	−	+
$\boldsymbol{\Theta}$	electric quadrupole moment	+	+
$\mathbf{B}$	magnetic field	+	−
$\dot{\mathbf{B}}$	time derivative of magnetic field	+	+
$\boldsymbol{m}$	magnetic dipole moment	+	−

The behaviour of the property tensors under time reversal and space inversion (see Table 2) can be inferred from Eqns (6) and (7) with the help of Table 1; e.g. the parity-even, time-odd magnetic dipole $\boldsymbol{m}$ induced by a parity-odd, time-odd time derivative of the electric field $\dot{\mathbf{E}}$ requires a parity-odd, time-even molecular response, here the optical rotation tensor, $G'_{\beta\alpha}$.

A molecule that is distinct from its mirror image is considered to be chiral [5]. Parity interconverts a chiral molecule into its mirror image. A chirally sensitive response thus requires a parity-odd property tensor; and since the isotropic part of any tensor is a scalar it follows that pseudoscalars (independent of the choice of coordinate axes, and of opposite sign for enantiomers) are associated with chirally

Table 2: Behaviour of various molecular response tensors under space inversion (parity, $\hat{P}$) and time reversal ($\hat{T}$).

response tensor		operators	$\hat{P}$	$\hat{T}$	isotropic part	chirally sensitive
$\alpha_{\alpha\beta}$	symmetric polarizability	$\hat{\mu}\hat{\mu}$	+	+	✓	
$\alpha'_{\alpha\beta}$	antisymmetric polarizability	$i\,\hat{\mu}\hat{\mu}$	+	−	✓	
$G'_{\alpha\beta}$	optical rotation tensor	$i\,\hat{m}\hat{\mu}$	−	+	✓	✓
$A_{\alpha,\beta\gamma}$	dipole-quadrupole polarizability	$\hat{\mu}\hat{\Theta}$	−	+		
$C_{\alpha\beta,\gamma\delta}$	quadrupole polarizability	$\hat{\Theta}\hat{\Theta}$	+	+	✓	
$\chi_{\alpha\beta} \quad = \chi^{dia.}_{\alpha\beta} + \chi^{para.}_{\alpha\beta}$	magnetizability	$\hat{m}\hat{m}$	+	+	✓	
$\alpha'_{\alpha\beta\gamma}$	Faraday rotation	$i\,\hat{\mu}\hat{\mu}\hat{m}$	+	+	✓	
$G'_{\alpha\beta\gamma}$	electric field on opt. rotation tensor	$i\,\hat{m}\hat{\mu}\hat{\mu}$	+	+	✓	
$\beta_{\alpha\beta\gamma}$	first hyperpolarizability	$\hat{\mu}\hat{\mu}\hat{\mu}$	−	+	✓	✓
$\gamma_{\alpha\beta\gamma\delta}$	second hyperpolarizability	$\hat{\mu}\hat{\mu}\hat{\mu}\hat{\mu}$	+	+	✓	

sensitive observables. Further, as a stationary fluid medium in the absence of an external magnetic field is invariant under time-reversal, we seek a time-even (and parity-odd) response tensor that has a non-vanishing isotropic part. We refer to Barron [3, 6] for a more complete account of the symmetry properties of optical activity phenomena.

In Table 2 it is seen that the optical rotation tensor $G'_{\beta\alpha}$ and the first hyperpolarizability $\beta_{\alpha\beta\gamma}$ are both odd under parity and even under time reversal. As required, the isotropic parts associated with $G'_{\beta\alpha}$ and $\beta_{\alpha\beta\gamma}$ are therefore time-even pseudoscalars. For a review of the optical activity phenomena associated with the optical rotation tensor, and in particluar vibrational and Raman optical activity, we refer to the references [3, 7, 8]. Henceforth, we will concetrate on $\overline{\beta}$ and discuss the nonlinear optical effects it may give rise to in a chiral liquid.

Sum-over-states expression of the isotropic part of the first hyperpolarizability

Second-harmonic generation (SHG) $2\omega = \omega + \omega$, sum-frequency generation (SFG) $\omega_3 = \omega_1 + \omega_2$, and difference-frequency generation (DFG) $\omega_3 = \omega_1 - \omega_2$ and the Pockels effect $\omega = \omega + 0$, are all described by the first hyperpolarizability $\beta_{\alpha\beta\gamma}$. Time-dependent perturbation theory may be used to obtain a sum-over-states expression for $\beta_{\alpha\beta\gamma}$. The isotropic component $\overline{\beta}$ of the sum-frequency hyperpolarizability is given by [9]:

$$\bar{\beta}(-\omega_3;\omega_1,\omega_2) = \frac{1}{6\,\hbar^2} \sum_{kj} \left(\boldsymbol{\mu}_{gk} \cdot [\boldsymbol{\mu}_{kj} \times \boldsymbol{\mu}_{jg}] \right)$$

$$\times \Big\{ \; (\tilde{\omega}_{kg} - \omega_3)^{-1} \left((\tilde{\omega}_{jg} - \omega_2)^{-1} - (\tilde{\omega}_{jg} - \omega_1)^{-1} \right)$$
$$+ (\tilde{\omega}^*_{kg} + \omega_1)^{-1} \left((\tilde{\omega}^*_{jg} + \omega_3)^{-1} - (\tilde{\omega}_{jg} - \omega_2)^{-1} \right)$$
$$+ (\tilde{\omega}^*_{kg} + \omega_2)^{-1} \left((\tilde{\omega}_{jg} - \omega_1)^{-1} - (\tilde{\omega}^*_{jg} + \omega_3)^{-1} \right) \Big\} \qquad (11)$$

The summation is over all excited states k and j, and the electric-dipole transition matrix element $\langle g|\hat{\boldsymbol{\mu}}|k\rangle \equiv \boldsymbol{\mu}_{gk}$. By allowing the transition frequency to be the complex quantity $\tilde{\omega}_{kg} \equiv \omega_{kg} - (\mathrm{i}/2)\,\Gamma_{kg}$, where ω_{kg} is the real transition frequency and Γ_{kg} the width at half the maximum height of the linear absorption spectral line from the ground state g to the the upper level k. Eqn. (11) is valid near resonance. The corresponding expression for DFG follows with the substitutions $\omega_2 \to -\omega_2$. The pseudoscalar $\bar{\beta}$ changes sign with the enantiomer and consequently averages to zero in the bulk of a racemic mixture (and in any achiral liquid). It may be seen that $\bar{\beta}$ vanishes for second-harmonic generation.

We now consider electro-optic effects in chiral liquids. The sum-over-states expression for the isotropic part of the Pockels tensor is [10]:

$$\bar{\beta}(-\omega;\omega,0) = \frac{1}{6\,\hbar^2} \sum_{kj} \left(\boldsymbol{\mu}_{gk} \cdot [\boldsymbol{\mu}_{kj} \times \boldsymbol{\mu}_{jg}] \right)$$

$$\times \Big\{ \; (\tilde{\omega}_{kg} - \omega)^{-1} \left(\omega^{-1}_{jg} - (\tilde{\omega}_{jg} - \omega)^{-1} \right)$$
$$+ (\tilde{\omega}^*_{kg} + \omega)^{-1} \left((\tilde{\omega}^*_{jg} + \omega)^{-1} - \omega^{-1}_{jg} \right)$$
$$+ \omega^{-1}_{kg} \left((\tilde{\omega}_{jg} - \omega)^{-1} - (\tilde{\omega}^*_{jg} + \omega)^{-1} \right) \Big\} \qquad (12)$$

The terms in brackets $\{\}$ in Eqn (12) are symmetric in k and j, and for real wavefunctions (as is appropriate in the absence of an external magnetic field) the triple vector product of transition dipole matrix elements is anti-symmetric, and hence the isotropic part of the Pockels tensor is zero. Eqn (12) may be deduced from Eqn (11) by setting an optical frequency *and* the associated complex damping term to zero [10]. This does not seem to have been appreciated in a number of recent publications [11-16]. Inconsistencies in the signs of the damping terms in sum-over-states expressions cause the Pockels tensor to have a non-vanishing isotropic component. We refer to reference [10] for a more detailed discussion.

In summary, the first hyperpolarizability has a nonvanishing isotropic component in a chiral liquid for sum- and difference-frequency generation, and this is discussed further in [17]. Second-harmonic generation and the Pockels effect are forbidden by symmetry in any liquid.

In the next section we show that the Pockels effect (linear change of the refractive index due to a static electric field) is also absent when the chiral liquid is dipolar and the interaction between the permanent molecular dipole moment and the electrostatic field is considered.

Orientational averages in the presence of a static field

If a molecule possesses a permanent electric dipole moment $\boldsymbol{\mu}^{(0)}$, then a static electric field, $\mathbf{F}$, will exert a torque on the molecules, such that they orient. The orientational average of a perturbed molecular tensor component $\mathbf{X}$ in space-fixed axes in the presence of a static electric field can be found by taking a classical Boltzmann average [3, 18]:

$$< \mathbf{X}(\Omega) >_{\mathbf{F}} = \frac{\int \mathbf{X}(\Omega)\, e^{-V(\Omega, \mathbf{F})/kT}\, d\Omega}{\int e^{-V(\Omega, \mathbf{F})/kT}\, d\Omega} \tag{13}$$

The potential energy V is a function of the static field and the molecule's orientation Ω in the field:

$$V(\Omega, \mathbf{F}) = V(\Omega) - \mu_\alpha^{(0)} F_\alpha - \frac{1}{2}\alpha_{\alpha\beta}^{(0)} F_\alpha F_\beta - \dots \tag{14}$$

where $\alpha_{\alpha\beta}^{(0)}$ is the static polarizability. Assuming that the sample is initially randomly oriented, and that $V(\Omega, \mathbf{F}) - V(\Omega) \ll kT$ we use the expansion

$$
\begin{aligned}
< \mathbf{X}(\Omega) >_{\mathbf{F}} \;=\;\; & < \mathbf{X}(\Omega) > + \frac{F_\alpha}{kT} < \mathbf{X}(\Omega)\mu_\alpha^{(0)} > \\
& + \frac{F_\alpha F_\beta}{k^2 T^2}\left[\frac{1}{2} < \mathbf{X}(\Omega)\mu_\alpha^{(0)}\mu_\beta^{(0)} > - \frac{1}{2} < \mathbf{X}(\Omega) >< \mu_\alpha^{(0)}\mu_\beta^{(0)} > \right. \\
& \left. + \frac{kT}{2}\left\{< \mathbf{X}(\Omega)\alpha_{\alpha\beta}^{(0)} > - < \mathbf{X}(\Omega) >< \alpha_{\alpha\beta}^{(0)} >\right\}\right] + \dots
\end{aligned}
\tag{15}
$$

where the brackets $< \dots >$ denote an isotropic average. If the duration of the applied static field is long compared to a molecular rotational time, the tensors $X_{\alpha\beta\gamma}$ and $X_{\alpha\beta\gamma\delta}$ in (8) should include linear terms $X_{\alpha\beta}\mu_\gamma^{(0)}/(kT)$ and $X_{\alpha\beta\gamma}\mu_\delta^{(0)}/(kT)$ respectively.

Considering the polarizability perturbed by the electrostatic field in Eqn (15), it is seen that the static field F_γ does not make a linear contribution to the hyperpolarizability $\beta_{\alpha\beta\gamma}(-\omega; \omega, 0)$, since the isotropic part of $\alpha_{\alpha\beta}(-\omega; \omega)\mu_\gamma^{(0)}/(kT)$

vanishes, and $\alpha'_{\alpha\beta}(-\omega;\omega)\mu_\gamma^{(0)}/(kT)$ is zero in a liquid in the absence of a magnetic field.

However, an electrostatic field may have a linear effect on the intensity of a sum-frequency signal in a chiral liquid which is discussed next.

A linear electro-optic effect in chiral liquids

We consider a coherent sum-frequency generation process in a chiral liquid in an electrostatic field. We include the second hyperpolarizability in the expression for the induced dipole oscillating at $\omega_3 = \omega_1 + \omega_2$ (see Eqn 7) and obtain

$$\Delta\mu_\alpha(-\omega_3;\omega_1,\omega_2,0) =$$

$$\frac{1}{2}\beta_{\alpha\beta\gamma}(-\omega_3;\omega_1,\omega_2)\,E_{1\beta}^{(0)}\,E_{2\gamma}^{(0)}\,\cos(\omega_1+\omega_2)t \tag{16}$$

$$+\frac{1}{3}\gamma_{\alpha\beta\gamma\delta}(-\omega_3;\omega_1,\omega_2,0)\,E_{1\beta}^{(0)}\,E_{2\gamma}^{(0)}\,F_\delta\,\cos(\omega_1+\omega_2)t$$

where we only consider the time-even hyperpolarizabilities. The second hyperpolarizability $\gamma_{\alpha\beta\gamma\delta}$ describes the response to third order in the applied fields and exists for all matter, even a spherical atom.

The scattering power at the sum-frequency is proportional to the square of the induced dipole moment $|\Delta\mu_\alpha|^2$ and consists of terms quadratic in the applied optical fields, one of which is linear in the applied static or low-frequency field, F [19]. The term linear in the electrostatic field arises from interference of the sum-frequency generator $\beta_{\alpha\beta\gamma}(-\omega_3;\omega_1,\omega_2)$ and the electric-field-induced sum-frequency generator $\gamma_{\alpha\beta\gamma\delta}(-\omega_3;\omega_1,\omega_2,0)\,F_\delta$ [19]. Their contributions to the scattering power can be distinguished.

If we choose the ω_1 beam to travel along the z direction and have its electric field vector oscillating along y and the ω_2 beam to be plane polarized in the yz plane, then the amplitude of the induced-dipole along x oscillating at the sum-frequency is

$$|\Delta\mu_x|^2 = \left|\frac{1}{2}\overline{\beta}\,E_{1_y}\,E_{2_z} + \frac{1}{3}\overline{\gamma}_3\,E_{1_y}\,E_{2_y}\,F_x\right|^2, \tag{17}$$

where the electrostatic field is taken along x. For an order-of-magnitude estimate we neglect the intensity component quadratic in the d.c. field and determine a fraction of the total signal that is linear in the d.c. field of

128

$$\frac{4\,|\overline{\gamma}_3\,F_x|}{3\,|\overline{\beta}|} \sim 10^{-2} \tag{18}$$

where we have taken $|\overline{\beta}|$ (at transparent frequencies) to be ~ 0.03 au [20], $|\overline{\gamma}|$ to be ~ 100 au, and $F_x \sim 2 \times 10^{-6}$ au $\approx 10^6$ V/m.

Conclusions

Symmetry arguments show that parity odd, time even molecular property tensors that have nonvanishing isotropic parts may give rise to chirally sensitive signals in liquids. Such tensors are the optical rotation tensor $G'_{\alpha\beta}$ in linear optics and the electric-dipolar first hyperpolarizability $\beta_{\alpha\beta\gamma}$ in nonlinear optics. Optical rotation, vibrational and Raman optical activity are all described by $G'_{\alpha\beta}$ (and its interference with the linear polarizability $\alpha_{\alpha\beta}$). Comparatively little is however known about the nonlinear optical phenomena in isotropic chiral media, such as sum- and difference-frequency generation [20]. It is expected that the hyperpolarizability $\overline{\beta}$ will be more sensitive than $\overline{\alpha}$ to environmental influences in the liquid [21].

The signal from a sum-frequency generation process in a chiral liquid in the presence of a static electric field is predicted to include an intensity component linear in the static field. Observation of the effect will require an appreciable isotropic component of the first hyperpolarizability.

Acknowledgments

A.D.B. and P.F. gratefully acknowledge funding by the Leverhulme Trust.

References

1. Buckingham, A. D. *Adv. Chem. Phys.* **1967**. *12*, 107.
2. Buckingham, A. D.; Raab, R. E. *Proc. R. Soc. Lond. A* **1975**. *345*, 365.
3. Barron, L. D. *Molecular Light Scattering and Optical Activity*. CUP, Cambridge, 1982.

4. Jeffreys, H. *Cartesian Tensors*. CUP, Cambridge, 1974.

5. Lord Kelvin *Baltimore Lectures*. C.J. Clay & Sons, London, 1904.

6. Barron, L. D. *Chem. Phys. Letters* **1986**. *123*, 423.

7. Barron, L. D.; Vrbancich, J. *Topics in Current Chemistry* **1984**. *123*, 151.

8. Nafie, L. A. *Vibrational Structure and Spectra, Vol. 10*, ed. Durig, J. R., Elsevier, Amsterdam, 1981 p. 153

9. Fischer, P. *Nonlinear Optical Properties of Chiral Media*. Ph.D. thesis, University of Cambridge, **1999**.

10. Buckingham, A. D.; Fischer, P. *Phys. Rev. A* **2000**. *61*, 035801.

11. Andrews, D. L.; Naguleswaran, S.; Stedman, G. E. *Phys. Rev. A* **1998**. *57*, 4925.

12. Koroteev, N. I. *JETP Letters* **1997**. *66*, 549.

13. Koroteev, N. I. *Frontiers in Nonlinear Optics*, eds. Walther, H.; Koroteev, N.; Scully, M. O., IOP, Bristol, 1993 p. 228

14. Zawodny, R.; Woźniak, S.; Wagnière, G. *Optics Comm.* **1996**. *130*, 163.

15. Woźniak, S. *Mol. Phys.* **1997**. *90*, 917.

16. Beljonne, D.; Shuai, Z.; Brédas, J. L.; Kauranen, M.;Verbiest, T.;Persoons, A. *J. Chem. Phys.* **1998**. *108*, 1301.

17. Kirkwood, J.; Albrecht, A. C.; Fischer, P.; Buckingham, A. D. *chapter in this ACS series*.

18. Buckingham, A. D.; Pople, J. A. *Proc. Phys. Soc. London Ser. A* **1955**. *68*, 905.

19. Buckingham, A. D.; Fischer, P. *Chem. Phys. Letters* **1998**. *297*, 239.

20. Fischer, P.; Wiersma, D. S.; Righini, R.; Champagne, B. C.; Buckingham, A. D. *Phys. Rev. Letters* **2000**. *85*, 4253.

21. Buckingham, A. D.; Concannon, E. P.; Hands, I. D. *J. Phys. Chem.* **1994**. *98*, 10455.

Chapter 10

Sum-Frequency Generation at Second Order in Isotropic Chiral Systems: The Microscopic View and the Surprising Fragility of the Signal

Jason Kirkwood[1,3], A. C. Albrecht[1,*], Peer Fischer[2], and A. D. Buckingham[2]

[1]Department of Chemistry, Cornell University, Ithaca, NY 14853
[2]Department of Chemistry, University of Cambridge, Lensfield Road, Cambridge CB2 1EW, United Kingdom
[3]Department of Chemistry, University of Rochester, Rochester, NY 14627–0216

The rotationally invariant component of the second order hyperpolarizability tensor is examined. Sum-frequency generation (SFG) arises in a chiral system only when at least two excited states are involved. No DC component and exceptionally sensitive resonance effects appear in this 3-wave mixing (3-WM) process. These are entirely unlike those seen in the more familiar even-WM spectroscopies. Both electronic (UV) and vibrational (IR) resonances are explored. Causes for the apparent fragility of SFG in optically active solutions are discussed and ways to circumvent them are suggested.

Introduction

The many nonlinear electric-field based spectroscopies, once treated perturbatively, are naturally classified by order. The sth order spectroscopies involve s incident field actions upon a material. The outcome of this action is a macroscopic, sth order polarization (the induced dipole density). The scalar product of its time derivative and the total field serves as the integrand in the calculation of the cycle-averaged rate of energy exchange between the incident fields and the material. Such net energy exchange must take place in all (one-photon or multiphoton) absorption and emission spectroscopies and, provided all incident fields are oscillating (none are DC), these so-called "Class I" spectroscopies can appear *only when s is odd* (*1*).

The induced, oscillating, sth order polarization also serves as a source term in Maxwell's equation for the electric field to produce a new field — the $(s+1)$th field in an overall process called $(s+1)$-wave mixing $((s+1)$-WM). The intensity of the new field may be detected in self-quadrature (homodyned) or placed into quadrature with an idler field (heterodyned). Unlike the Class I spectroscopies, here phase-matching becomes an important feature since the signal requires the successful macroscopic build-up of the new field. These "coherent" nonlinear spectroscopies need not have light/matter energy exchange that survives cycle averaging. They have been termed the "Class II" nonlinear spectroscopies and, unlike those of Class I, may appear at *all orders in s*.

However the macroscopic material response to the incident fields is built of contributions from each of the constituent molecules of the sample. This entails an ensemble average of the microscopic (molecular) induced polarization. At sth order such an average involves $(s+1)$ direction cosines that project the s actions of the incident field and the consequent induced dipole within the molecular frame to the macroscopic polarization of the sample in the laboratory frame. The ensemble average amounts to an average of the product of the $(s+1)$ direction cosines over the orientational distribution of the molecules. Normally samples consisting of microscopically randomly oriented molecules such as gases, liquids, and microscopically disordered solids, are symmetric with respect to inversion in the macroscopic frame. However each direction cosine is odd with respect to such inversion. Thus Class II spectroscopies for such randomly oriented systems should *not exist at even s* (an odd number of direction cosines). Indeed Class II spectroscopies at even order are unknown for such randomly oriented systems. However there is one exception ! The formal requirement for the microscopic susceptibility at any order to survive unweighted rotational averaging is that the tensor possess a rotationally invariant component. It was pointed out long ago (*2*) how at second order the microscopic polarizability tensor, beta, does in fact have a rotationally invariant component — it is a pseudoscalar. Though clearly vanishing for ordinary systems, it was noted how for any material lacking inversion symmetry on the macroscopic scale, this special component of the beta tensor should survive. An optically active system satisfies this requirement. So a solution of randomly oriented, optically resolved, chiral centers may give rise to sum-frequency generation (SFG). This suggested an exciting new nonlinear spectroscopy that would promise a background-free way to examine only the chiral centers in any complex system. Chiral centers are of general interest, but since they are ubiquitous in both proteins and nucleic acids such a new method to study them in the absence of any background might be of considerable use in biophysics.

Two reports of SFG from optically resolved arabinose have appeared (*3, 4*). However just last year an extensive and careful re-examination of SFG from these systems suggests a likely artifactual nature of the reported signals and that in fact SFG from such optically active systems is yet to be demonstrated (*5, 6*). Nevertheless an encouraging positive experiment first reported at this Symposium (*7*) shows an SFG signal from randomly oriented limonene in which an infrared resonance plays an enhancing role.

The present contribution reports results from a detailed study (*8*) of the rotationally invariant microscopic beta tensor element. The aim is to search for reasons behind the evident fragility of the SFG signal and thereby to suggest ways for its optimization. The exploration is cast in the frequency domain so it is the dispersive properties of SFG that are examined. To our surprise we have discovered that the dispersion of such *odd*-WM from randomly oriented systems is qualitatively different from that seen in the well-known, ubiquitous, *even*-WM spectroscopies.

We begin by presenting the analytic form for the rotationally invariant part of the beta tensor. Then a minimal three-state model is introduced for exploring the dispersive properties of SFG. Both electronic and vibrational resonances are included. These properties are displayed in the form of six figures in which both the amplitude (the modulus) and the phase of the response are shown, with the two incident frequencies and the energy level spacings as parameters. The dispersive behaviour of the SFG intensity (if homodyned) is not shown. But this is proportional to the square of the plotted amplitude. A brief concluding section with some speculation ends this contribution.

Theory

General

Let the sth order microscopic electrical susceptibility be called $\phi^{(s)}$ (where in more familiar terms $\phi^{(1)} \equiv \alpha$, the linear polarizability; $\phi^{(2)} \equiv \beta$, the first hyperpolarizability, and $\phi^{(3)} \equiv \gamma$, the second hyperpolarizability). It is an $(s+1)$-rank tensor having $3^{(s+1)}$ Cartesian components. In the density matrix formalism the analytic expression for $\phi^{(s)}$ is built from contributions from all possible "Liouville paths" that trace the evolution of states of the molecule as it becomes (nonlinearly) polarized. Each path defines a particular time ordering of the actions on the molecule of the incident fields and whether a given action is on the bra side or the ket side of the basis set used to describe the molecule. A dual Feynman diagram, or equivalently, a wave-mixing energy level diagram (WMEL), is used to depict uniquely any given path. Provided no incident field acts more than once (this is the case of fully nondegenerate wavemixing) there are altogether $s!2^s$ paths to consider at sth order (2^s for the choice of bra or ket action at each perturbing event and $s!$ for the ways of time ordering the actions of s distinct fields). The contribution of each such path to the analytic expression for $\phi^{(s)}$ consists of two factors. One is the $(s+1)$th order transition dipole tensor; the other consists of the s energy denominator factors characterizing the perturbative approach at sth order. These two factors are linked by s independent sums over the basis set (provided the unperturbed molecule is in its ground state). Let the transition dipole tensor for the Lth Liouville path be called $\mu_L^{(s+1)}(\{s\})$ and the energy factor $EF_L^{(s)}(\{s\})$ in which $\{s\}$ acknowledges how s different dummy indices of the basis set are to appear. $EF_L^{(s)}(\{s\})$ is responsible for the dispersion of $\phi^{(s)}$ and carries the various resonances possible in any $(s+1)$-WM process. The $\mu_L^{(s+1)}(\{s\})$ are responsible for how the Lth Liouville path responds to excitation with variously polarized exciting fields. And for real wavefunctions the $EF_L^{(s)}(\{s\})$ also is entirely responsible for its contribution to the phase (and its dispersion) of $\phi^{(s)}$. (However the overall sign of the contribution from path L is influenced by $\mu_L^{(s+1)}(\{s\})$ even when it is pure real.)

Summing over all Liouville paths and with the s independent sums over the basis set we write $\phi^{(s)} = \sum_L \sum_{\{s\}} \mu_L^{(s+1)}(\{s\}) EF_L^{(s)}(\{s\})$. Each of the $3^{(s+1)}$ components of the macroscopic susceptibility tensor, $\chi^{(s)}$, is obtained by summing over the contribution from all of the molecules in a unit volume or, equivalently, by taking the orientationally averaged projection of all relevant components of $\phi^{(s)}$ onto the appropriate laboratory Cartesian axes, I, J, K, and multiplying by the number density, N. Thus $\chi_{IJK\cdots}^{(s)} \propto N \sum_L \sum_{\{s\}} \left\langle \mu_L^{(s+1)}(\{s\}) \right\rangle_{IJK\cdots} EF_L^{(s)}(\{s\})$.

In the present three-wave mixing problem, $s=2$, so there is a double sum m,n over all virtual states and $\chi_{IJK}^{(2)} \propto N \sum_L \sum_{m,n} \left\langle \mu_L^{(3)}(m,n) \right\rangle_{IJK} EF_L^{(2)}(m,n)$. For randomly oriented molecules the orientational averaging, formally involving 729 $\left(\left(3^2 \right)^3 \right)$ terms, leaves only the pseudoscalar of the beta tensor contributing at most only to those six tensor components of $\chi^{(2)}$ for which $I \neq J \neq K$. This alone has serious experimental implications for it requires that the two incident laser fields be orthogonally polarized (field 1 along J, field 2 along K) and that the signal field be analyzed with the polarization perpendicular to each (along I).

At $s=2$ there are 8 Liouville paths to consider. It turns out that for the orientational averaging over the randomly oriented molecules the *magnitude* of the rotationally averaged transition dipole tensor is the same for all Liouville paths. Only the sign changes as the time ordering of incident fields 1 and 2 is reversed. Thus the sum over L is a sum just over $EF_L^{(2)}(m,n)$ *provided* the sign change is incorporated in four of its terms. We shall call this sum on L (with the sign change) simply $EF^{(2)}(m,n)$.

With i, j, k defining a molecule-based Cartesian frame, we have for the common factor $\left\langle \mu_L^{(3)}(m,n) \right\rangle_{IJK} \equiv \mu_{RotInv}^{(3)}(m,n)$, where

$$\mu_{RotInv}^{(3)}(m,n) = \frac{1}{6} \times \begin{pmatrix} (\mu_i)_{mg}(\mu_j)_{mn}(\mu_k)_{ng} - (\mu_i)_{mg}(\mu_k)_{mn}(\mu_j)_{ng} \\ +(\mu_j)_{mg}(\mu_k)_{mn}(\mu_i)_{ng} - (\mu_j)_{mg}(\mu_i)_{mn}(\mu_k)_{ng} \\ +(\mu_k)_{mg}(\mu_i)_{mn}(\mu_j)_{ng} - (\mu_k)_{mg}(\mu_j)_{mn}(\mu_i)_{ng} \end{pmatrix} \qquad (1)$$

the same pseudoscalar form that would follow directly from the vector algebra $\vec{\mu}_{mg} \cdot \left(\vec{\mu}_{mn} \times \vec{\mu}_{ng} \right)$. Similarly we can write $\beta_{RotInv} = \sum_P (-1)^P \beta_{ijk}$ in which P is the pairwise permutation operator on the molecule-based Cartesian indices. p is the permutation count (starting at zero), and the sum is over the six such permutations. This pseudoscalar component of the beta tensor must change sign upon changing from one enantiomer to the other and must therefore vanish for an achiral system.

Thus in a system of randomly oriented molecules we can write for the beta pseudoscalar

$$\beta_{RotInv} = \sum_{m,n} \mu_{RotInv}^{(3)}(m,n) EF^{(2)}(m.n) \qquad (2)$$

where the sum is twice independently over the entire basis set.

Because the two time orderings of incident fields 1 and 2 are *exactly opposite in sign* $EF^{(2)}(m,n)$ reduces to an expression containing only four terms multiplied by the color difference between the two incident fields, $(\bar{\nu}_2 - \bar{\nu}_1)$. We obtain:

$$EF^{(2)}(m,n) = \frac{\left(\bar{v}_2 - \bar{v}_1\right)}{(hc)^2}$$

$$\times \left(\begin{array}{c} \dfrac{1}{\left(i\gamma_{mg} + \bar{v}_1 - \bar{v}_{mg}\right)\left(i\gamma_{mg} + \bar{v}_2 - \bar{v}_{mg}\right)\left(i\gamma_{mn} + \bar{v}_2 + \bar{v}_1 - \bar{v}_{mn}\right)} \\[4pt] + \\[4pt] \dfrac{1}{\left(i\gamma_{mg} + \bar{v}_1 + \bar{v}_{mg}\right)\left(i\gamma_{mg} + \bar{v}_2 + \bar{v}_{mg}\right)\left(i\gamma_{mn} + \bar{v}_2 + \bar{v}_1 + \bar{v}_{mn}\right)} \\[4pt] + \\[4pt] \dfrac{1}{\left(i\gamma_{mg} + \bar{v}_1 - \bar{v}_{mg}\right)\left(i\gamma_{mg} + \bar{v}_2 - \bar{v}_{mg}\right)\left(i\gamma_{ng} + \bar{v}_2 + \bar{v}_1 - \bar{v}_{ng}\right)} \\[4pt] + \\[4pt] \dfrac{1}{\left(i\gamma_{mg} + \bar{v}_1 + \bar{v}_{mg}\right)\left(i\gamma_{mg} + \bar{v}_2 + \bar{v}_{mg}\right)\left(i\gamma_{ng} + \bar{v}_2 + \bar{v}_1 + \bar{v}_{ng}\right)} \end{array} \right) \tag{3}$$

This factoring immediately shows how *i) degenerate frequency summing (SHG) cannot occur in isotropic systems even if of chiral molecules.* It follows that *no purely DC* $\chi^{(2)}$ can exist (for this would be a special case of degeneracy — both frequencies zero). This means that the dispersion spectrum of SFG from random systems starts with zero amplitude in the DC limit. This is entirely unlike what is seen in the *even*-WM spectroscopies.

Now another important observation may be made. We note from Eq. (1) how the pseudoscalar is such that $\mu_{RotInv}^{(3)}(m,n) = -\mu_{RotInv}^{(3)}(n,m)$. This means that *all* $m = n$ *terms* in the sum over states *must vanish! ii) Only the* $m \neq n$ *terms can contribute to* β_{RotInv}. This anti-Hermitian behaviour also allows one to convert the double sum over states (in Eq. (2)) into a one-sided double sum: $\beta_{RotInv} = \sum_{n>m} \mu_{RotInv}^{(3)}(m,n)\left(EF^{(2)}(m,n) - EF^{(2)}(n,m)\right)$. Clearly, *SFG from a random distribution of (chiral) molecules cannot exist in a simple two-state (ground and a single excited state (m=n)) model.* Again this is unlike all of the familiar *even*-WM spectroscopies seen from randomly oriented systems. For example in 4-WM single state resonances (as in (Class I) resonance Raman spectroscopy) are ubiquitous. Thus *iii) For SFG to occur through a rotationally invariant second order susceptibility a three-state model is the minimum possible.*

Finally we must note that: *iv) SFG in normally dispersive media cannot be phase-matched.* The k-vector of the sum frequency will always exceed the sum of the k-vectors of the incident fields even when they are collinear (5). This important feature for enhancing the Class II spectroscopies is absent in this case. Resort to samples made anomalously disperse may overcome this disadvantage.

The three-state model

We now introduce a simple three-state model, the minimum model capable of describing SFG in a chiral, disordered, system. in order to explore the dispersion of the amplitude and phase of β_{RotInv} as the frequencies of the two incident light fields are changed and as the energy level system and the damping parameters are varied.

The three-state system shall consist of the ground state g, an excited state a and another state above that, b. The energies (in cm^{-1}) of a above g shall be called simply A, of b above g, simply B. Incident laser field 1 will be at L_1, field 2 at L_2, and the SFG at $S = L_1 + L_2$ (all in cm^{-1}). It is also convenient to define the color difference $\Delta_{21} = L_2 - L_1$. The damping constants are chosen depending on the context, though for electronic dephasing we shall pick 1000 (or 1500) and for vibrational dephasing 10 (all in cm^{-1}). The large electronic dephasing constants are intended to simulate (in part) the normally broad Franck-Condon vibronic widths expected for electronic transitions of polyatomic molecules in solution.

The simple three level model is achieved by selecting just the $m=a$ and $n=b$ term of the one-sided double $\beta_{RotInv} = \sum_{n>m} \mu^{(3)}_{RotInv}(m,n)\left(EF^{(2)}(m,n) - EF^{(2)}(n,m)\right)$.

(the dipole operators are defined to exclude the ground state dipole moment). Thus our working expression for the three-level model is simply:

$$\beta_{RotInv}(A/B) = \mu^{(3)}_{RotInv}(a,b) \times \left(EF^{(2)}(a,b) - EF^{(2)}(b,a)\right) \qquad (4)$$

with $\mu^{(3)}_{RotInv}(a,b)$ obtained from Eq. (1). As we shall see the requirement that $EF^{(2)}(a,b) \neq EF^{(2)}(b,a)$ for β_{RotInv} not to vanish exposes a remarkable fragility of the signal. For this three-state model the difference between $EF^{(2)}(a,b)$ and $EF^{(2)}(b,a)$ is given by:

$$EF^{(2)}(a,b) - EF^{(2)}(b,a) = \frac{(L_2 - L_1)}{(hc)^2}$$

$$\times \left(\begin{array}{c}
\dfrac{1}{\left(i\gamma_{ag} + L_1 - A\right)\left(i\gamma_{ag} + L_2 - A\right)\left(i\gamma_{ab} + L_2 + L_1 - A + B\right)} \\[2mm]
- \\[2mm]
\dfrac{1}{\left(i\gamma_{bg} + L_1 - B\right)\left(i\gamma_{bg} + L_2 - B\right)\left(i\gamma_{ab} + L_2 + L_1 - B + A\right)} \\[2mm]
+ \\[2mm]
\dfrac{1}{\left(i\gamma_{ag} + L_1 + A\right)\left(i\gamma_{ag} + L_2 + A\right)\left(i\gamma_{ab} + L_2 + L_1 + A - B\right)} \\[2mm]
- \\[2mm]
\dfrac{1}{\left(i\gamma_{bg} + L_1 + B\right)\left(i\gamma_{bg} + L_2 + B\right)\left(i\gamma_{ab} + L_2 + L_1 + B - A\right)} \\[2mm]
+ \\[2mm]
\dfrac{1}{\left(i\gamma_{ag} + L_1 - A\right)\left(i\gamma_{ag} + L_2 - A\right)\left(i\gamma_{bg} + L_2 + L_1 - B\right)} \\[2mm]
- \\[2mm]
\dfrac{1}{\left(i\gamma_{bg} + L_1 - B\right)\left(i\gamma_{bg} + L_2 - B\right)\left(i\gamma_{ab} + L_2 + L_1 - A\right)} \\[2mm]
+ \\[2mm]
\dfrac{1}{\left(i\gamma_{ag} + L_1 + A\right)\left(i\gamma_{ag} + L_2 + A\right)\left(i\gamma_{bg} + L_2 + L_1 + B\right)} \\[2mm]
- \\[2mm]
\dfrac{1}{\left(i\gamma_{bg} + L_1 + B\right)\left(i\gamma_{bg} + L_2 + B\right)\left(i\gamma_{ag} + L_2 + L_1 + A\right)}
\end{array} \right) \qquad (5)$$

The three-level model explored

We proceed to seek out the properties of $\beta_{RotInv}(A/B)$ as being qualitatively representative of any such paired states in the double summation of Eq. (2). In fact since we know almost nothing about $\mu_{RotInv}^{(3)}(a,b)$, the appropriate variables for exploration are the amplitude and phase of $EF^{(2)}(a,b) - EF^{(2)}(b,a)$ (Eq. (5)), the former being just $\left| EF^{(2)}(a,b) - EF^{(2)}(b,a) \right|$. Atomic units (a.u.) are used in the presentation[1]. Should a value be available for $\left| \mu_{RotInv}^{(3)}(a,b) \right|$ one may immediately scale the Figures to $\left| \beta_{RotInv}(A/B) \right|$.

States a and b in the first Figures 1-3 are considered to be electronic states. In Figures 4-6 a is a vibrational state while b is electronic. Accordingly it is expected that these two very different situations should possess very different $\mu_{RotInv}^{(3)}(a,b)$.

Depending on the parameters, normally only small subsets of the Liouville paths contribute significantly to the signal at a given order. Here at second order with only 8 paths it is not difficult to always use the full expression even after summing over two states (Eq. (5)). There is no practical need to isolate the strong contributors as one usually does for the higher order polarizabilities. Finally we note that the (homodyned) SFG signal is proportional to the *square* of the plotted amplitude.

In each Figure the amplitude of $EF^{(2)}(a,b) - EF^{(2)}(b,a)$ (in a. u.) is shown as a solid curve, its phase (in radians) as a dashed curve.

Figure 1 explores the dependence of the amplitude on the color difference away from any resonance. The key feature is how the signal vanishes as L_2 and L_1 approach each other ($\Delta_{21} = 0$, color degeneracy). In the plot we have excluded the $0 < L_1 < 5,000$ cm^{-1} region to avoid infra-red resonances and also the important hybrid dynamic/static problem when L_1 becomes a DC field at $L_1 = 0$. The maximum of the amplitude appears approximately when $L_2/L_1 \approx 5$.

Figure 2, still away from resonance, shows how the amplitude varies with the shift of B to the blue of A. When the two states are degenerate there can be no signal. Thus one encounters the unusual effect that, as B recedes to the blue, the amplitude *grows* only then to decline somewhat as B is pushed further into

---

[1] The atomic units for the elements of the β tensor are $\left(4\pi\varepsilon_0\right)^2 a_0^5 / e$; for $\mu_{RotInv}^{(3)}$ they are $\left(ea_0\right)^3$; and for $EF^{(2)}$ they are $\left(4\pi\varepsilon_0 a_0 / e^2\right)^2$.

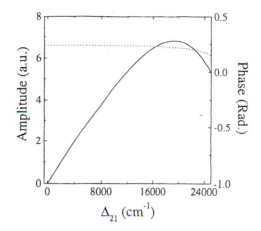

Figure 1. Off-resonance. (γ_{ag}=1,000 cm^{-1}, γ_{bg}=1,000 cm^{-1}; γ_{ab}=1,500 cm^{-1}). S = L_2 + L_1 = 30,000 cm^{-1}; A = 40,000 cm^{-1}; B = 56,000 cm^{-1}. Δ_{21} = L_2 − L_1 is varied. A maximum occurs at 20,000 (L_1 = 5,000 cm^{-1} and L_2 = 25,000 cm^{-1}). Δ_{21} = 10,000 cm^{-1} corresponds to excitation with harmonics, namely L_2/L_1 = 2.

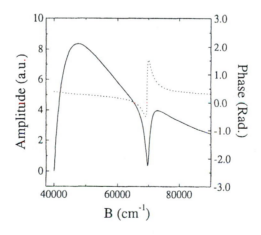

Figure 2. Off-resonance. (γ_{ag}=1,000 cm^{-1}, γ_{bg}=1,000 cm^{-1}; γ_{ab}=1,500 cm^{-1}). S = 30,000 cm^{-1}; Δ_{21} = 20,000 cm^{-1}; A = 40,000 cm^{-1}. State B is tuned to the blue from A to >>>A. A maximum occurs at 47,500 cm^{-1}, and a sharp dip is seen at B = 70,000 cm^{-1}. The former is a resonance of L_2 with B-A, while the dip appears when S = B-A.

the vacuum UV. The dip at $B = 70,000$ cm^{-1} corresponds to an anti-resonance when $S = B - A$ [2].

In Figure 3 the various resonances are exposed as the SFG signal is tuned from zero into the deep vacuum UV. The strongly dispersive behaviour of the phase similarly exposes the four resonances – first S with A, then L_2 with A, next S with B, and finally L_2 with B. Also there is a nearly three orders of magnitude (6 orders of magnitude in intensity) increase of amplitude compared to that in Figure 1.

Next, in Figure 4, a vibrational resonance is explored in the SFG process by regarding state A as vibrational (at 1,000 cm^{-1}) and L_1 is tuned through the resonance. The SFG amplitude is strongly enhanced by the narrow vibrational resonance as the phase reveals the expected anomalous dispersion. This models closely the positive experimental results on limonene (7).

Next, in Figure 5, state B is moved across the SFG signal from below as L_1 remains resonant with vibrational level A. Basically one encounters the electronic resonance as B passes through S. And the amplitude shows considerable enhancement over that seen in electronically off-resonant Figure 4.

Finally in tuning S (with a fixed Δ_{21}) across the full spectrum, Figure 6 shows how the vibrational resonance as L_1 passes through A is so much stronger than the subsequent electronic resonances of S with B, then L_2 with B. This most surely reflects the difference in linewidths taken for the two types of transitions.

Conclusions

This examination of the microscopic β_{RotInv} reveals new properties not seen in *even*-WM susceptibilities. Probably first and foremost is the finding that $\mu^{(3)}_{RotInv}(m,n) = -\mu^{(3)}_{RotInv}(n,m)$ forcing all diagonal $(m = n)$ terms of the double sum-over-states to vanish. One must resort to excited state *pairs* to obtain any contribution. At this point this antisymmetry ($\mu^{(3)}_{RotInv}(m,n) = -\mu^{(3)}_{RotInv}(n,m)$) for *any* state pair lends considerable fragility to the strength of the SFG signal requiring that the two excited states be separated in energy (i.e. not be degenerate), yet not be too separated to depart from the advantages of pre-resonance or resonance opportunities. This delicate behavior must underlie the exceptionally strong resonance enhancement seen in the three-level model even when we have used what we regard as conservatively large damping constants. We note how the electronically resonant peak for the amplitude seen in Figure 3

---

[2] This particular feature is unique to the density matrix approach. While conventional perturbation theory essentially reproduces all of the plots shown, it cannot exhibit this dephasing induced extra resonance.

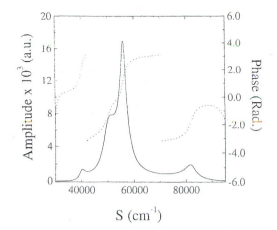

Figure 3. Exploring electronic resonances. ($\gamma_{ag} = 1,000$ cm^{-1}, $\gamma_{bg} = 1,000$ cm^{-1}; $\gamma_{ab} = 1,500$ cm^{-1}). Δ_{21} is fixed at 30,000 cm^{-1}. A = 40,000 cm^{-1}; B = 56,000 cm^{-1}. S is tuned from 30,000 cm^{-1} to 95,000 cm^{-1}. Resonances are seen at S = 40,000 cm^{-1} (L_2 = 43,000 cm^{-1}, L_1 = 13,000 cm^{-1}); at S = 50,000 cm^{-1} (L_2 = 40,000 cm^{-1}, L_1 = 10,000 cm^{-1}); at S = 56,000 cm^{-1} (L_2 = 30,000 cm^{-1}, L_1 = 26,000 cm^{-1}); and at S = 82,000 cm^{-1} (L_2 = 56,000 cm^{-1}, L_1 = 26,000 cm^{-1}). As expected, the resonances occur when either S or a single laser frequency (L_1 or L_2) matches an electronic transition.

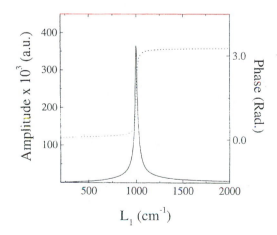

Figure 4. Exploring the IR resonance. A is fixed at 1,000 cm^{-1}; B is fixed at 40,000 cm^{-1}. ($\gamma_{ag} = 10$ cm^{-1}, $\gamma_{bg} = 1,000$ cm^{-1}; $\gamma_{ab} = 1,000$ cm^{-1}). L_2 = 28,000 cm^{-1} and L_1 is tuned.

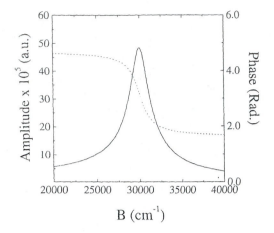

Figure 5. Electronically enhanced IR resonance. Same parameters as in Figure 4 except now $L_2 = 29,000 \ cm^{-1}$ and L_1 is fixed at the peak of the IR resonance. As the location of B is tuned from 20,000 cm^{-1} to 40,000 cm^{-1} the IR peak is enhanced by the resonance as B passes through S at 30,000 cm^{-1}.

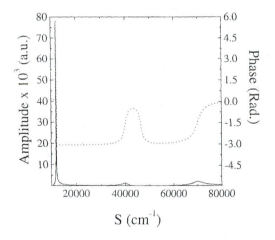

Figure 6. A comparison of the vibrationally resonant IR peak with pure electronic resonances. Parameters are the same as in Figure 4, except now Δ_{21} is fixed at 10,000 cm^{-1} and S is tuned from 10,000 cm^{-1} to 80,000 cm^{-1}. At S = 12,000 cm^{-1} ($L_2 = 11,000 \ cm^{-1}$, $L_1 = 1,000 \ cm^{-1}$) the purely IR resonance takes place (as in Figure 4). At S = 40,000 cm^{-1} ($L_2 = 25,000 \ cm^{-1}$, $L_1 = 15,000 \ cm^{-1}$) the purely electronic resonance of S with B occurs. And at S = 70,000 cm^{-1} ($L_2 = 40,000 \ cm^{-1}$, $L_1 = 30,000 \ cm^{-1}$) the purely electronic resonance of L_2 with B occurs. The relative weakness of the electronic resonances is a manifestation of the vibrational damping parameter being two orders of magnitude smaller than the electronic damping.

amounts to a signal intensity that is nearly 8 orders of magnitude stronger than what is seen for the off resonance experiment (Figure 1). Perhaps more to the point one may compare the fall-off of about 3 orders of magnitude shown upon detuning a Lorentzian peaking at 56,000 cm^{-1} to 30,000 cm^{-1} with the behavior of the SFG amplitude *squared* - which declines by approximately 8 orders of magnitude as S is moved from 56,000 cm^{-1} to 30,000 cm^{-1}. Clearly this second order process from a randomly oriented system of (chiral) molecules exhibits a very special kind of resonance not seen in the *even*-WM spectroscopies.

Such dramatic changes of signal strength with reasonable variations in experimental parameters is also a consequence of the fact that the signal must vanish as the sum frequency vanishes. It is likely that the extant experiments on arabinose lie somewhere within this region where the signal changes extremely rapidly and their artifactual nature (6) may reflect far from optimum excitation conditions.

Figures 4 to 6 reveal interesting vibrationally resonant SFG (see also (7)). Though the amplitude is strong we must be careful in comparing purely electronic SFG with the vibrationally resonant version for the unknown $\mu^{(3)}_{RotInv}(a,b)$ will be quite different for the two cases. Nevertheless the interesting case of the electronically enhanced IR resonance shown in Figure 5 is a valid one since here $\mu^{(3)}_{RotInv}(a,b)$ is the same (mixed electronic/vibrational) across the whole plot.

Clearly one would like to design a chiral molecule and choose laser sources to optimally offer a specific two-excited state A, B system resonantly optimized for the behavior shown in Figure 3. At this point the remaining distant state pairs in the full sum-over-states, because of the exceptional sensitivity of the amplitude to the distance in state energies, should not play a significant role in producing SFG.

On the other hand, given a suitable excited state pair energetic situation for a given choice of laser colors, one must also seriously attend to the non-vanishing of $\mu^{(3)}_{RotInv}(a,b)$. One can suggest some constraints needed for this purpose. Without apparent loss of generality we may choose the local Cartesian system such that axis i is exactly along $(\mu_i)_{ag}$. Thus $(\mu_j)_{ag} = (\mu_k)_{ag} = 0$ and

Eq. (1) reduces simply to $\mu^{(3)}_{RotInv}(a,b) = (\mu_i)_{ag} \left((\mu_j)_{ab} (\mu_k)_{bg} - (\mu_k)_{ab} (\mu_j)_{bg} \right)$.

For this to be strong it is not only necessary that the transition moments themselves be strong but also that one, and only one, of the Cartesian components (j or k) be unequal or even *change sign* depending on whether it appears in the a,b matrix element or in the b,g element. We hope to explore this issue to see whether this requirement is a particularly challenging one or one that is readily observed.

Finally it will be fascinating to explore the microscopic view of 5-WM at fourth order to see whether similarly dramatic resonance effects appear in the rotationally invariant part of $\phi^{(4)}$. At least phase matching is possible for such processes (4).

Acknowledgements

ACA wishes to express gratitude for the most helpful discussions with Professor Koroteev on this topic at ICORS-'98 in Cape Town. ACA and JK are grateful for support under NSF Grant CHE-9616635 as are ADB and PF for support by the Leverhulme Trust.

References

1. A discussion of the present outline for classifying the nonlinear spectroscopies is to be found in: Kirkwood, Jason C.; Ulness, Darin J.; Albrecht, A. C. *J. Phys. Chem. A*, **2000**, *104*, 4167-4173.
2. Giordmaine, J. A. *Phys. Rev.* **1965**, *138*, 1599.
3. Rentzepis, P. M.; Giordmaine, J. A.; Wecht, K.W. *Phys. Rev. Lett.* **1966**, *16*, 792-794.
4. Shkurinov, A. P.; Dubrovskii, A. V.; Koroteev, N. İ. *Phys. Rev. Lett* **1993**, *70*, 1085-1088.
5. Fischer, Peer Ph.D. thesis, University of Cambridge, Cambridge, UK, **1999**.
6. Fischer, P.; Wiersma, D. S.; Rhighini, R.; Champagne, B.; Buckingham, A. D.; *Phys. Rev. Lett.,* **2000**, *85*, No. 20.
7. Kulakov, T.; Belkin, M.; Yan, L.; Ernst, K.-H.; Shen, Y. R. Abstracts of papers of the American Chemical Society, **2000**, Vol. 219, pp. 449-PHYS. See also *Phys. Rev. Lett.,* **2000**, *85*, No. 21.
8. The full presentation is to be appear in the literature. Among other things this more complete treatment will examine difference frequency generation (DFG) in addition to SFG.
9. The expression for the second order macroscopic electric-dipolar susceptibility tensor element, $\chi_{IJK}^{(2)}$, and the double Feynman diagrams for the eight Liouville paths, in terms of the microscopic parameters (though not for a local Cartesian frame) are to be found in Shen, Y. R., *The Principles of Nonlinear Optics;* John Wiley and Sons, Inc.: New York, NY, 1984, pp 17-21.

Chapter 11

Application of Chiral Symmetries in Even-Order Nonlinear Optics

André Persoons[1], Thierry Verbiest[1], Sven Van Elshocht[1], and Martti Kauranen[2]

[1]Laboratory of Chemical and Biological Dynamics, University of Leuven, Celestijnenlaan 200 D, B–3001 Heverlee, Belgium
[2]Institute of Physics, Tampere University of Technology, P.O. Box 692, FIN–33101, Tampere, Finland

We give an overview of the role of chirality in second-order nonlinear optics. In particular, we describe the use of chiral materials to create highly symmetric materials and thin films for second-order nonlinear optics. We discuss the role of magnetic-dipole contributions to the second-order nonlinearity and the possible applications of chiral nonlinear optical materials.

Introduction

Even-order nonlinear optics (NLO) has several applications in the field of opto-electronics (1,2). Several of these nonlinear processes are straightforward to experimentally demonstrate but their application in devices has been hampered by the lack of appropriate materials. Necessary requirements for second-order nonlinear optical materials include the absence of centrosymmetry, stability (thermal and mechanical), low optical loss and large and fast nonlinearities (1).

Perhaps the most stringent requirement for second-order NLO is the absence of centrosymmetry. On a molecular level, this has been achieved by connecting electron donor and acceptor groups by a highly polarizable π-conjugated system, yielding strongly dipolar molecules (3). On a macroscopic level the centrosymmetry must be broken artificially by techniques such as aligning molecular dipoles in an external electric field (poling) or by depositing Langmuir-Blodgett films or self-assembled films (1). These techniques result in noncentrosymmetric media with polar order.

However, polar order is not required for a material to be noncentrosymmetric (4). For example, chiral materials comprised of enantiomerically pure chiral molecules are inherently noncentrosymmetric. Accordingly they give rise to second-order nonlinear optical effects even in the absence of polar order. In fact, frequency mixing has been observed in isotropic solutions of chiral sugar molecules and the electro-optic effect has been predicted to occur in isotropic media (5,6). A significant advantage of chiral isotropic media as compared to traditional polar materials is their inherent thermodynamic stability. Furthermore, we have recently shown that the susceptibility coefficients related to chirality can be quite large (7). In addition, the nonlinear optical properties of chiral molecules and polymers can be significantly enhanced by optimizing magnetic-dipole contributions to the nonlinearity, as was recently demonstrated for thin films of chiral poly(isocyanide)s (8). Therefore, we believe that chirality can be a valuable alternative in the search for new second-order NLO materials.

In this paper, we give an overview of the role of chirality in the field of second-order nonlinear optics and the potential applications of chiral materials in this field.

Theoretical Background

Consider the situation where several optical fields at different frequencies are incident on a nonlinear medium. In general, electric-dipole interactions as well as magnetic-dipole and electric-quadrupole interactions will contribute to the nonlinear response (9). However, in the remainder of this paper, we will only consider electric-dipole and magnetic-dipole contributions to the nonlinearity of the medium. One reason for this is that the magnetic-dipole interaction is actually somewhat stronger than the quadrupole interaction (10). Furthermore, the importance of magnetic interactions in chiral media is well established and one can argue that chirality favours the magnetic interactions (11-13). Furthermore, the general symmetry properties of magnetic-dipole and electric-quadrupole interactions are very similar in second-order nonlinear optics, and separation of the two is difficult. Hence, the quadrupole interaction can be implicitly included in the magnetic-dipole interaction. Then, up to first order in the magnetic-dipole contributions, the nonlinear polarization is given by (2)

$$P_i(\omega_p + \omega_q) = \sum_{jk} \sum_{(pq)} \chi_{ijk}^{eee}(\omega_p + \omega_q; \omega_p, \omega_q) E_j(\omega_p) E_k(\omega_q) +$$

$$\sum_{jk} \sum_{(pq)} \chi_{ijk}^{eem}(\omega_p + \omega_q; \omega_p, \omega_q) E_j(\omega_p) B_k(\omega_q)$$

(1)

where the subscripts refer to Cartesian coordinates in the laboratory (macroscopic) frame. In addition, the materials develops a nonlinear magnetization

$$M_i(\omega_p + \omega_q) = \sum_{jk\,(pq)} \chi_{ijk}^{mee}(\omega_p + \omega_q; \omega_p, \omega_q) E_j(\omega_p) E_k(\omega_q) \tag{2}$$

In eqs (1) and (2), the notation (pq) indicates that, in performing the summation over p and q, the sum $\omega_p + \omega_q$ is to be held fixed, although ω_p and ω_q are each allowed to vary. The superscripts in the susceptibility components associate the respective subscripts with electric-dipole (e) or magnetic-dipole (m) interactions. Both nonlinear polarization and magnetization act as sources of second-harmonic generation. The number of nonvanishing susceptibility components depends on the symmetry of the materials system. In the following paragraphs we will explicitly consider second-order nonlinear optical processes in several specific cases of materials symmetry.

Second-order Nonlinear Optical Processes in Centrosymmetric Isotropic Media

For a centrosymmetric isotropic medium, all electric tensor components χ_{ijk}^{eee} vanish. However, because of the different symmetry properties of magnetic quantities the tensors χ^{eem} and χ^{mee} have one independent component (13) :

$$\chi^{eem} = \chi_{xyz}^{eem} = \chi_{yzx}^{eem} = \chi_{zxy}^{eem} = -\chi_{xzy}^{eem} = -\chi_{yxz}^{eem} = -\chi_{zyx}^{eem}$$

and

$$\chi^{mee} = \chi_{xyz}^{mee} = \chi_{yzx}^{mee} = \chi_{zxy}^{mee} = -\chi_{xzy}^{mee} = -\chi_{yxz}^{mee} = -\chi_{zyx}^{mee}$$

Hence, we can rewrite eqs (1) and (2) as :

$$\begin{aligned}P(\omega_p + \omega_q) &= \chi^{eem}(\omega_p + \omega_q; \omega_p, \omega_q) E(\omega_p) \times B(\omega_q) + \\ &\quad \chi^{eem}(\omega_p + \omega_q; \omega_q, \omega_p) E(\omega_q) \times B(\omega_p)\end{aligned} \tag{3}$$

and

$$M(\omega_p + \omega_q) = 2\chi^{mee}(\omega_p + \omega_q; \omega_p, \omega_q) E(\omega_p) \times E(\omega_q) \tag{4}$$

The functional forms of eqs (3) and (4) imply that isotropic centrosymmetric materials cannot support nonlinear optical processes were the two input beams are copropagating. In that case, the polarization and magnetization would be directed along the wavevectors of the incident beams and cannot give rise to coherent radiation. However, processes that rely on input beams that cross each other are possible. For example, second-harmonic generation with two noncoplanar input beams is possible through the tensor χ^{eem} (χ^{mee} is zero for second-harmonic generation). As for the magnitude of the effect, one could argue that magnetic contributions to the nonlinearity are weak and that the effect would be difficult to detect. However, it has been shown that chirality enhances the strength of magnetic interactions (12). Hence, combining equal amounts of both enantiomers in an isotropic medium (for example a racemic solution of chiral molecules, which is isotropic and centrosymmetric) could therefore still lead to a nonlinear optical response. To the best of our knowledge, second-harmonic generation through χ^{eem} in an isotropic medium has never been observed. However, second-harmonic generation in centrosymmetric crystals of chiral molecules has been observed and attributed to magnetic-dipole contributions (12).

Chiral Isotropic Media

The highest possible symmetry that allows for a nonlinear optical process through the tensor χ^{eee} is an isotropic chiral medium. Such a medium is for example an isotropic solution of enantiomerically pure molecules. In that case, the nonvanishing tensor components are

$$\chi^{eee} = \chi^{eee}_{xyz} = \chi^{eee}_{yzx} = \chi^{eee}_{zxy} = -\chi^{eee}_{xzy} = -\chi^{eee}_{yxz} = -\chi^{eee}_{zyx},$$

$$\chi^{eem} = \chi^{eem}_{xyz} = \chi^{eem}_{yzx} = \chi^{eem}_{zxy} = -\chi^{eem}_{xzy} = -\chi^{eem}_{yxz} = -\chi^{eem}_{zyx},$$

$$\chi^{mee} = \chi^{mee}_{xyz} = \chi^{mee}_{yzx} = \chi^{mee}_{zxy} = -\chi^{mee}_{xzy} = -\chi^{mee}_{yxz} = -\chi^{mee}_{zyx}$$

Hence, eqs (1) and (2) transform into

$$\begin{aligned}
\mathbf{P}(\omega_p + \omega_q) = {} & 2\chi^{eee}(\omega_p + \omega_q; \omega_p, \omega_q)\mathbf{E}(\omega_p) \times \mathbf{E}(\omega_q) + \\
& \chi^{eem}(\omega_p + \omega_q; \omega_p, \omega_q)\mathbf{E}(\omega_p) \times \mathbf{B}(\omega_q) + \\
& \chi^{eem}(\omega_p + \omega_q; \omega_q, \omega_p)\mathbf{E}(\omega_q) \times \mathbf{B}(\omega_p)
\end{aligned} \tag{5}$$

and

$$\mathbf{M}(\omega_p + \omega_q) = 2\chi^{mee}(\omega_p + \omega_q; \omega_p, \omega_q)\mathbf{E}(\omega_p) \times \mathbf{E}(\omega_q) \tag{6}$$

It is clear from eqs (5) and (6) that, in order to observe any nonlinear optical effect, the input beams must not be copropagating. Furthermore, nonlinear optical effects through the tensor χ^{eee} requires two different input frequencies. For example, sum-frequency generation in isotropic solutions of chiral molecules through the tensor χ^{eee} has been experimentally observed (5). Recently, this technique has been proposed as a new tool to study chiral molecules in solution (14).

Another interesting effect is the possibility of the electro-optic effect in chiral isotropic media through the tensor χ^{eee} (6,15). If we neglect the weaker magnetic-dipole contributions, the source polarization is given by :

$$\mathbf{P}(\omega) = 2\chi^{eee}(\omega;\omega,0)\mathbf{E}(\omega) \times \mathbf{E}(0) \tag{7}$$

The tensor $\chi^{eee}(\omega;\omega,0)$ will be nonvanishing when material damping is properly accounted for.

For a plane wave propagating in the z-direction and a static electric field (with amplitude E_0) applied along z, the nonlinear polarization can be written as

$$\mathbf{P}\pm(\omega) = \pm i2\chi^{eee}E_0\mathbf{E}\pm(\omega) \tag{8}$$

Where + and – refer to the left- and right-hand circular polarization of the optical field, $\mathbf{E}_\pm(\omega) = E\pm(\hat{\mathbf{x}} \pm i\hat{\mathbf{y}})/\sqrt{2}$ with $E_\pm$ the scalar amplitude of the field. Important is that the overall imaginary factor in eq 8 reverses the role of the real and imaginary parts of the susceptibility. Therefore, in isotropic chiral media, index modulation is due to the imaginary part of χ, while electro-absorption is due to the real part of χ.

A time-dependent perturbative density-matrix formalism (16) yields the following expression for the susceptibility (5):

$$\chi^{eee}(\omega;\omega,0) = \frac{N}{6\hbar^2}\sum_{n,m}\left[-\frac{i2\gamma_{ng}}{\omega_{ng} - i\gamma_{ng}}\frac{\mathbf{d}_{gn}\cdot(\mathbf{d}_{nm}\times\mathbf{d}_{mg})}{(\omega_{mg} - \omega - i\gamma_{mg})(\omega_{ng} + i\gamma_{ng})} \right.$$

$$-\frac{i2\gamma_{mg}}{\omega_{mg} - i\gamma_{mg}}\frac{\mathbf{d}_{gn}\cdot(\mathbf{d}_{nm}\times\mathbf{d}_{mg})}{(\omega_{ng} + \omega + i\gamma_{ng})(\omega_{mg} - i\gamma_{mg})}$$

$$\tag{9}$$

$$+ iK_{nmg}\frac{\mathbf{d}_{gn}\cdot(\mathbf{d}_{nm}\times\mathbf{d}_{mg})}{(\omega_{mg} - \omega - i\gamma_{mg})(\omega_{ng} + i\gamma_{ng})}$$

$$\left. + iK_{nmg}\frac{\mathbf{d}_{gn}\cdot(\mathbf{d}_{nm}\times\mathbf{d}_{mg})}{(\omega_{ng} + \omega + i\gamma_{ng})(\omega_{mg} - i\gamma_{mg})} \right]$$

Where g is the ground state and m and n are intermediate states. $\mathbf{d}_{nm}$ are transition dipole moments, $\omega_{nm} = \omega_n - \omega_m$ is the transition frequency and γ_{nm} the damping rate between states n and m, and

$$K_{nmg} = \frac{\gamma_{mn} - \gamma_{mg} - \gamma_{ng}}{\omega_{mn} - \omega - i\gamma_{mn}} \tag{10}$$

The first two terms in eq 9 are only nonvanishing in the presence of damping. The last two terms have to be considered in the presence of dephasing, which leads to incomplete destructive interference between the quantum-mechanical pathways. This factor becomes resonant whenever the optical frequency is close to the energy difference between two unpopulated intermediate excited states.

A close inspection of eq 9 leads to several surprising results. When the optical frequency is detuned far from any material resonances, the damping rates can be neglected in the denominators of eqs 9 and 10 and the susceptibility becomes purely imaginary. Eq 8 then implies that index modulation is possible through nonresonant interactions. However, when the optical frequency is close to the energy difference between intermediate excited states the dephasing-induced terms become dominant and essentially real. Eq 8 then implies that the optical field can experience gain due to the electro-optic response. We must note however, that the existence of the electro-optic effect in chiral isotropic media depends critically on how damping is treated. Recently, the validity of the density-matrix formalism in treating quantum-mechanical damping and, as a consequence the existence of the electro-optic effect in chiral isotropic media, has been challenged (16). Therefore, experimental work in this particular field could be used to test various approaches to the quantum-mechanical description of nonlinear optical phenomena.

Chiral Thin Films With In-plane Isotropy

We have observed significant magnetic contributions to second-harmonic generation of Langmuir-Blodgett (LB) films of chiral poly(isocyanides) (8). Such films belong to the C_∞ symmetry group and their nonlinearity can be described by four independent components of χ^{eee}, seven independent components of χ^{eem} and four independent components of χ^{mee}. To obtain an idea about the relative magnitude of the various susceptibilities, we performed an extensive analysis of the second-order nonlinearity of thin films of a chiral chromophore-functionalized poly(isocyanide) (Fig. 1), using a recently developed polarization technique based on second-harmonic generation (8, 17). The relative magnitudes of all components of χ^{eee}, χ^{eem}, and χ^{mee} are listed in Table I. The components that are nonvanishing for chiral surfaces only are referred to as chiral components. Susceptibility components that are non-vanishing for any isotropic (chiral or achiral) surface are called achiral components.

From table 1 it is clear that the largest magnetic susceptibility component is of the order of 20 % of the largest electric susceptibility components. If we use an absolute value of 9 pm/V for the zzz components of χ^{eee}, the magnitude of the largest magnetic contribution would be on the order of 2 pm/V. This value certainly indicates that magnetic-dipole contributions to the nonlinearity can be useful for nonlinear optical applications.

Table I. Relative magnitudes of the chiral and achiral susceptibility components of Langmuir-Blodgett films of a chiral chromophore-functionalized poly(isocyanide) (8).

Tensor	achiral components	chiral components	relative magnitude
χ^{eee}	zzz		1.00
	zxx=zyy		0.62
	xxz=xzx=yyz=yzy		0.60
		xyz=xzy=-yxz=-yzx	0.08
χ^{eem}		zzz	0.23
		zxx=zyy	0.12
		xxz=yyz	0.01
		xzx=yzy	0.05
	xyz=-yxz		0.06
	zxy=-zyx		0.13
	xzy=-yzx		0.06
χ^{mee}		zzz	0.15
		zxx=zyy	0.00
		xxz=xzx=yyz=yzy	0.00
	xyz=xzy=-yxz=-yzx		0.03

SOURCE: data from Reference 8

Chiral Thin Films With Twofold Symmetry

We have also investigated Langmuir-Blodgett films of the chiral helicenebisquinone derivative shown in Fig. 2 (7, 18). The enantiomerically pure material combines into large large helical stacks when incorporated into LB films. The large helical stacks further organize into bundles. Consequently, the samples have a C_2 symmetry. In Langmuir-Blodgett films of the racemic material, such strong organization does not occur.

The supramolecular organization in the nonracemic (*i.e.* prepared from enantiomerically pure material) LB films has a profound impact onto the second order nonlinear optical properties [7]. For example, a 30-layer thick LB-sample made from the pure enantiomer generates a second harmonic signal which is three orders of magnitude higher compared to the signal from a sample with the same number of layers but made from the racemic mixture. The reason is that the nonlinear optical response is dominated by the chiral (electric) tensor components. These components were found to be equal in magnitude for both enantiomers but opposite in sign. No evidence of magnetic contributions to the nonlinearity was found.

An important property of this material is that the anisotropy (and the supramolecular organisation) of the LB samples can be manipulated by external factors (19). Because of the intimate connection between film structure and second-order nonlinear optical response it should therefore be possible to optimize and fine-tune the nonlinear optical response.

We found that at least two experimentally controllable parameters have an influence on the anisotropy of the LB-samples. First, the dipping procedure (i.e. horizontal or vertical dipping) has a profound influence on the structure of the LB films (20). The symmetry of LB-films prepared by vertical and horizontal dipping was investigated by recording the SH intensity while rotating the samples around their surface normal. The rotation patterns obtained for a p-polarized fundamental beam and a s-polarized second-harmonic beam are shown in Fig. 3. Both patterns clearly reveal a twofold symmetry, but the vertically dipped samples are clearly more anisotropic. Furthermore, the maximum signal intensity is about 50 percent higher for the vertically dipped sample.

Second, we found that heating also influences the structure of the LB-samples and consequently the SH-intensity generated by the film. To study the effect of heating, horizontally dipped LB-samples were heated to 220 °C (the melting point of the bulk material is 211 °C) at an average rate of 2 degrees/min. The material was kept at that temperature for 5 minutes and than cooled at an average rate of 4 degrees/min. Next, SHG rotation patterns were recorded before and after heating. From Fig. 3 it is clear that heating significantly increases the in-plane anisotropy. Furthermore, the maximum amount of SHG is more than a factor of two higher for the heated sample.

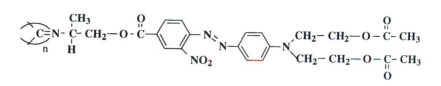

Figure 1. Chemical structure of the chromophore-functionalized poly(isocyanide)

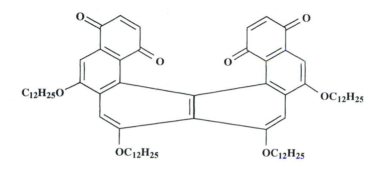

Figure 2. Chemical structure of the helicenebisquinone

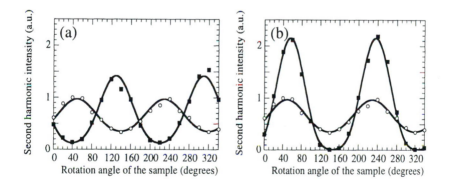

Figure 3. Second-harmonic generation versus the rotation angle of the sample for (a) horizontally (open circles) and vertically dipped (filled squares) sample, and (b) heated (filled squares) and unheated (open circles) horizontally dipped sample

154

The helicenebisquinone can also be used to achieve a new way of quasi-phase matching (21). Quasi-phase matching is important in frequency conversion processes, in which the phase relation between the nonlinear source polarization and the generated field can only be maintained over the distance of a coherence length. However, by reversing the sign of the nonlinearity after each coherence length, the phase relation between the source and the generated field will be restored, allowing the nonlinear signal to grow quasi continuously. In systems with polar order, for example electric field poled materials, this can be achieved by periodically poling the material. In chiral materials, the second-order nonlinear optical coefficients associated with chirality differ in sign for the two enantiomers. Therefore, quasi-phase matched frequency conversion in periodically alternating stacks of the enantiomers of a chiral material should be possible. Since the nonlinearity in the LB- films of the helicenebisquinone is dominated by chirality, this material is an excellent candidate to demonstrate this new type of quasi-phase matching.

We first tested the compatibility of both enantiomers by constructing films whose units were four layers of a single enantiomer, either (P)-(+)- or the (M)-(-). The films were composed of 32 layers comprised of eight identical (P/P/P...) or alternating (P/M/P...) units. Second-harmonic light generated in the samples was detected in transmission, where the coherence length is much larger than the thickness of the samples. In that case, the nonlinear response of each unit can be added. Hence, because of the chirality of the helicene, the second-harmonic intensity from films composed of a single enantiomer should increase quadratically with the total number of layers. On the hand, films composed of alternating stacks of the two enantiomers should exhibit no SHG. Such behaviour was experimentally observed. The results are shown in Table II.

Table II. SH-intensity in transmission of LB-films composed of stacks of the (P)-(+)- and (M)-(-)-enantiomer of the chiral helicenebisquinone.

Number of layers	(P)-(+)-enantiomer SH-intensity (arb. u.)	(P)-(+)-enantiomer alternated With (M)-(-)-enantiomer SH-intensity (arb. u.)
0	0	-
4	1	-
8	3	0.1
12	7	0.3
16	9	0.1
20	21	0.5
24	22	0.1
28	35	0.6
32	46	0.1

Subsequently, the coherence length for the material was determined. Measurements were done in reflection, where the coherence length is much shorter, to avoid having to deposit a too large number of layers. Films were made consisting of two stacks (one for each enantiomer) with a varying number of layers. Such a procedure avoids possible interference of the SH-signal that is generated in transmission and reflected by the film/substrate and substrate/air interfaces. In addition, any interference of linear optical activity of the material is avoided. For an angle of incidence of 42.6°, the coherence length was determined to be approximately 48 layers (21).

As a final step, LB-films were constructed with stacks of 48 layers, alternating both enantiomers. Table III clearly demonstrates that there is a continuous build-up of the second harmonic signal, thus demonstrating quasi-phase-matching in chiral materials.

Table III. SH-intensity in reflection of LB-films composed of stacks of 48 layers of the (P)-(+)- and (M)-(-)-enantiomer of the chiral helicenebisquinone.

Number of layers	SH-intensity (arb. u.)	stacks of 48 layers
48	1	P
96	4	P/M
144	9	P/M/P
192	14	P/M/P/M
288	32	P/M/P/M/P/M

Conclusions

We have given an overview of how chirality can be used to advantage in second-order nonlinear optics. Since chiral materials are inherently noncentrosymmetric, second-order nonlinear optical effects can be observed in highly symmetric media. Furthermore, magnetic-dipole contributions are enhanced by chirality and can significantly increase the nonlinear optical response. An important application of chiral materials in second-order nonlinear optics is the possibility of achieving quasi-phase-matching in structures consisting of alternating stacks of the different enantiomers of the chiral material.

Acknowledgements

We acknowledge the support of the Belgian Government (IUAP P4/11) and the Fund for Scientific Research-Flanders (FWOV G.0338.98, 9.0407.98). T.V. is a postdoctoral fellow of the Fund for Scientific Research-Flanders.

156

References

1. Prasad, P.N.; Williams, D.J. *Introduction to Nonlinear Optical Effects in Molecules & Polymers*; Wiley Interscience: New York, 1990
2. Boyd, R.W. *Nonlinear Optics*; Academic, San Diego, 1992
3. Verbiest, T.; Houbrechts, S.; Kauranen, M.; Clays, K.; Persoons, A. *J. Mater. Chem.* **1997**, *7*, 2175
4. Giordmaine, J.A. *Phys. Rev.* **1965**, *138*, A1599
5. Rentzepis, P.M.; Giordmaine, J.A.; Wecht, K.W. *Phys. Rev. Lett.* **1966**, *16*, 792
6. Kauranen, M.; Persoons, A. *Nonlinear Optics* **1999**, *19*, 309
7. Verbiest, T.; Van Elshocht, S.; Kauranen, M.; Hellemans, L.; Snauwaert, J.; Nuckolls, C.; Katz, T.J.; Persoons, A. *Science* **1998**, *282*, 913
8. Kauranen, M.; Maki, J.J.; Verbiest, T.; Van Elshocht, S.; Persoons, A. *Phys. Rev. B* **1997**, *55*, R1985
9. Wagniére, G. *Linear and Nonlinear Optical Properties of Molecules*; VCH, Weinheim, 1993
10. Louden, R. *The Quantum Theory of Light, 2nd ed.*; Oxford, Oxford, 1983
11. Shuai, Z.; Brédas, J.L. *Adv. Mater.* **1994**, *6*, 486
12. Meijer, E.W.; Havinga, E.E. *Phys. Rev. Lett.* **1990**, *65*, 37
13. Kauranen, M.; Verbiest, T.; Persoons, A. *Journal of Nonlinear Optical Physics & Materials* **1999**, *8*, 171
14. Yang, P.K.; Huang, J.Y. *J. Opt. Soc. Am. B* **1998**, *15*, 1698
15. Beljonne, D.; Shuai, Z.; Brédas, J.L.; Kauranen, M.; Verbiest, T.; Persoons, A. *J. Chem. Phys.* **1998**, *108*, 1301
16. Bloembergen, N. *Nonlinear Optics 4th ed.*; Benjamin, Reading, 1982
17. Maki, J.J.; Kauranen, M.; Persoons, A. *Phys. Rev. B* **1995**, *51*, 1425
18. Nuckolls, C.; Katz, T.J.; Van Elshocht, S.; Verbiest, T.; Kuball, H.-G.; Kiesewalter, S.; Lovinger, A.J.; Persoons, A. *J. Am. Chem. Soc.* **1999**, *121*, 3453
19. Van Elshocht, S.; Verbiest, T.; Katz, T.J.; Nuckolls, C.; Busson, B.; Kauranen, M.; Persoons, A. *accepted for publication in Nonlinear Optics*
20. Van Elshocht, S.; Verbiest, T.; de Schaetzen, G.; Hellemans, L.; Phillips, K.E.S.; Nuckolls, C.; Katz, T.J.; Persoons, A. *accepted for publication in Chem. Phys. Lett.*
21. Busson, B.; Kauranen, M.; Nuckolls, C.; Katz, T.J.; Persoons, A. *Phys. Rev. Lett.* **2000**, *84*, 79

Chapter 12

X-ray Natural Circular Dichroism: Introduction and First Results

Robert D. Peacock

Department of Chemistry, University of Glasgow, Glasgow G12 8QQ, Scotland, United Kingdom

We have extended the measurement of Natural Circular Dichroism to the X-ray region for the first time by using the high circular polarization rates available from the helical undulator source Helios II at ESRF (Grenoble, France). XNCD spectra for two enantiomeric pairs of uniaxial single crystals have been measured: $Na_3Nd(dig)_3.2NaBF_4.6H_2O$ and $2[Co(en)_3Cl_3].NaCl.6H_2O$. The XANES parts of the Nd $L_{2,3}$ and Co K edge X-ray absorptions show circular dichroism corresponding to chiral multiple scattering paths of the photoelectron. In addition both compounds show quadrupole allowed pre-edge features ($2p \rightarrow 4f$ for Nd and $1s \rightarrow 3d$ for Co) which have exceptionally large Kuhn dissymmetry factors.

[1]Dedicated to the memory of Agnes Louise Davis Peacock (1912-1999)

Introduction

In 1895, the same year that Röntgen reported the discovery of X-rays (*1*), Cotton coined (*2*) the term **circular dichroism** to describe the differential absorption of right and left circularly polarized light by a chiral substance He had previously measured the ellipticty induced in plane polarized light after it had passed through basic copper and chromium *d*-tartarate solutions and subsequently looked at the absorption of left and right circularly polarized sodium light by the same solutions. Since ellipticity and circular dichroism are simply related, and since, until relatively recently, it was much easier to measure ellipticity than circular dichroism, most CD experiments prior to the 1960's were made in this way

The adoption of the Pockel's Cell in 1960 (*3*) and the photoelastic modulator in 1966 (*4*), together with developments in signal processing and photomultiplier technology, led to commercial CD instruments which operate between 1400 and 180 nm. This energy range expanded in the 1970's with the development of CD in the infra-red region (*5, 6*) and the vacuum ultraviolet region (*7*). At much the same time the complementary spectroscopies of Raman CID (*8*) and Circularly Polarized Luminescence (*9*) were developed.

Before we began our excursions into the X-ray region, CD could be measured between roughly 1000 cm^{-1} in the infra-red region and 75,500 cm^{-1} in the vacuum uv region. Put another way, CD could be measured in vibrational spectra and in electronic transitions within valence states and between valence and Rydberg orbitals.

Magnetic circular dichroism (MCD) was developed in the 1970's (*10*) and was first measured in the X-ray region in 1987 (*11*). It is worth asking the question why MCD was measured twelve years ago and is now considered a routine X-ray spectroscopy, while the measurement of Natural CD had to wait till 1997. There are two main reasons: firstly MCD is generally a large effect - the dissymmetry factor (g = $\Delta A/A$) often being between 10^{-1} and 10^{-2}. Estimates of the likely value of g for XNCD were variously put at between 10^{-3} and 10^{-8} (see below). Secondly the magnetic experiment has the considerable advantage that the sign changes with the direction of the magnetic field (and is zero in the absence of the field). This enables the experiment to be performed with a single hand of circularly polarized light and field reversal. NCD requires both hands of circularly polarized light, and both enantiomers and a racemate if (at least for the first experiments) a good baseline is to be obtained.

The synchrotron radiation obtained from bending magnets varies from horizontal polarization at the centre of the beam to high rates of circular polarization at the top and bottom of the beam. However the beam intensity decreases from the centre to the edges, so it is only possible to obtain relatively high rates of circular polarization with low flux. To measure Natural CD, either the sample must be physically moved up and down relative to the beam centre, or a moveable slit placed in the beam path to enable the top or bottom parts of the beam to be focussed on the sample. Clearly neither approach is suited to making the first measurements of what was considered to be a weak phenomenon. The helical undulator (*12*) built at ESRF in 1994-6 (*13*) produces a beam of right or left circularly polarized X-rays which is switched by altering the relative phase of the periodic magnetic arrays. Thus the position and focus of the beam on the sample is the same for both hands of light. This presented a real possibility of measuring XNCD.

One previous attempt at measuring natural optical activity in the X-ray region should be mentioned. In 1990 Siddons *et al* measured (*14*) the rotation of plane polarized light at the Co K-edge for powdered samples of $Co(en)_3Br_3$. Enantiomorphous spectra in the pre-edge region were obtained for the enantiomeric complexes but there is a considerable problem with interpreting the data (discussed in more detail below).

X-Ray Spectroscopy

Figure 1 shows a schematic energy level diagram appropriate to the K-edge spectrum of a first row transition metal ion and the $L_{2,3}$-edge spectrum of a lanthanide ion.

In the lanthanide case, the principal absorption is the electric dipole allowed transition from the 2p orbital to the unoccupied continuum εd and εs states. Some 10 eV to low energy of the electric dipole allowed transitions ("white line") is an electric quadrupole allowed transition from the 2p orbital to the bound 4f orbitals. This weak (pre-edge) transition has not been directly observed in absorption. It has been observed in the XMCD spectrum of Gd^{3+} (*15*) and in the resonant inelastic X-ray scattering of Yb metal (*16*) some 5 - 10 eV to low energy of the L_3 edge. In the case of 3d transition metal K-edge spectra, the electric dipole allowed absorption is 1s $\rightarrow$ εp and the electric quadrupole 1s $\rightarrow$ 3d pre-edge feature, although weak, is well resolved from the edge and can be observed directly in the X-ray spectra of transition metal compounds with partially filled d orbitals, eg Co(III).

In addition to the atomic type transitions described above, X-ray spectra show considerable fine structure for several hundred eV above the edge. The region extending to around 50 - 100 eV above the edge is called XANES (X-ray Absorption Near Edge Structure) while that to higher energy is called EXAFS (Extended X-ray Absorption Fine Structure). In both cases the fine structure is caused by the ejected photo-electron being back-scattered by neighbouring atoms. The difference between EXAFS and XANES comes from the relation between the photoelectron de Broglie wavelength and the distance travelled by the photoelectron between scattering events.

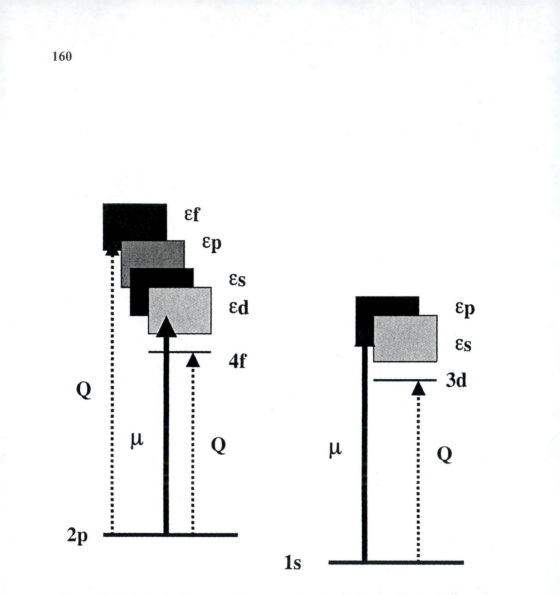

Figure 1: Schematic diagram of the energy levels of a lanthanide ion (left) and a 3d transition metal ion (right). The allowed electric dipole transitions (μ) are shown as bold arrows; the allowed electric quadrupole transitions (Q) as dotted arrows

The de Broglie wavelength decreases from around 2Å in the middle of the XANES region to less than around 0.5Å by 500 eV after the edge. The result of this is that EXAFS is dominated by single scattering events (ie the photoelectron is scattered by a single neighbour) while XANES has a much larger contribution from multiple scattering paths.

The Origins of X-ray CD

The Rosenfeld equation relates the circular dichroism of a transition i → f in a randomly oriented sample to the scalar product of its electric and magnetic dipole moments (E1-M1 mechanism)

$$R = Im\langle i |\mu| f \rangle.\langle f |\mathbf{m}| i \rangle$$

In the case of an achiral chromophore in a chiral environment, such as a tris-chelated metal complex, first order perturbation theory may be employed. The zero order electric moment and magnetic moment arise from different transitions and are mixed by the chiral perturbation ($\mathbf{V}$). Considering the case of the Nd^{3+} ion , and taking the electric dipole transition as 2p → nd,

$$R \approx Im\langle 2p |\mu| nd \rangle\langle nd |\mathbf{V}| n' p \rangle\langle n' p |\mathbf{m}| 2p \rangle$$

Unfortunately there is a considerable problem for the E1-M1 mechanism in that the magnetic dipole transition is forbidden . The magnetic dipole selection rule $|\Delta\ell| = 0$, allows the transition from 2p to the np and continuum εp states. Unfortunately $\mathbf{m}$ is a pure angular operator and cannot connect states which are radially orthogonal. This results in the $|\Delta n| = 0$ selection rule for bound states and also clearly forbids 2p → εp, except *via* core-hole relaxation. An estimate can be made (*16*) of the contribution of core-hole relaxation from the relevant radial overlap integral, <2p|4p>, which can be calculated to be approximately equal to 10^{-2} from the relevant hydrogenic wavefunctions. Remembering that typical values for the dissymmetry factor, $|\Delta\varepsilon/\varepsilon|$, are 10^{-2} for a magnetic dipole allowed transition (for example the $^1A_1 → {}^1E({}^1T_1)$ transition of $[Co(en)_3]^{3+}$ or the n → π* transition of a ketone) or 10^{-3} for an electric dipole allowed transition (for example a charge transfer transition) we can estimate that the maximum value of the dissymmetry factor for the 2p → εd transition from the E1-M1 mechanism is likely to be around 10^{-5} for a metal complex. This is far too small to be measured using the circular polarization rates available above and below the orbit plane of radiation coming from conventional bending magnets, and indeed is on the limit of what might be measurable using radiation from a helical undulator.

There is an additional problem with the E1-M1 mechanism for K-edge spectra or L_1 edge spectra. Magnetic dipole transitions are forbidden from s orbitals so the only possible source of magnetic dipole intensity involves 1s - np orbital mixing.

Two attempts at calculating the E1-M1 contribution to NXCD have been published. Alagna $et\ al$ (18) calculated the dissymmetry factors for the carbon K-edge spectra of propylene oxide to be of the order of 10^{-3}. This relatively large value can be attributed to the closeness of the 1s and 2p orbitals in carbon compared to the 1s and 4p orbitals in a transition metal ion. Goulon $et\ al$ (19) calculated the sulfur K and L edge XNCD for the chiral conformation of H_2S_2. Dissymmetry factors were found to be around 10^{-8} for the K edge and 10^{-4} for the $L_{2,3}$ edges.

Fortunately, for oriented systems, there is an additional source of circular dichroism intensity via the electric dipole-electric quadrupole (E1-E2) mechanism. This mechanism was first developed by Chiu (20), elaborated by Buckingham and Dunn (21) and applied by Barron (22) to explain the CD of the $^1A_1 \rightarrow {}^1E$ ($^1T_2$) transition in the axial crystal spectrum of $[Co(en)_3]^{3+}$. This transition, which is magnetic dipole forbidden and electric quadrupole allowed was the only case, until the present work, where the E1-E2 mechanism had been used to account for CD.

Applying the E1-E2 mechanism to Nd^{3+} we have, for the pre-edge 2p → 4f, electric quadrupole allowed, transition, and assuming the principal source of electric dipole intensity is the 2p → nd transitions:

$$R \approx \langle 2p|\mathbf{Q}|4f\rangle\langle 4f|\mathbf{V}|nd\rangle\langle nd|\mathbf{\mu}|2p\rangle$$

Conversely the CD of the main, electric dipole allowed transition 2p → nd is given by

$$R \approx \langle 2p|\mathbf{\mu}|nd\rangle\langle nd|\mathbf{V}|4f\rangle\langle 4f|\mathbf{Q}|2p\rangle$$

Choice of samples

Based on the experience of visible/uv CD and the expectation that the E1-E2 mechanism, which is only applicable to oriented species, might be significant, we decide to perform our first measurements on oriented single crystals. The choice of our first sample, $Na_3Nd(digly)_3.2NaBF_4.6H_2O$, was dictated by the energy at which ESRF beamline ID12A had the maximum circular polarization. In fact, as will be seen later, the choice was an excellent one The crystal structure of $Na_3Nd(digly)_3.2NaBF_4.6H_2O$ had not been determined (although that of the [ClO₄]⁻ analogue had been) and so we determined (*23*) the crystal structure. This is shown in Figure 2.

The complex is a tris-chelate in which the Nd^{3+} ion is coordinated by 6 carboxylate and 3 ether oxygen atoms. Included in the picture are the 6 nearest neighbour water molecules which are connected to the anion by hydrogen bonds. (The chirality of a tris chelate is designated Λ if the three chelate ligands form a left handed screw or Δ if they form a right handed screw.)

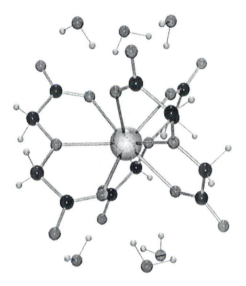

Figure 2: View of the Λ-[Nd(digly)₃]³⁻ anion (and associated water molecules) perpendicular to the C_3 axis

Having established the viability of measuring XNCD we chose 2[Co(en)₃Cl₃].NaCl.6H₂O as our second subject of investigation. The [Co(en)₃]³⁺ ion has been of historic importance in the development of transition metal optical activity. A crystal of 2[Λ-Co(en)₃Cl₃].NaCl.6H₂O provided the first transition metal complex to have its absolute configuration determined by the anomalous scattering of X-rays (*24*). The same compound was the first transition metal complex to have its electronic CD spectrum measured in both the solid state and in solution (*25, 26*) and as such has been the complex of choice for testing the various theories of transition metal CD in the visible region (*27 - 30*). In addition, the presence of a well-resolved pre-edge (1s → 3d) feature some 18 eV to low energy of the K edge absorption and the expectation that the E1-E2 mechanism would be dominant, made [Co(en)₃]³⁺ a natural choice. The structure of 2[Λ-Co(en)₃Cl₃].NaCl.6H₂O was originally determined by 2-D

164

crystallography some 40 years ago. Accordingly we have redetermined (*31*) the structure.

Results and Discussion

$Na_3Nd(digly)_3.2NaBF_4.6H_2O$

Figure 3 shows the Nd L_3 edge absorption spectrum and the XNCD spectra (multiplied by 100) of enantiomorphic single crystals of $Na_3Nd(digly)_3.2NaBF_4.6H_2O$ (*32*). The spectra clearly show that we have measured CD in the X-ray region of the spectrum. As expected, the enantiomorphic single crystals have mirror image circular dichroism spectra. The intensity of the CD is much larger than that expected from the E1-M1 mechanism, the dissymmetry factor in the XANES region being ~ 3 x 10^{-3}, and suggests that the intensity comes from the E1-E2 term. This is confirmed by the disappearance of the CD in the powder spectra as required by this mechanism

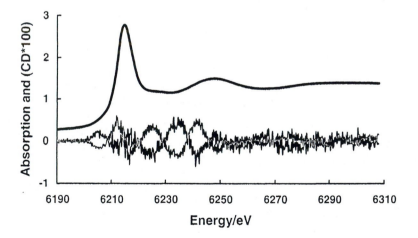

Figure 3: Axial absorption and CD*100 spectra for single crystals of Λ- (dark) and Δ- (light) $Na_3Nd(digly)_3.2NaBF_4.6H_2O$

The XNCD spectra are best divided into three regions for discussion: the quadrupole allowed 2p → 4f pre-edge transition, the white line/ XANES region and the EXAFS region.

A major feature of the spectra is a strong dichroism (g is at least 7×10^{-3}, and is probably considerably larger, since the transition cannot be independently detected in the absorption spectrum) exactly where the predicted electric quadrupole-allowed 2p → 4f transitions are expected. The position of the transition some 9 eV to low energy of the white line is similar to that of equivalent transitions in other rare earths (*15, 16*). Enantiomeric features with g ~ 10^{-3} are also seen under the white line and in the XANES region of the spectrum with weaker features evident in the EXAFS. The XANES region of the spectrum has a significant contribution from multiple scattering. Single scattering paths cannot be chiral by definition and therefore cannot contribute to the dichroism (the theory of chiral multiple scattering is described in detail in the following chapter). The CD spectrum, therefore, singles out multiple scattering pathways, which explains why the dichroism is relatively stronger in XANES than in EXAFS. We have calculated the CD spectrum using the

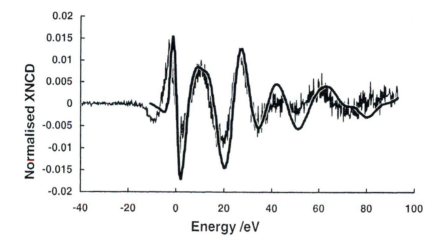

Figure 4: Comparison between the experimental and calculated XNCD signals, normalised to the respective atomic absorptions. The experimental curve has been multiplied by a factor of 4.5.

multiple scattering formalism. A comparison of experiment and calculation is shown in Figure 4. The result is an excellent agreement. A Fourier transform of the XANES CD gives distances of 7 and 8.2 Å while the equivalent analysis of the absorption XANES gives peaks at about 2.5 and 3.5Å . Thus while the absorption has a major contribution from the single scattering paths involving the Nd and the ligating oxygen atoms (2.433 , 2.535Å) and the Nd and the first carbon neighbours (3.51Å) the CD is entirely due to multiple scattering. The

166

two chiral paths which best fit the data are Nd - O(carboxyl) - C(carboxyl) (7.05 Å) and Nd - O(carboxyl) - O(carboxyl adjacent ligand) (8.25 Å).

$2[Co(en)_3Cl_3].NaCl.6H_2O$

Figure 5 shows the absorption and XNCD spectra of axial single crystals of $2[\Lambda$- and Δ-$Co(en)_3Cl_3].NaCl.6H_2O$ (*33*) . The most noticeable feature of the spectra is the *spectacular size* of the 1s → 3d CD observed below the Co *K* edge. The Kuhn dissymmetry factor, g = ($\Delta A/A$), is 12.5% if the raw CD and absorption are used. As expected, a racemic crystal gave no XNCD signal. To attempt to explain this dissymmetry factor, which is comparable to that of the magnetic dipole allowed $^1A_1 \rightarrow {}^1A_2(T_{1g})$ and $^1E(T_{1g})$ d ↔ d transitions, we

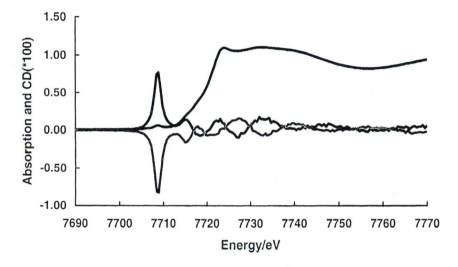

Figure 5: Axial absorption and XNCD spectra of
$2[Co(en)_3Cl_3].NaCl.6H_2O$ - the dark curve is the CD of the Δ crystal, the light curve that of the Λ crystal

extended the *ab initio* approach to the calculation of core-valence CD for the first time. In order to investigate the efficacy of the E1-E2 mechanism for the 1s → 3d transition, we have performed both frozen core and relaxed core HF calculations in a Gaussian orbital basis (*33*). Such calculations of the absorption spectra of transition metal complexes by *ab initio* methods are rare (*34*) and there is only one example of an *ab initio* calculation of transition metal d ↔ d

NCD (*35*). We used as our model complex the Λ-$[Co(en)_3]^{3+}$ ion in a D_3 geometry optimization, starting from the crystal structure (*31*) of $2[\Lambda$-$Co(en)_3Cl_3].NaCl.6H_2O$. As a check, the NCD of the magnetic dipole allowed $^1A_1 \rightarrow {}^1A_2(T_{1g})$ and $^1E(T_{1g})$ valence transitions were reproduced satisfactorily, with similar agreement to experiment to that found in references 30 and 35. The results confirm the importance of the quadrupole-dipole interference term in the mechanism of the XNCD in this oriented crystal. The sign of the 1s $\rightarrow$ 3d pre-edge signal is correctly reproduced as is the order of magnitude of the Kuhn dissymmetry factor ($g_{calc} = 0.11$, $g_{obs} = 0.125$; $R_{calc} = -7.8 \times 10^{-44}$ cgsu, $R_{obs} = -8.7 \times 10^{-44}$ cgsu). The source of electric dipole transition moment for the pre-edge CD is approximately 97% Co-based, as expected for a transition emanating from the 1s core orbital.

The large magnitude of the CD in the pre-edge excitation is satisfactorily explained by the E1-E2 mechanism which is particularly efficient in this system. The g factor for 1s $\rightarrow$ 3d is the same order of magnitude as that of the d $\leftrightarrow$ d excitations where the latter are dominated by the E1-M1 mechanism. As a comparison, the dissymmetry factor is approximately 22% for the 3d $\rightarrow$ 3d $^1A_1 \rightarrow {}^1E(T_{1g})$ transition which is electric- and magnetic-dipole allowed. The 1s $\rightarrow$ 3d transition is essentially pure quadrupole in character, with about 10% electric dipole activity. Almost all of this borrowed transition moment is effective in the CD. The additional frequency factor in the E1-E2 mechanism provides an approximately 3000-fold enhancement compared to the same transition moments if they were operating in the visible spectral region.

Returning to the optical rotation measurements of Siddons *et al* (*14*), we can see why the data present a problem. We have shown both experimentally and theoretically that the E1-E2 mechanism accounts for essentially all the CD intensity in the X-ray spectra. However that mechanism gives zero intensity in random orientation, the condition under which the ellipticity experiment was performed. The only suggestion at present is that there must have been some residual alignment of the crystallites of the $[Co(en)_3]Br_3$ samples.

α-LiIO$_3$

Shortly after the observation of XNCD in $Na_3Nd(digly)_3.2NaBF_4.6H_2O$, Goulon *et al* reported (*36*) the XNCD spectra of the uniaxial gyrotropic crystal α-LiIO$_3$ at the L_1, L_2 and L_3 Iodine edges. A particularly interesting feature of this system is that the individual $[IO_3]^-$ ions are achiral (C_{3v} symmetry) and the chirality is provided by the helical arrangement of the ions in the crystal. Thus all chiral scattering paths must involve at least second shell neighbours.

The spectra at the L_2 and L_3 edges were found to have the same general appearance and sign, while that of the L_1 edge was of opposite sign and somewhat more intense. Once again multiple scattering calculations involving the E1-E2 mechanism satisfactorily reproduced the experimental results.

Experimental

The XCD spectra were carried out at the ESRF beamline ID12A which is dedicated to polarization-dependent XAFS studies (*13*). Circularly or elliptically polarized X-ray photons were generated with a helical undulator Helios II (*12*) which has the optional capability to flip the helicity of the emitted photons. The fixed exit monochromator was equipped with a pair of Si(111) crystals cooled down to its optimum temperature of -140 C. Given the ultra-low emittance of the source, the energy resolution was close to the theoretical value, 0.9 eV at the Nd L_3 edge, while the core-hole life time broadening is 3.65 eV (*37*). Polarimetry, using a 1/4-wave plate (*38*), allowed the circular polarization rate of the monochromatic beam to be determined to be of the order of 78%. The typical beam size at the sample (i.e. at 52 meters from the source point) was 300x300 μm^2 due to the horizontal focussing capability of a horizontally deflecting SiC mirror. All the spectra were obtained by monitoring the fluorescence yield as a function of the energy. The crystals were inserted in a stainless-steel high vacuum compatible fluorescence chamber equipped with photodiodes operated in a backscattering geometry

Hexagonal crystals of Λ- and Δ- $Na_3Nd(digly)_3.2NaBF_4.6H_2O$ (digly = the dianion of diglycolic acid) and crystals of $2[\Lambda$-$Co(en)_3Cl_3].NaCl.6H_2O$ and $2[\Delta$-$Co(en)_3Cl_3].NaCl.6H_2O$ were grown from aqueous solution by slow evaporation. The unique axes was identified by optical microscopy and confirmed by measuring the axial CD spectrum in the visible region. For each compound two enantiomeric crystals were selected and mounted in the sample compartment of ID12A. Spectra of $Na_3Nd(digly)_3.2NaBF_4.6H_2O$ were measured at ambient temperature while $2[\Lambda$-$Co(en)_3Cl_3].NaCl.6H_2O$ was cooled to a nominal 100K. In order to compensate for possible spurious signals due to crystal decomposition or drift we used a simple data acquisition program which changed the polarization of the light in the sequence 1(+), 2(-), 3(-), 4(+). Defining the dichroic signal D = (-) - (+), any spurious signal can be eliminated and the true dichroic signal is given by D = [-1(+) + 2(-) +3(-) -4(+)]/2. The crystals of $Na_3Nd(digly)_3.2NaBF_4.6H_2O$ were severely dehydrated in the high vacuum of the beam line. After the experiment the CD of thin sections of the crystals were measured. The axial CD in the visible region was found to be unchanged thus indicating that the dehydration had not changed the molecular chirality of the sample nor had the orientation of the $[Nd(digly)_3]^{3-}$ ion relative to the crystal axes been altered.

Acknowledgements

The measurement and understanding of Natural X-ray CD has been a team effort. The team members are Lucilla Alagna, Tommaso Prosperi and Stefano Turchini from ICMAT-CNR, Rome, Calogero Natoli from INFN-Laboratori di Frascati, Chantal Goulon-Ginet from Joseph Fourier University, Grenoble, José Goulon and Andrei Rogalev from ESRF, Brian Stewart from the University of Paisley and myself. It goes without saying that without the excellent beamline produced by José and Andrei, the project would have been impossible

References

1. Röntgen, W. C. **1895**, Communication to the Würzburg Physical and Medical Society.
2. Cotton; A. *Compt. Rend. Paris*, **1895**, *120*, 989, 1044.
3. Grosjean, M.; Legrand, M. *Compt. Rend. Paris*, **1960**, *251*, 2150
4. Billardon, M.; Badoz, J. *Compt. Rend. Paris*, **1966, 262,** 1672.
5. Dudley, R. J.; Mason, S. F.; Peacock, R. D. *J. Chem. Soc. Chem . Comm.*, **1972**, 1084 - 1085.
6. Chabay, I.; Hsu, E. C.; Holzwarth, G. *Chem. Phys. Lett.*, **1972**, *15*, 211 - 214
7. Schnepp, O.; Allen, S.; Pearson, E. F. *Rev. Sci. Instum.*, **1970**, *41*, 1136.
8. Barron, L. D.; Bogaard, M. P.; Buckingham, A. D. *J. Amer. Chem. Soc.* **1973**, *95*, 603 - 605.
9. Richardson, F. S.; Riehl, J. P. Chem. Rev., **1977**, *77*, 773 - 792.
10. Schatz, P. N.; McCaffery, A. D. *Quart. Rev. Chem. Soc*, **1969**, *23*, 552 - 584.
11. Schutz, G.; Wagner, W.; Wilhelm, W.; Kienle, P.; Zeller, R.; Frahm, R.; Materlik, G. *Phys. Rev. Lett.*, **1987**, *58*, 737 - 740.
12. Elleaume, P. *J. Synchrotron Rad.* , **1994**, *1*, 19 - 26.
13. Goulon J.; Rogalev A.; Gauthier C.; GoulonGinet C.; Paste S.; Signorato R.; Neumann C.; Varga L.; Malgrange C.; *J. Synchrotron Radiation*, **1998**, *5*, 232 - 238
14. Siddons, D.P., Hart, M., Amemiya, Y. & Hastings, J.B. *Phys. Rev. Lett.*, **1990**, *64*, 1967 - .1970
15. Giorgetti, C.; Dartyge, E.; Brouder, C.; Baudelet, F.; Meyer, C.; Pizzini, S.; Fontaine, A.; Galera, R. M. *Phys. Rev. Lett.*, 1995, **75**, 3186 - 3189.
16. Krisch, M. H.; Kao, C. C.; Calieke, W. A.; Hamalainen, K.; Hastings, J. B. *Phys Rev Lett.*, 1995, **74**, 4931 - 4934.
17. Stewart, B. J. *J de Physique IV*, colloque C3, supplément au J de Physique III, **1994**, *4*, C9

18. Alagna, L.; Di Fonzo, S.; Prosperi, T.; Turchini, S.; Lazeretti, P.; Malagoli, M.; Zanazi, R.; Natoli, C. R.; Stephens, P. J. *Chem. Phys. Letts.*, **1994**, *223*, 402 - 410.

19. Goulon, J.; Sette, F.; Moise, C.; Fontaine, A.; Perey, D.; Rudolf, P.; Baudelet, F. *Jpn. J. Appl. Phys.*, **1993**, Suppl. *32-2*, 284 - 289.

20. Chiu, Y.N. *J. Chem. Phys.*, **1970**, *52*, 1042 - 1053.

21. Buckingham, A. D.; Dunn, M. B. *J. Chem. Soc.* (A), **1971**, 1988 - 1991.

22. Barron, L. D. *Molec. Phys.*, **1971**, *21*, 241 - 246.

23. Farrugia, L. J.; Peacock, R. D.; Stewart, B. *Acta Cryst. C*, submitted.

24. Saito, Y.; Nakatsu, K.; Shiro, M.; Kuroya, H. *Acta Cryst.* **1954**, *7*, 636

25. Mathieu, J-P. *C.R Acad. Sci. Paris*, **1953**, *236*, 2395-2397

26. McCaffery, A. J.; Mason, S. F. *Molec .Phys.*, **1963**, *6*, 359-371.

27. Moffit, W. J. *J. Chem. Phys.*, **1956**, *25*, 1189-1198.

28. Mason, S. F; Seal, R. H. *Molec. Phys*, **1976**, *31*, 755-775.

29. Strickland, R.W.; Richardson, F. S. *Inorg. Chem.*, **1973**, *12*, 1025-1036

30. Evans, R. S.; Schriener, A. F.; Hauser, P. J. *Inorg. Chem.*, **1974**, *13*, 2185-2192.

31. Farrugia, L. J.; Peacock, R. D.; Stewart, B *Acta Cryst C,* **2000**, *56C*, 149 - 151.

32. Alagna, L.; Prosperi, T.; Turchini, S.; Goulon, J.; Rogalev, A.; Goulon-Ginet, C.; Natoli, C. R.; Peacock, R. D.; Stewart, B. *Phys. Rev. Letters*, **1998**, *80*, 4799-4802.

33. Stewart, B.; Peacock, R. D.; Alagna, L.; Prosperi, T.; Turchini, S.; Goulon, J.; Rogalev, A.; Goulon-Ginet, C.. *J. Amer. Chem. Soc.*, **1999**, *121*, 10233 - 10234.

34. Vanquickenborne, L. G.; Coussens, B.; Ceulmans, A.; Pierloot, K. *Inorg. Chem.* **1991**, *30*, 2978-2986.

35. Ernst, M. C.; Royer, D. J. *Inorg Chem.*, **1993**, *32*, 1226-1232.

36. Goulon, J.; Goulon-Ginet, C.; Rogalev, A.; Gotte, V.; Malgrange, C.; Brouder, C.; Natoli, C. R. *J. Chem. Phys.*, **1998**, *108*, 6394 - 6403.

37. Krause, M. O.; Olivier, J. H. *J. Phys. Chem. Ref. Data*, **1979**, *8*, 329 - 338.

38. Goulon, J.; Malgrange, C.; Giles, C.;. Neumann, C.; Rogalev, A.;. Moguilline, E.; de Bergevine, F.; Vettier, C. *J. Synchrotron Rad.*, **1996**, *3*, 272 - 281.

Chapter 13

X-ray Natural Circular Dichroism: Theory and Recent Developments

Brian Stewart

Department of Chemistry and Chemical Engineering,
University of Paisley, Paisley PA1 2BE, Scotland, United Kingdom

A theoretical model is described for the origins of natural circular dichroism (XNCD) in X-ray pre-edge and near-edge (XANES) photoabsorption. The predominant contribution results from the interference between the electric dipole and electric quadrupole transition moments (E1-E2 mechanism) and gives rise to orientation-dependent CD. For the photoelectron continuum absorption, a chiral multiple-scattering approach successfully acounts for the XANES CD of rare-earth L edge absorption in enantiomorphic single crystals. The same approach successfully treats gyrotropic single crystal XNCD where individually achiral units are arranged helically. Ab-initio Gaussian orbital calculations have been extended to treat core-valence CD and are able to account for the large pre-K edge CD seen in $2[Co(en)_3Cl_3].NaCl.6H_2O$.

Introduction

Since the days of Pasteur, the importance of mirror image properties of matter have been recognized as playing a key role in the structure of biomolecules and their discriminatory interactions. Physical techniques giving molecular chirality information have been developed using the differential interaction of circularly polarized radiation with chromophores in the visible, near and vacuum ultraviolet, the near infrared and the vibrational infrared spectral regions. Recently these techniques have been extended to the X-ray region (*1,2*) due to the development of high-brilliance synchrotron sources with insertion devices capable of delivering well-defined and high circular polarization rates. This article describes some of the developments in the theory of natural CD in the X-ray region where, by extending the well-developed analytical tools for X-ray near-edge absorption features, element-specific information about local chirality may be obtained.

The extremes of the electromagnetic spectrum have been less accessible for instrumental reasons (see preceeding chapter). Additionally however, there are sum rules and dimensional arguments which suggest that the CD effect is expected to decrease in magnitude for very long or very short wavelengths in comparison to the scale of a chiral molecular structure. More accurately, the characteristic excitation length should be considered. This is especially relevant in the vibrational IR where wavelengths are very large in comparison to the characteristic length of localized vibrational excitations. The Condon-Eyring sum rule for optical activity:

$$\sum_f \mathbf{R}_{if} = Im\left(\sum_f \langle i|\boldsymbol{\mu}| f \rangle \cdot \langle f |\mathbf{m}| i \rangle \right) = 0$$

implies that optical rotation vanishes at very low and very high energies relative to the excitation energies of the electronic system. However, this does not mean that circular dichroism is necessarily small at these extremes. In particular, in the X-ray region, the interaction with quadrupole transition moments is enhanced when the dimensions of the electronic displacement are comparable to the wavelength of the radiation (breakdown of the dipole approximation). This allows substantial CD to be derived from the E1-E2 mechanism (see below). This situation pertains to the near-edge absorption (XANES) where multiple scattering of the photoelectron by near neighbour atoms is largely responsible for the structure of the absorption profile. In the present work, it is also relevant to pre-edge excitations with effectively pure quadrupole character (eg 1s-3d in the transition metals and 2p-4f in the lanthanides).

Photon–molecule interactions

Absorption

The quantum mechanical approach to light absorption (*3*) involves evaluating matrix elements of a radiation-molecule interaction operator between initial and final states, subject to the conservation of energy. The cross section for a photon absorption process in which there is an electronic transition from an initial state $\langle i|$ to a final state $|f\rangle$ is given by (*4*)

$$\sigma = (4\pi^2 \, \alpha_o \hbar \, / e^2 \omega) \, |\langle i| \, H_{int}|f\rangle|^2 \, \delta(E_f - E_i - \hbar\omega)$$

for radiation of angular frequency ω. α_o is the fine structure constant and the δ-function takes care of energy conservation. The interaction operator here is essentially the scalar product of the electron momentum $\mathbf{p}$ and the space part of the vector potential $\mathbf{A}$ of the radiation field:

$$H_{int} = (e/m) \, \mathbf{p} \cdot \mathbf{A}, \qquad \mathbf{A} = \varepsilon \, e^{i\mathbf{k}\cdot\mathbf{r}}$$

for photons with wavevector $\mathbf{k}$ and polarization vector ε interacting with an electron at position $\mathbf{r}$.

The vector potential may be expanded in a truncated Taylor series,

$$e^{i\mathbf{k}\cdot\mathbf{r}} = 1 + i \, \mathbf{k}\cdot\mathbf{r} \quad (\text{provided that } \mathbf{k}\cdot\mathbf{r} \ll 1)$$

which when substituted into the cross-section expression gives:

$$\sigma = (4\pi^2 \, \alpha_o \, \hbar/e^2\omega) \, |\langle i| \, \varepsilon\cdot\mathbf{p} + i \, (\varepsilon\cdot\mathbf{p})(\mathbf{k}\cdot\mathbf{r}) \, |f\rangle|^2 \, \delta(E_f - E_i - \hbar\omega)$$

The $|\langle i| \, \varepsilon\cdot\mathbf{p} \, |f\rangle|^2$ contribution gives rise to the familiar electric dipole absorption responsible for a large part of electronic spectroscopy. The second term in the matrix element may be divided into a symmetric part, identifiable with the electric quadrupole interaction, and an antisymmetric part which is the magnetic dipole operator:

$$(\varepsilon\cdot\mathbf{p})(\mathbf{k}\cdot\mathbf{r}) = \tfrac{1}{2}[\, (\varepsilon\cdot\mathbf{p})(\mathbf{k}\cdot\mathbf{r}) + (\varepsilon\cdot\mathbf{r})(\mathbf{k}\cdot\mathbf{p}) \,] + \tfrac{1}{2}[\, (\varepsilon\cdot\mathbf{p})(\mathbf{k}\cdot\mathbf{r}) - (\varepsilon\cdot\mathbf{r})(\mathbf{k}\cdot\mathbf{p}) \,]$$
$$= \quad \text{electric quadrupole} \quad + \quad \text{magnetic dipole}$$

The presence of these higher order terms in the expansion of the transition operator results in further contributions to the absorption cross section. Due to their dependence on powers of $\mathbf{k \cdot r}$, these are expected to be smaller than the electric dipole contribution in many circumstances. For example, consider visible light (λ = 500 nm) for which $|\mathbf{k}| = 2\pi/\lambda \sim 2 \times 10^6$ m^{-1}. Since $|\mathbf{r}| \approx 10^{-10}$ m for a valence electron, $\mathbf{k \cdot r} \approx 2 \times 10^{-4}$.

At X-ray wavelengths however, $|\mathbf{k}| \sim 5 \times 10^9$ m^{-1} and $\mathbf{k \cdot r} \approx 5 \times 10^{-1}$. Quadrupole activity is thus expected to be considerably enhanced in the X-ray region compared with the visible.

Optical activity

In a general treatment of higher order contributions to optical activity, Chiu (5) showed that, when the transition operator is written as a series with electric and magnetic multipole contributions[1]:

T = E1 + E2 + E3 + ... + M1 + M2 + M3 + ...

  = electric dipole + electric quadrupole + electric octupole + ...
   + magnetic dipole + magnetic quadrupole + magnetic octupole + ...

then contributions to the total transition probability due to cross terms between transition moment operators of opposite parity are responsible for the differential absorption of circularly polarized light. The leading term is the pseudoscalar product between the electric (μ) and the magnetic (**m**) dipole operators which occurs in the famous Rosenfeld-Condon equation for the rotational strength in isotropic systems:

$$R = Im\{\langle i|\mu|f\rangle \cdot \langle f|\mathbf{m}|i\rangle\}$$

---

[1] The multipoles behave as tensors of increasing rank (k) and alternating parity given by $(-1)^k$ for electric moments and $(-1)^{k-1}$ for magnetic moments. The isotropic absorption probability is a scalar invariant and is given by the sum of scalar product terms arising from each multipole :

$$T^{(0)} \sim E1 \cdot E1 \ + \ E2 \cdot E2 + ... + \ M1 \cdot M1 \ + \ M2 \cdot M2 \ + ...$$

For oriented systems of less than cubic symmetry, the anisotropy in absorption is related to corresponding tensor products, eg $[E1 \otimes E1]^{(2)}$.

This will be referred to as the E1-M1 mechanism. A pseudoscalar quantity is a rotational invariant but, unlike a true scalar, it has odd parity and therefore inverts its sign between enantiomeric systems.

Chiu's investigation of optical activity arising from higher-order cross terms recognized that in many cases the effects would be orientation dependent. Pseudoscalar terms are the only ones which survive in random orientation (molecules in solution or liquid phase). But at the same order of perturbation as E1-M1 there is a product of the electric dipole and electric quadrupole transition operators (E1-E2 mechanism). Since this product involves tensors of unequal rank, the result cannot be a pseudoscalar and this term would not therefore contribute in random orientation. This term, which gives rise to orientation-dependent optical activity, was developed by Buckingham & Dunn (6) and recognized by Barron (7) as a potential contribution to the visible CD in oriented crystals containing the $[Co(en)_3]^{3+}$ ion. Until the present work this was the only case where E1-E2 optical activity was thought to be significant. Work in XNCD has found that the predominant contribution to X-ray optical activity is from the E1-E2 mechanism. The reason for this is that the E1-M1 contribution depends on the possibility of a significant magnetic dipole transition probability and this is strongly forbidden in core excitations due to the radial orthogonality of core with valence and continuum states. This orthogonality is partially removed due to relaxation of the core-hole excited state, but this is not very effective and in the cases studied so far there is no definite evidence of pseudoscalar XNCD. Calculations (1) indicate the E1-M1 contribution to be ca. $1/1000^{th}$ of the E1-E2 part.

For light propagating along a 3-fold or higher axis (the z axis) in an oriented sample, the rotational strength, including E1-M1 and E1-E2 parts, is given in cartesian components by (8):

$$R_{if} = -\omega_{if}\, Re\, \{\langle i|\, \mu_x|f\rangle\langle f|Q_{yz}|i\rangle - \langle i|\mu_y|f\rangle\langle f|Q_{xz}|i\rangle\} \qquad \text{(E1-E2)}$$
$$+ Im\, \{\langle i|\mu_x|f\rangle\langle f|m_x|i\rangle + \langle i|\mu_y|f\rangle\langle f|m_y|i\rangle\} \qquad \text{(E1-M1)}$$

The E1-E2 part corresponds to a 2nd rank odd parity tensor. In terms of spherical irreducible tensors, this E1-E2 optical acivity derives from the following irreducible tensor product:

$$[\mu^{(1)}Q^{(2)}]_p^{(2)}$$

For 3-fold or higher uniaxial symmetry, only the cylindrical component (p=0) does not vanish. The rotational behaviour of such a 2nd rank tensor is determined by the order parameter,

$$S_0^{(2)} = \tfrac{1}{2}(3\cos^2\theta - 1)$$

where θ is the angle between the molecular or crystal z axis and the wavevector of the radiation. For the case of random orientation, $<\cos^2\theta> = 1/3$, $S^{(2)}_0 = 0$ and the E1-E2 contibution vanishes.

X-ray Absorption Spectroscopy

An important underlying feature of XAS is the core nature of the one-electron initial state. As a consequence, the transition matrix elements involve integrals which are expected to be dominated by the region close to the photabsorbing atom. The major part of an X-ray edge absorpion is therefore atomic in nature and of no interest to structural chemists. Of course it is the small modulations of the edge absorption due to the presence of neighbouring atoms that has to be analysed for structural information (EXAFS)(9). The common approach here is to use scattering theoretical methods (10,11) to describe the propagation of the photoelectron in the neighbourhood of the absorber. Conventional EXAFS analysis is applied to energies above ~100eV from the ionisation threshold where the de Broglie wavelength of the photoelectron is short in comparison to interatomic spacings and the effect of neighbour atoms is largely accounted for by single-scattering. Below ~100eV (XANES region), the photoelectron wavelength is comparable to the near-neighbour distances and multiple scattering processes are of significant importance. The analysis of multiple scattering (10-12) is complicated but it is known to provide 3D structural information about the neighbourhood of a photoabsorber whereas the single scattering interpretation of EXAFS gives only a radial distribution function. For this last reason it should be anticipated that multiple scattering analysis will be necessary to account for the effects of chirality in XAS.

In the context of chirality, the above situation is reminiscent of the model of an intrinsically achiral chromophore in a dissymmetric environment. The sources of transition amplitude are confined to the photoabsorber atom and the chiral arrangement of neighbouring atoms provides a perturbation able to couple those transition moments in such a way as to the produce optical activity. Since, as seen above, the E1-E2 coupling is the most significant for XNCD, we should look for multiple scattering processes that are capable of coupling an electric dipole with an electric quadrupole excitation.

In the scattering interpretation, the absorption process is separated into a sequence of photon interaction, electron propagation and scattering events. Rearranging the absorption cross-section expression,

$$\sigma = (4\pi^2\,\alpha_o\,\hbar/e^2\omega)\sum_f \langle i|H_{int}|f\rangle\delta(E_f - E_i - \hbar\omega)\langle f|H_{int}|i\rangle$$

the structure

$$\sum_f |f\rangle\delta(E_f - E_i - \hbar\omega)\langle f| = -\frac{1}{\pi} Im[G(E_i + \hbar\omega)]$$

represents the electron propagation and scattering processes and may be identified with the imaginary part of the Green's function for the photoelectron. Introducing the atomic transition operator, $T^{\pm}$, with circular polarization components (T^+ corresponds to left circular photon polarization in the sense of Born & Wolf (*13*)):

$$\sigma^{\mp} = 4\pi^2 \, \alpha_o \, \hbar\omega \, \langle \, \phi_c(r) \, | \, T^{\mp} \, ImG(r,r'; (\hbar\omega\text{-}I_c)) \, T^{\pm} \, | \, \phi_c(r') \, \rangle$$

$$T^{\pm} = E1^{\pm} + M1^{\pm} + E2^{\pm} + \ldots\ldots = r^{\pm} \pm i(\alpha/2)(r\times\nabla)^{\pm} + i(\alpha/4) \, \omega \, z \, r^{\pm} + \ldots\ldots$$

for light propagating along the Z axis. The Green's function of the system may be written exactly with the use of muffin-tin potentials for each atomic site and may be expanded in a multiple scattering series (see Figure 1).

This series, which has been shown to converge (*14*) for photoelectron energies above ~20 eV, describes single, double and higher scatterings by neighbours of the photoabsorber. Each term in the multiple scattering series possesses the symmetry of the complete cross-section so that a particular scattering path contributes to XNCD only if the geometry of that path fulfills the usual symmetry requirements of natural optical activity. For the E1-M1 contribution these are well-known. In the E1-E2 case, optical activity vanishes

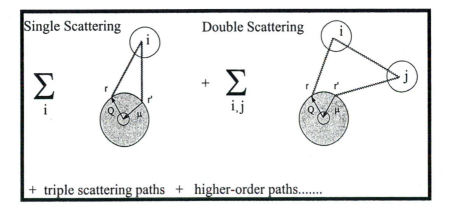

Figure 1. The Multiple Scattering Series and XNCD. Q and μ represent the atomic matrix elements of the quadrupole and dipole operators on the photoabsorber

for random orientation and the orientation of paths in relation to the radiation wavevector must be considered. Symmetry rules for the selection of active paths are given in the next section.

Multiple Scattering Contributions in Xanes CD

The Green's function approach, together with the muffin-tin potential description, allows the use of powerful angular momentum coupling algebra. The XNCD cross-section may be written (15):

$$\sigma^- - \sigma^+ = -4\pi^2 \alpha_0 \hbar \omega \sqrt{\tfrac{2\pi}{15}} (2\ell_0 + 1) \sum_{\ell, \ell'} \begin{pmatrix} \ell & 1 & \ell_0 \\ 0 & 0 & 0 \end{pmatrix} \begin{pmatrix} \ell' & 2 & \ell_0 \\ 0 & 0 & 0 \end{pmatrix} \begin{Bmatrix} 2 & \ell' & \ell \\ \ell_0 & 1 & 2 \end{Bmatrix}$$

$$\times \sqrt{(2\ell + 1)(2\ell' + 1)} D_\ell Q_{\ell'} Re \left[\sum_p Y_p^{(2)*}(k) \tau^{00}(\ell\ell';2p) \right]$$

where D_ℓ and $Q_{\ell'}$ are radial matrix elements of the dipole and quadrupole operators between the core level (angular momentum ℓ_0) and the components in the final photoelectron state with angular momenta ℓ and ℓ' respectively. $Y^{(2)}{}_p(k)$ is a spherical harmonic function of the wavevector direction and is the equivalent to the order parameter previously introduced. $\tau^{00}(\ell,\ell';2p)$ is an irreducibly coupled form of the scattering path operator which, like the full XNCD cross-section, behaves like an odd-parity 2nd rank tensor. For trigonal or higher axial symmetry crystals, p=0 only and the symmetry criterion for the contribution of a scattering path to XNCD is that $\tau^{00}(\ell,\ell';20)$ is invariant (totally symmetric) in the point group of the photoabsorber site.

For a K or L_1 edge, $\ell_0 = 0$, the transition operators restrict ℓ,ℓ' to be 1 and 2 respectively and the XNCD cross-section depends on the interference of p and d character final state components. For L_2 or L_3 edges, $\ell_0 = 1$, $\ell = 0,2$ and $\ell' = 1,3$. The 6-j symbol forbids any contribution from s character final state components and the XNCD cross-section contains contributions from d-p and d-f interferences. The quadrupole matrix elements generally favour the final f state component and this may be seen in Figure 2. which shows the calculated XNCD (16) arising from 2p→εp and 2p→εf quadrupoles in the [Nd(digly)$_3$] system. Figure 3 illustrates the relatively insignificant XNCD arising from 2p→εp magnetic dipole excitations (E1-M1 mechanism).

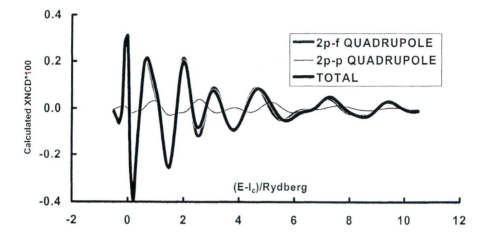

Figure 2. Multiple Scattering XANES CD calculation showing contributions from 2p→εp, 2p→εf quadrupoles.

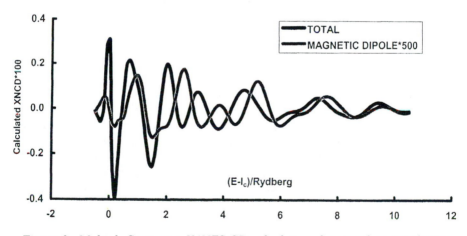

Figure 3. Multiple Scattering XANES CD calculation showing the contribution from the 2p→εp magnetic dipole

Symmetry Considerations

Since the complete XNCD experiment conserves parity, it is invariant to a mirror reflection. The effect of a mirror reflection is to interchange enantiomers and to interchange the hands of circular polarisation. Thus any circular dichroism is unchanged by this operation. To test for the contribution of a MS path, apply a symmetry operation to the molecule/crystal which:

1. interconverts enantiomers/enantiomorphs and
2. preserves the orientation of the radiation wave vector in the molecular/crystal axis system.[2]

Such operations constitute the secondary operations of the symmetry group $D_{\infty h}$ for radiation propagating along the C_∞ axis: ie mirror planes ($\infty \sigma_v$) including the z axis (3-fold axis of Nd(digly)$_3$) and σ_h.

Since these operations interconvert enantiomers, any CD contribution must change sign. Therefore, if the effect of the operation is to transform a MS path into itself, such a path must be intrinsically achiral and contributes zero to the CD. If, however, the path is converted into an equivalent path then the set of such paths has a net zero contribution.

Significantly, all **single scattering paths** which originate and return to the origin are transformed into themselves by vertical mirror planes. Thus they are intrinsically achiral and do not contribute to CD. Considering double scattering paths, Figure 4 shows the two major contributors in the Nd(digly)$_3$ system, as found from a sine Fourier transfrom analysis of the experimental XANES CD. Figure 5 shows an achiral double-scattering path (transformed into itself by a vertical mirror plane) of comparable length.

Longer and higher-order MS paths are expected to be less important because of the decay of the photoelectron due to various loss processes. However, for pseudoscalar optical activity, since the CD must be invariant to rotation by any angle about any axis, all contributing paths must be *intrinsically chiral*, ie at least triple scattering (4 atom) paths.

In summary, we must recognise that in a chiral molecular system, MS paths may possess a symmetry greater than that of the full system and in particular may possess an achiral symmetry. In the latter case the paths do not contribute to the optical activity. The paths may be intrinsically achiral. Alternatively, if paths occur as enantiomorphically-related equivalent sets then such sets have no net contribution.

---

[2] condition **2.** is necessary for the pseudotensor part of the optical activity (quadrupole-dipole mechanism but also the orientation-dependant part of the El-M1 mechanism) in which the CD depends on the orientation of the molecule through the $S^{(2)}_0$ order parameter. This condition is relaxed in considering pseudoscalar CD.

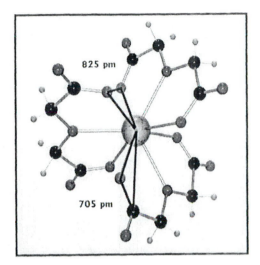

Figure 4. The Two Main Multiple Scattering Paths Contributing to XANES CD in the [Nd(dig)₃] cluster.

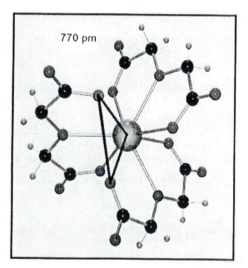

Figure 5. An Achiral Double-Scattering Path in the [Nd(dig)₃] cluster.

The XNCD experiment thus selects out only multiple scattering contributions from an absorption which is dominated by atomic and single-scattering processes. Additionally, it is selective for local chirality around the photoabsorber site in an extended structure.

Ab initio Calculations of $[Co(en)_3]^{3+}$ 1s→3d XNCD

Gaussian orbital HF-SCF calculations (2) are able to account for the large XNCD seen in the pre-K-edge feature of Λ-2$[Co(en)_3Cl_3]$.NaCl.6H$_2$O. Table I compares experiment and theory for the axially oriented crystal. The E1-E2 contributions to the visible d-d CD were also calculated for the first time and support the original suggestion of Barron (7) concerning the $^1T_{2g}$ band.

Table I. Calculated and Experimental Rotational Strengths ($/10^{-40}$ cgsu) for Λ-2$[Co(en)_3Cl_3]$.NaCl.6H$_2$O

Orbital excitation	Transition	Experiment[a]	Calculated[b]	
			E1-E2	E1-M1
(1s → 3d)	$^1A \rightarrow {}^1E$	-8.7×10^{-4}	-7.8×10^{-4}	$< 10^{-10}$
(3d → 3d)	$^1A \rightarrow {}^1E\ (^1T_{1g})$	+62.9	-0.26	+41.5
	$^1A \rightarrow {}^1A_2(^1T_{1g})$	-58.6	$< 10^{-10}$	-42.9
(3d → 3d)	$^1A \rightarrow {}^1E\ (^1T_{2g})$	+2.0	+0.66	+0.56

[a] Axial crystal Λ-2$[Co(en)_3Cl_3]$.NaCl.6H$_2$O

[b] Ab initio Gaussian orbital SCF for 1s-3d, SCF + RPA for d-d using Gamess-UK (17); Wachters Co basis with Goddard's d exponents + extra diffuse s functions, 189 basis functions in total.

Transition operator expectation values calculated using the MAGOPS program (18)

The source of electric dipole transition moment for the pre-edge CD is ca. 97% Co-based, as expected for a transition emanating from the 1s core orbital.

Conclusions

The XNCD derived from the E1-E2 mechanism vanishes for random orientation. Whilst this may be seen as a limitation, the orientation sensitivity may be exploitable, eg in studies of chiral surface adsorbates. Most measurements to date have used a uniaxial crystal orientation but it has been shown (*19*) that XNCD can be extracted from measurements on biaxial faces using the fluorescence detection method. Although XNCD is contaminated by linear dichroism (LD) in such cases, the LD and birefringeance appear to be negligible for the emitted X-ray photons which travel in a transparent region. This may be important for a wider application of the technique to obtain chirality information on more common non-uniaxial crystals.

Thanks

The author would like to acknowledge many stimulating interactions with coworkers over the several years of this project. He especially thanks Dr Tommaso Prosperi, ICMAT-CNR(Roma), Dr Calogero R Natoli and INFN(Frascati) for their hospitality and an introduction to the art of multiple scattering.

References

1. Alagna, L.; Prosperi, T.; Turchini, S.; Goulon, J.; Rogalev, A.; Goulon-Ginet, C.; Natoli, C. R.; Peacock, R. D.; Stewart, B. *Phys. Rev. Letters*, **1998**, *80*,4799-4802.
2. Stewart, B.; Peacock, R. D.; Alagna, L.; Prosperi, T.; Turchini, S.; Goulon, J.; Rogalev, A.; Goulon-Ginet, C.. *J. Amer. Chem. Soc.*, **1999**, *121*, 10233 - 10234.
3. *Atoms and Molecules;* Weissbluth, M; Academic Press: New York and London, 1978; chapter 22.
4. Chiu, Y.N. *J. Chem. Phys.*, **1969**, *50*, 5336-5349; ibid **1970**, *52*, 1042-1053.
5. Brouder, Ch. *J. Phys.: Condens. Matter,* **1990**, *2*, 701-738.
6. Buckingham, A. D.; Dunn, M. B. *J. Chem. Soc.* (A), **1971**, 1988 - 1991.
7. Barron, L. D. *Molec. Phys.*, **1971**, *21*, 241.
8. Barron, L. D., *Molecular Light Scattering and Optical Activity*; Cambridge University Press: Cambridge, 1982: μ_x, μ_y; m_x, m_y; Q_{xz}, Q_{yz} are components

of, respectively, the electric dipole, magnetic dipole and electric quadrupole operators.

9. Schaich, W. L., *Phys. Rev.* **1973**, *B8,* 4028-4032.
10. Lee, P. A. and Pendry, J. B. *Phys. Rev.* **1975**, *B11,* 2795.
11. Brouder, C., Ruiz Lopez, M. F., Pettifer, R. F., Benfatto, M. and Natoli, C. R., *Phys. Rev.* **1989**, *B39,* 1488-1500.
12. T. A. Tyson, K. 0. Hodgson, C. R. Natoli, and M. Benfatto, *Phys. Rev.* **1992**, *B46*, 5997.
13. Born, M. and Wolf, E. *Principles of Optics*, 6th edition, Pergamon Press 1985, p23ff.
14. Benfatto, M. and Natoli, C. R., *J. Phys. (Paris) Colloq.* **1986,** *C8,* 11-23.
15. Natoli, C. R.; Brouder, Ch.; Sainctavit, Ph.; Goulon, J.; Goulon-Ginet, C.; Rogalev, A. *Eur. Phys. J.,* **1998**, *B4,* 1-11.
16. XANES CD calculations used a modified version of the 'Continuum' code of Natoli and coworkers (see also *11*).
17. GAMESS-UK program suite (Guest, M. F; Sherwood, P. ,EPSRC Daresbury laboratory, 1995)
18. Dykstra, C. E.; Augspurger, J. D., MAGOPS: one-electron operator program, Quantum Chemistry Program Exchange, program no. 585.
19. Goulon, J., Goulon-Ginet, C., Rogalev, A., Gotte, V., Brouder, C. and Malgrange, C., *Eur. Phys. J.* **1999,** *12*, 373-385.

New Approaches in Chiral Recognition

Chapter 14

Aspects of Molecular Chirality Observed by Scanning Probe Microscopy

B. A. Hermann, U. Hubler, and H.-J. Güntherodt

Institute of Physics, University of Basel, Klingelbergstrasse 82, CH–4056
Basel, Switzerland

The rich variety of molecular chirality ultimately comes down to local properties of individual stereocenters in organic molecules. A local technique like scanning probe microscopy (SPM) is an excellent tool for investigating and understanding chiral properties at a (sub-) molecular level. This article reviews results of scanning tunneling microscopy (STM) and atomic force microscopy (AFM) studies of chiral molecules in air and in vacuum. These results clearly indicate that SPM is useful for studying many aspects of chiral molecules – ranging from analysis of order phenomena in chiral molecular films to imaging and even the controlled repositioning of individual molecules.

When Pasteur discovered the existence of enantiomers by applying optical microscopy to salt crystals of a "racemic acid" (*1*), he helped forming the first basic picture of molecular structure. Since then, researchers have been fascinated by indirect structural influences of pure enantiomers, coexisting enantiomorphous domains of racemates, and spontaneous formation of chiral domains by achiral molecules. The fact that achiral molecules in the gas phase can form chiral molecular assemblies on a surface underlines the importance of investigating monolayers and films of chiral molecules or chiral supramolecular assemblies.

Nowadays, the size scale for observing chirality in experiments has approached the size of molecules themselves or even submolecular units.

Shortly after its invention by G. Binnig and H. Rohrer in 1982 (2), the scanning tunneling microscope (STM) was used to investigate layers of molecules and also single molecules adsorbed on surfaces (3, 4). Since then, the STM has become a well-established tool for imaging surfaces and adsorbates with a resolution down to the atomic scale. Even the structure of single molecular orbitals (HOMO, LUMO) is accessible through STM experiments. The atomic force microscope (AFM) (5) has become a standard characterization method when analyzing long range order phenomena in molecular films. The full potential of this method, however, comes to light when individual molecules are accessed by a chemically functionalized scanning tip (6).

Due to the operating principle of scanning probe microscopy, the molecules to be studied have to be deposited on a well defined, flat (in the case of STM an electrically conducting) substrate. These samples can then be studied in a wide variety of environments – ranging from "dry conditions" (like ambient conditions, in vacuum, and even at low temperature) to "wet environments" (where sample and tip are immersed in a liquid). The latter method is also called STM at the solid liquid interface and is discussed by Flynn et al. This article will mainly focus on the investigation of chiral molecules under dry conditions, including layers as well as macromolecular assemblies and analysis of single molecules.

Chirality exists in nature in great abundance. Many "hot topics" like DNA or carbon nanotubes fall into the category of chiral substances. In the case of single walled nanotubes, the electrical conductance varies from metallic to semiconducting depending on their chirality. Although they provide good examples of the importance of molecular chirality, a description of the results in these fields is beyond the scope of this article. Rather than providing a complete overview, this article intends to show the versatility of scanning probe techniques for the investigation of the diverse aspects of molecular chirality.

Introducing the Principle of STM and AFM

Scanning Tunneling Microscopy (STM)

Consider the quantum mechanical tunneling effect resulting when a sharp metallic tip is brought to close vicinity of an electrically conducting substrate, and a voltage (10 mV...10 V) is applied between tip and substrate. Then a small

_nt (1pA...1nA) flows even before the tip actually touches the surface due to the overlap of the electronic wave functions of tip and sample. Electrons "tunnel" across the gap between tip and substrate forming the so called tunneling current. Not only the voltage applied across the gap (bias voltage), but also the width of the tunneling gap d and the electronic properties of both tip and sample determine the tunneling current. This is the basic principle of scanning tunneling microscopy. Assuming that both the tip and sample are metallic, the tunneling current is given in WKB[1] approximation by the equation

$$I_{tunnel} \propto U_{bias} \cdot \exp(-d\sqrt{2m_e\Phi}/\hbar) \, .$$

Here, d is the tip-sample distance (width of the tunneling gap), and Φ is the energy needed to extract an electron of mass m_e from the surface (the barrier height or work function). An increase of only 1 Å in the tip-sample distance can easily be detected by a STM, since the tunneling current then drops about an order of magnitude (assuming $\Phi \approx 4eV$ for two metals). This strong dependence of the tunneling current on d is the reason for the extremely high resolution achieved in STM experiments. Topographic effects (through d) as well as the electronic structure (through local variations of Φ) affect a STM image. More generally speaking, the tunneling current is not only determined by the sample-tip distance, but also by the electronic density of states near the Fermi level of both, the tip and sample. This effect is crucial when investigating molecules on surfaces, because the electronic properties of the molecule itself as well as the electronic interactions between the molecule and substrate strongly influence STM images – the tunneling probability can either be reduced or enhanced by the presence of a molecule on the substrate. Examples will be discussed in greater detail below.

With piezoelectric elements, the tip is scanned line by line parallel to the surface (x,y-direction, typical resolution 0.1 Å) and precisely moved in z-direction (typical resolution 0.01 Å). In the most common STM operation mode, an electronic feedback system is used to keep the tunneling current at a constant value by lowering and raising the scanning tip while scanning across the surface ("constant current mode"[2]). The voltage applied to the piezoelectric element by the feedback loop (to attain the vertical movement of the tip) is recorded and processed by a computer to give an image of the sample surface. As long as the electronic density of states remains constant across the surface, the image corresponds perfectly to the surface topography. When investigating more complex samples (such as molecules adsorbed onto a metal substrate), the convolution of electronic effects with topography must be taken into account. For instance, an

---

[1] WKB: Wentzel-Kramers-Brillouin.

[2] In the "constant height mode" the tip is scanned at a certain z-height while the spatial variations of the tunneling current are recorded. This operation mode is appropriate only on very flat surfaces and is often applied when measuring at the solid liquid interface.

adsorbed molecule may appear as a depression due to reduced tunneling probability, see "STM on Molecules" below.

A more detailed and comprehensive discussion of STM principles can be found e.g. in (7).

Atomic Force Microscopy (AFM)

To investigate non-conducting samples, the more widely known AFM can be used. The scanning tip (usually a microfabricated Si or Si_3N_4 tip) is attached to a cantilever. While the tip is scanning across the surface in the so-called contact mode[3], the force between tip and sample results in a measurable bending of the cantilever[4]. Similar to STM, either the z-movement of the scan piezo needed to keep the cantilever deflection constant (constant force mode) or the deflection of the cantilever while scanning at a constant height (constant height mode) is measured. In most cases, the line followed by the scanning tip is interpreted as the sample topography, however, this assumption is not always valid when imaging soft materials. The forces responsible for the deformation of the cantilever, however, can originate from a variety of interactions, such as Pauli and nuclei repulsion (topography in contact mode), van der Waals, chemical, electrostatic, magnetic or frictional forces.[5] Recently, with sophisticated, home-built AFMs operated in UHV[6] "true atomic resolution" similar to STM was achieved (8, 9).

Molecules Investigated by STM

Usually, organic molecules are "insulators", i.e. there is a large gap in the electronic band structure between the highest occupied (HOMO) and the lowest unoccupied molecular orbital (LUMO). As STM relies on the detection of an electronic current, a monolayer or sub-monolayer coverage of the surface with molecules is desirable. A molecular layer of greater thickness will most probably be penetrated by the tunneling tip resulting in an unwanted mechanical interaction between scanning tip and molecules. Secondly, STM is sensitive to elec-

---

[3] In contrast to non-contact (10) or tapping mode, where oscillating cantilevers are used.

[4] Various methods can measure the bending of the cantilever: deflection of a laser beam by the cantilever back, interferometric or capacitive methods or the use of piezo-resistive cantilevers.

[5] To study frictional forces the torsion of the cantilever rather than its bending is measured.

[6] Ultra high vacuum, base pressure $< 10^{-9}$ mbar.

tronic states near the Fermi level. The adsorbed, insulating molecule presents an intermediate state for the tunneling electrons. The tunneling current therefore results from a combination of the direct interaction between tip and substrate through space and the interaction mediated by the adsorbate. The (complex) sum of these two parts yields a signal which depends on the lateral tip position as well as the tip-to-sample distance (11).

The electronic structure of the molecules, charge transfer from the substrate to the molecule as well as mechanical interactions may produce a variety of effects:

- The electronic interaction between adsorbed molecule and substrate may alter the apparent shape of molecules. STM images of molecules may even vary with the molecule's adsorption site, as has been demonstrated for benzene on Pt(111) by Weiss et al. (12).

- Most strikingly, a molecule can even appear as a depression of the surface due to reduced tunneling probability. This is an effect often observed on silicon surfaces, where the dangling bonds of the Si surface lead to a high tunneling current. When these dangling bonds are saturated by an adsorbate, they no longer contribute to the tunneling current and a "depression" in the surface is observed (13).

- The imaged electronic states (or orbitals) can be "selected" by the applied bias voltage. In some cases, by reversing the bias voltage and setting it to an appropriate value, either the HOMO or LUMO may be imaged resulting in different appearances of the same molecule (14).

- Due to the same reason, when looking at complex molecules, often only certain parts of a molecule (e.g. π-orbitals of a phenyl ring within the molecular structure) contribute to the image contrast while other parts remain "invisible" (15).

- Molecules are also subject to mechanical interactions with the scanning tip. This interaction may be used for controlled repositioning of single molecules. Molecules may be picked up or moved around (pushed, pulled or levitated) by the scanning tip (16).

- This mechanical interaction can also lead to a distortion of the molecular shape, especially when large and flexible molecules are studied. The pressure[7] exerted by the STM tip can lead to a significant compression of the molecule.

---

[7] Measurements and calculations show that on graphite, the force between the STM tip and the sample amounts to approx. 10^{-6} N (U_{bias}=50mv; I_t=1nA) (17, 18). Assuming a rather blunt tip with an area of 10^4 nm^2, the resulting pressure can be estimated to be of the order 1 kbar (0.1 GPa).

If all these effects are taken into account in a thorough analysis of STM data, STM proves to be a powerful, reliable and attractive tool for the investigation of molecules and molecular layers. Examples are discussed in the following sections.

Spontaneous Formation of Enantiomers or Enantiomorphous Domains Starting from Achiral Molecules

Depositing achiral molecules from the gas phase or liquid onto a substrate can induce a two-dimensional chirality due to the adsorbing configuration and possibly restricted orientations. By a molecular self-assembly process, it is even possible that these molecules form two-dimensional chiral crystals. Enantiomorphous domains originating from achiral molecules were observed by STM (*19, 20*) as well as AFM (*21*).

Pasteur´s Experiment at the Nanometer Scale

The spontaneous formation of chiral macromolecular assemblies from achiral molecules was recently reported by Böhringer et al. (*22, 23*). Their experiment was carried out in UHV with 1-nitronaphthalene (NN) adsorbed onto a Au(111) surface. Since the NN molecules lie flat on the gold surface and are asymmetric, a two-dimensional chirality is induced upon adsorption, even though NN is achiral in the gas phase. While this chirality is not directly evident from STM images, surprisingly molecular clusters of ten NN-molecules each were detected on the surface, see Figure 1. Not only is the configuration of these decamers stable, but they exist in two different orientations forming two enantiomers. The authors claim to have observed the synthesis of a two-dimensional chiral supramolecular assembly forming a racemic mixture on the surface. By bringing the STM tip closer to the surface[8], the authors were able to move decamers in a controlled way, thereby separating *(S)*- and *(R)*-oriented decamers. One can look upon this as the first equivalent to Pasteur's experiment performed at the nanoscale.

---

[8] This is achieved by reducing the tunneling resistance by three orders of magnitude.

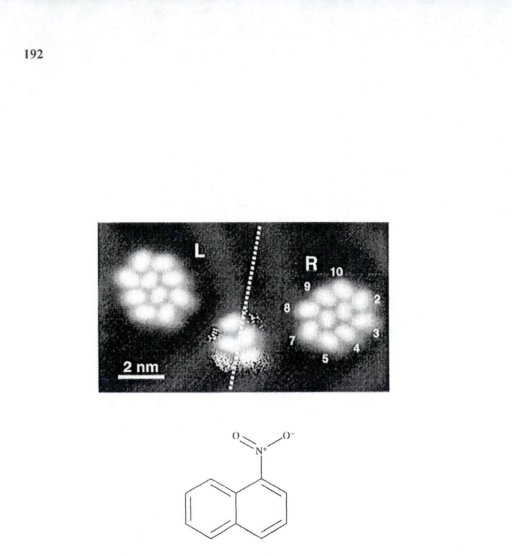

Figure 1. STM image of two-dimensional chiral decamers of 1-nitronaphthaline on Au(111) taken in UHV at 50 K (U_{Bias}=-150 mV, I= 10 pA). (Reproduced from reference 22.)

Chiral Ordering of Achiral Molecules on Non-Chiral Surfaces

Recently, F. Charra and J. Cousty (*24*) observed the emergence of a chiral order when increasing the triangular aspect ratio of a achiral molecule by tuning the length of the alkyloxy-chains by STM at the solid-liquid interface[9]. Although neither the molecule nor the surface (HOPG) present any chirality, the hexakis-9,11-alkyloxytriphenylene (H-9,11-T) order in beautiful chiral arrangements on the surface as opposed to the H-2,3,6,7-T. The authors interpret this mechanism, which involves steric hindrance, superlattice formation and conformational mobility, in the framework of a simple frustrated triangular Ising net.

Discriminating Enantiomers and Enantiomorphous Domains of Racemates

The formation of chiral crystalline monolayer domains in Langmuir-Blodgett films of enantiomerically pure lipids was first discovered in 1984 by Weis and McConnel using epifluorescence optical microscopy (*25*). Some years later, chiral symmetry breaking in Langmuir-Blodgett films prepared from racemic molecules was observed by AFM (*26*), and recently racemic mixtures of 1,1´-binaphthalene-2,2´-dithiol were investigated by Othani et al. (*27*).

When enantiomorphous domains are formed from a racemic mixture, there are two possible mechanisms: either the racemic mixture spontaneously separates into two-dimensional domains of enantiomerically pure material, or different domains can be composed of racemates.[10]

Domains of Opposite Chirality Forming in Racemic Mixtures

In 1996, Stevens et al. reported STM investigations of a racemic mixture as well as enantiomerically pure monolayers of liquid crystals deposited on a graphite[11] surface (*28*). The authors attributed the STM contrast mainly to the aromatic parts of the biphenylbenzoates. In the case of pure enantiomers, highly ordered monolayers were observed with molecular resolution. The chirality of these films could be resolved and directly related to the chirality of the adsor-

---

[9] These results are mentioned for the sake of comprehensiveness, even though they were obtained at the solid-liquid interface.

[10] Since there is no way to physically separate the domains, it is usually impossible to distinguish the two cases by measuring the optical activity in solution.

[11] Highly oriented pyrolytic graphite (HOPG) is easily cleaved to give a clean, atomically flat surface.

bate. On the other hand, when a racemic mixture was deposited on the surface, domains of opposite chirality formed and domain walls appeared. After thorough comparison of this data with STM images of the pure enantiomers, the authors concluded that the domains were enantiomerically pure. This work provides an example where a racemic mixture of biphenylbenzoates forms clearly separated enantionerically pure domains.

A more detailed overview of STM studies of chiral liquid crystal molecules, up to 1996, can be found in (*29, 30*).

Determining Absolute Chirality

Recently, Lopinski et al. reported the determination of the absolute chirality of individual molecules by STM (*31*). When a *trans-2*-butene is adsorbed on a Si(100) surface, its two C atoms bind to a Si-dimer on the substrate thereby determining the orientation of the adsorbate relative to the silicon surface. Because the unsaturated bond in *trans-2*-butene has two reactive faces the molecule can adsorb in two chirally different ways due to the geometrical arrangement of the methyl groups and hydrogens. Although these enantiomers are indistinguishable by spectroscopic techniques, STM allows the determination of the position as well as the orientation of the methyl groups[12] relative to the substrate and thus permits the absolute chirality of the adsorbed *trans-2*-butene molecules to be determined.

Pure Enantiomers: Structural and Conformational Influences of Chiral Centers

Numerous structural and conformational influences can result from chiral centers of molecules. Some of the most striking are discussed below: spirals, with the same handedness as the chiral molecules they consist of, on macroscopic spherulites of crystallized films, an enantiospecific substrate/molecule combination or a surprising case of diastereoselectivity. Only a limited insight can be given here, as many interesting AFM/STM investigations exist.

---

[12] The authors point out, that in contrast to most adsorbates on silicon, methyl groups are imaged as protrusions in the STM image instead of depressions.

Molecular Film Growth Influenced by Pure Enantiomers

In the field of crystal engineering in two dimensions, AFM is regularly used to determine the long-range order of Langmuir-Blodgett (LB) films after their transfer from aqueous to solid phase on a scale of a few microns. For example, Eckardt et al. (*32*) showed that the chiral amphiphile tetracyclic acid (TCA), though as rigid as the achiral trinorboranecarboxylic acid (TNBC), transfers with less disruptive results onto mica.

An STM study (*33*) of the liquid crystal 7CB (4-heptyl-4′-cyanobiphenyl) revealed unexpected domain angles in LB-films on HOPG. Unlike the widely studied 8CB, the 7CB monolayers do not favor the expected angle of 60° between domain boundaries. Also the 13° angle expected between domain boundaries of different chirality meeting on a HOPG terrace is not found. With a novel "rotated row" model the authors can explain the angle of 17°, often observed between domains of opposite chirality. Even small changes in the molecular shape or functionality can influence the angle between domains of different chirality underlining the importance of structural examination by SPM techniques.

The defect mobility in chiral smectic thin films was investigated in a study of biphenylyl benzoate liquid crystals on HOPG with STM. Surprisingly, the authors find a greater mobility across rows in the 2-D lamellar crystals than within the rows (*34*).

Molecular Chirality Influencing Gross Morphology

An interesting example where molecular chirality is responsible for structural features on a scale of a few 100 μm is found in (*35*). In this work AFM investigations (see Figure 2) of melt-crystallized spherulites of the optically active *(R)* and *(S)* forms of poly(epichlorohydrin) show that the molecular geometry of the isochiral helices is translated beyond the lamellar level to macroscopic spherulites (diameter of up to several 100μm). The direction of the spiral on the surface of the spherulites depends on the handedness of the constituent poly-enantiomer. In the case of the optically active poly(epichlorohydrin), the asymmetrical units in the backbone of the optically pure enantiomer affect the development of the spherulite and thus transmit structural properties to the gross morphology.

Single Chiral Molecules and Assemblies Thereof

Designing a surface capable of chiral recognition is a goal envisaged for a long time. The system *trans*-2-butene on Si(100) discussed below seems to be a step in this direction.

Figure 2. AFM image of enantionmerically pure melt crystallized (80°C) poly(epichlorohydrin). Image recorded in constant force mode, image size 80 μm × 80 μm. (Reproduced from reference 35.)

Enantiospecifity

The adsorption of individual *trans-2*-butene molecules on Si(100) leading to equal numbers of both chiral faces of the molecule (an overall racemic mixture) was described above. An enantiospecific system with 1S(+)-3-carene on Si(100) was found by Lopinski et al. (*36*). Following the idea of introducing steric hindrance at one of the faces, the naturally occurring bicyclic alkene (carene, 3,7,7-trimethylbicyclo[4.1.0]hept-3-ene) was introduced to a clean Si surface under UHV conditions. Two orientations of the molecule, rotated by 180°, were found, and in one of the orientations the molecules appeared tailored. Both orientations had the same chirality, even though carene molecules can initially form two enantiomers when double bonding to a single Si dimer. Interpreting their STM images with gradient corrected density functional methods, the authors conclude that one enantiomer undergoes a further reaction, where the cyclopropyl ring on the molecule opens and leads to two additional bonds to the silicon surface (bridging configuration). As the tailored molecules conform to such bridging molecules, only this enantiomer is observed. In this way, an enantiospecific molecule-surface system could be demonstrated.

A thorough study of the growth of *R*, *R*-tartaric acid on Cu(110) in view of enantioselective heterogeneous catalysis has been reported by Ortega Lorenzo et al. (*37*).

Investigating Chirality of Complex Molecules

One complex and large macromolecule, where the chirality can even influence the chemical behavior (*38*), is a $C_{180}H_{266}O_{39}$ dendrimer[13]. The conformation of most of the larger dendrimers, especially the chiral ones, is unknown, as they do not crystallize. "Classical" analysis methods (like e.g. x-ray diffraction) cannot be applied to investigate the conformation leaving only SPM techniques as available tools for investigating the conformation of these molecules on surfaces. However, in the case of these very large, complex and flexible molecules, molecule-surface interactions as well as interactions with the tunneling tip become crucial, often prohibiting a straight forward analysis of STM data.

---

[13] Dendrimers are macromolecules consisting of branches reaching out from a central core. They have a complex, perfectly branched structure with radial symmetry, a high degree of self-similarity, and an unique molecular weight.

A variety of dendrimers have been synthesized during the past years. Besides their beauty, the dendrimers proved useful for many applications ranging from highly functional and specialized catalysts to "dendritic boxes" in medical applications (drug release) (*39*). It is clear that the complex, three-dimensional conformation of the dendrimers has a significant influence on their properties.

Diastereoselectivity

An unexpected case of diastereoselectivity was found by Murer and Seebach (*38*) while synthesizing chiral 2nd generation dendrimers ($C_{180}H_{266}O_{39}$) with a chiral triol center and chiral building blocks at all branching points. They found, when a branch bromide with *(S)* configuration at the four benzylic centers is used, the reaction proceeds smoothly to the compact dendrimer **2** with three branches. With the corresponding *(R)*-configured bromid, however, the reaction stops at the doubly coupled product **1a** (see Figure 3). Due to the long distance of the chiral center from the reaction site, an indirect influence of the chirality on the conformation of the branches is proposed to be responsible for this behavior.

Three different samples of dendrimers, deposited on Pt(100), were investigated by STM (*40*): the three-branched product **1** ($C_{180}H_{266}O_{39}$) as well as the *(R)*- and *(S)*-configured two-branched dendrimers **2a** and **2b**.[14] STM images of dendrimer **1** show a quasi-periodic arrangement with a distance of 6 Å between neighboring protrusions with height differences of only approximately 1 Å. Two conclusions are drawn from this observation: (a) the pressure exerted on the dendrimers by the STM tip at the chosen tunneling parameters surprisingly forces the dendrimers into an almost planar and rather regular configuration at the surface, and (b) the distance between the protrusions corresponds to the distance of two neighboring phenyl rings inside the dendrimer structure. In the context of this paper it is less surprising that the contrast of these STM images most likely originates from the aromatic portions of the molecules.[15]

---

[14] For the *(S)*-configured two-branched dendrimer, the reaction was stopped at the intermediate stage by a deficiency of the branch bromide during the reaction and chromatographic separation. No structural differences of these two products are evident from NMR measurements.

[15] This assumption is supported by STM reference measurements of benzene on Pt(100) and by additional structural features observed at low temperature (77K). To the best of our knowledge, these are the first STM images of dendrimers with submolecular resolution.

While it is difficult to distinguish individual dendrimers inside a molecular layer, in a comparison of dendrimers **1** and **2a,** a layer of dendrimer **2a** shows a significantly reduced symmetry in accordance with the lower symmetry of the molecule. More interestingly, there are two main differences observed in STM images of layers of enantiomers **2a** and **b** in Figure 4: Dendrimer **2b,** reacting to the fully coupled product, arranges slightly more densely on the surface, while for dendrimer **2a,** a feature (encircled in Figure 4) hinting towards a branch folded back is often observed. Both observations are consistent with the assumption, that the conformation is prohibiting the full reaction of enantiomer **2a:** the folded back branch may cover the reaction site – while the less dense arrangement may suggest that the third branch is unable to fully approach the reaction site.

Conclusions

The aim of this article was to give some insight into the diversity of STM and AFM investigations on chiral molecules and molecular arrangements under dry conditions. AFM is most often used to characterize long range order phenomena in molecular monolayers and films. STM not only allows the chirality of single molecules to be studied and in some cases permits the identification of their absolute chirality, but also allows the manipulation of individual molecules (even decamers). An enantiospecific substrate/molecule system was identified and the conformational properties of large dendrimers showing diastereoselectivity were further understood with help of STM images. The chosen examples clearly show that scanning probe methods are valuable tools in investigating many aspects of molecular chirality and provide a manifold picture of chirality on length scales ranging from micro- to nanometer.

Acknowledgements

The authors would like to thank Dr. H. P. Lang and Dr. M. Lantz for proofreading the manuscript and for valuable discussions. We profited from fruitful discussions with Prof. Dr. D. Seebach, Dr. T. Sifferlen, Dr. G. Greiveldinger, Dr. P. B. Rheiner, Prof. J. P. Rabe, Dr. P. Samorí, and Prof. Dr. E. C. Constable. B. A. H. acknowledges financial support from the German foundation of the chemical industries and the BMBF.

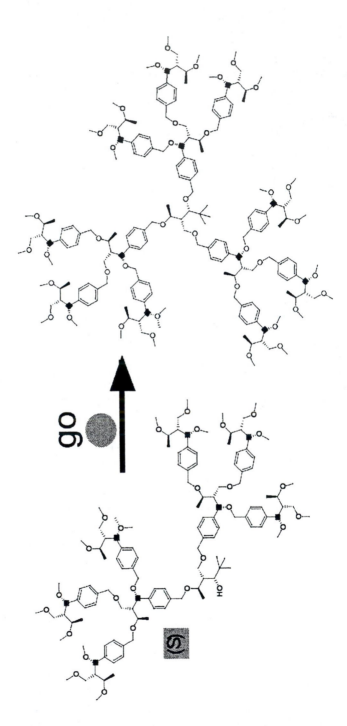

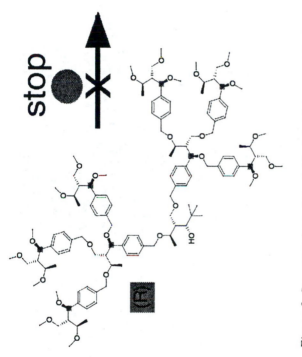

Figure 3. Structure formulae of chiral dendrimers exhibiting diastereoselectivity.

202

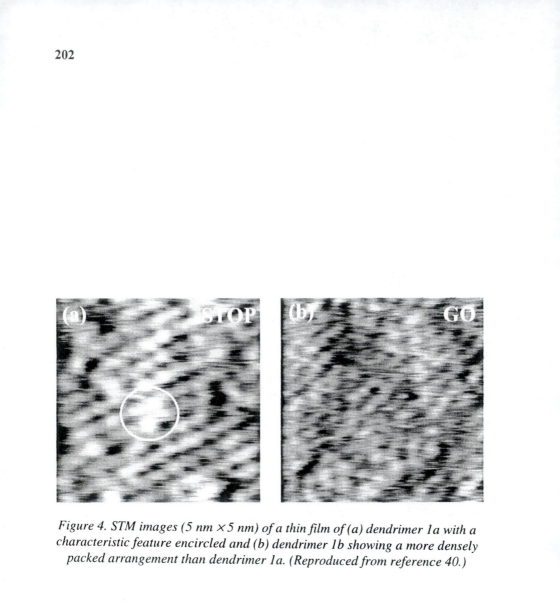

Figure 4. STM images (5 nm × 5 nm) of a thin film of (a) dendrimer 1a with a characteristic feature encircled and (b) dendrimer 1b showing a more densely packed arrangement than dendrimer 1a. (Reproduced from reference 40.)

References

1. Pasteur, L. C. R. *Hebd. Séances Acad. Sci.* **1848**, *26*, 535-539.
2. Binnig, G.; Rohrer, H.; Gerber, Ch.; Weibel, E. *Phys. Rev. Lett.* **1982**, *49*, 37.
3. Ohtani, H.; Wilson, R. J.; Chiang, S.; Mate, C. M. *Phys. Rev. Lett.* **1988**, *60*, 2398.
4. Lippel, P. H.; Wilson, R. J.; Miller, M. D.; Wöll, Ch.; Chiang, S. *Phys. Rev. Lett.*, **1989**, *62*, 171.
5. Binning, G.; Quate, C. F.; Gerber, Ch. *Phys. Rev. Lett.* **1986**, *56*, 930.
6. McKendry, R.; Theoclitou, M.-E.; Abell, C.; Rayment, T. *Jpn. J. Appl. Phys,* **1999**, *38*, 3901-3907.
7. *Scanning Tunneling Microscopy I;* Wiesendanger, R.; Güntherodt, H.-J., Eds.; Springer: Berlin Heidelberg New York, 1994.
8. Giessibl, F.J. *Science* **1995**, *267*, 68.
9. Bammerlin, M.; Lüthi R.; Meyer E.; Baratoff A.; Lü, J.; Guggisberg, M.; Gerber Ch.; Howald, L.; Güntherodt, H.-J. *Probe Microscopy* **1997**, *1*, 3.
10. Guggisberg, M; Bammerlin, M.; Loppacher, Ch.; Pfeiffer, O.; Abdurixit, A.; Barwich, V.; Bennewitz, R.; Baratoff, A.; Meyer, E.;Güntherodt, H.-J. *Phys. Rev. B,* **2000**, *61*, 11151.
11. Sautet, P.; Joachim C. *Chem. Phys. Lett.* **1988**, *153*, 511.
12. Weiss, P. S.; Eigler, D. M. *Phys. Rev. Lett.* **1993**, *71*, 3139.
13. Wolkow, R. A.; Avouris Ph. *Phys. Rev. Lett.* **1988**, *60*, 1049.
14. Lazzaroni, R.; Calderone, A.; Brédas, J. L.; Rabe, J. P. *J. Chem. Phys.* **1997**, *58*, 269.
15. Rabe, J. P.; Buchholz, S. *Phys. Rev. Lett.* **1991**, *66*, 2096.
16. Meyer, G.; Repp, J.; Zöphel, S.; Braun, K.-F.; Hla, S. W.; Fölsch, S.; Bartels, L.; Moresco, F.; Rieder, K. H. to appear in *Single Molecules*, **2000**,*1*.
17. Mamin, H. J.; Ganz, E.; Abraham, D. W.; Thomson, R. E.; Clarke, *J. Phys. Rev.* **1986**, *B34*, 9015.
18. Mate, C. M.; Erlandsson, R.; McClelland, G. M.; Chiang, S. *Surf. Sci.* **1989**, *208*, 473.
19. Smith, D. P. E. *J. Vac. Sci. Technol. B* **1991**, *9*, 1119.
20. Sowerby, S. J.; Heckel, W. M.; Petersen, G. B. *J. Mol. Evol.* **1996**, *43*, 419.
21. Viswanathan, R.; Zasadzinksi, J. A.; Schwartz, D.K. *Nature* **1994**, *368*, 440.
22. Böhringer, M.; Morgenstern, K.; Schneider, W.-D.; Berndt, R. *Angew. Chem. Int. Ed.* **1999**, 38, *821*.
23. Böhringer, M.; Morgenstern, K.; Schneider, W.-D.; Berndt, R.; Mauri, F.; De Vita, A.; Car, R. *Phys. Rev. Lett.* **1999**, 83, *324*.
24. Charra, F.; Cousty, J. *Phys. Rev. Let.* **1998**, *80*, 1682.
25. Weis, R. M.; McConnel, H. M. *Nature* **1984**, *310*, 47.

26. Eckhardt, C. J.; Peachy, N. M.; Swanson, D. R.; Takacs J. M.; Khan, M. A.; Gong, X.; Kim, J.-H.; Wang, J.; Uphaus, R. A. *Nature* **1993,** *362,* 614.
27. Othani, B.; Shintani A.; Uosaki, K. *J. Am. Chem. Soc.* **1999,** *121,* 6515.
28. Stevens, F.; Dyer, D. J.; Walba, D. M. *Angew. Chem. Int. Ed. Engl.* **1996,** *35,* 900.
29. Walba, D. M.; Stevens, F.; Clark, N. A.; Parks, D. C. *Acc. Chem. Res.* **1996,** *29, 591.*
30. Walba, D. M.; Stevens, F.; Dyer D. J.; Parks, D. C.; Clark, N. A.; Wand, M. D. *Enantiomer* **1996,** *1,* 267.
31. Lopinski, G. P.; Moffatt, D. J.; Wayner, D. D. M.; Wolkow R. A. *Nature* **1998** *392,* 909.
32. Eckardt, C. J.; Peachey, N. M.; Takes, J. M. *Thin Solid Films,* 1994, *242,* 67.
33. Schulze, J.; Stevens, F.; Beebe, Jr., T. B. *J. Phys. Chem.* **1998,** *102,* 5298.
34. F. Stevens, D. J. Dyer, D. M. Walba, R. Shao and N. A. Clark, *Liq. Cryst.* **1997,** *22,* 531.
35. Singfield, K. L.; Klass, J. M.; Brown, G. R. *Macromolecules* **1995,** *28,* 8006.
36. Lopinski, G. P.; Moffatt, D. J.; Wayner, D. D. M.; Zgierski, M. Z.; Wolkow, R. A. *J. Am. Cem. Soc.,* **1999,** *121,* 4532.
37. Ortega Lorenzo, M.; Haq, S.; Bertrams, T.; Murray, P.;Raval, R. and Baddeley, C. J., *J. Phys. Chem.* **1999,** *103,* 10661.
38. Murer, P.; Seebach, D. *Angew. Chem. Int. Ed.* **1995,** 34, *2116.*
39. Tomalia, D. A.; Naylor, A. M.; Goddard III, W. A. *Angew. Chem.* **1990,** *102,* 119.
40. Hermann, B. A.; Hubler, U.; Jess, P.; Lang, H. P.; Güntherodt, H.-J.; Greiveldinger, G.; Rheiner, P. B.; Murer, P.; Sifferlen, T.; Seebach, D. *Surf. Interface Anal.* **1999,** *27,* 507.

Chapter 15

Raising Flags: Chemical Marker Group STM Probes of Self-Assembly and Chirality at Liquid–Solid Interfaces

Dalia G. Yablon[1], Hongbin Fang[1],
Leanna C. Giancarlo[2], and George W. Flynn[1],*

[1]Department of Chemistry, Columbia University, New York, NY 10027
[2]Department of Chemistry, Mary Washington University,
Fredericksburg, VA 22401

STM has been used to study racemic 2-bromohexadecanoic acid and a mixture of this chiral molecule with achiral hexadecanoic acid physisorbed at a liquid-solid interface. Racemic 2-bromohexadecanoic acid segregates into enantiomerically pure domains on a graphite surface. The absolute chirality of individual molecules can be directly determined from STM images. Achiral hexadecanoic acid self-assembles into two enantiomorphous domains and, in a mixture with 2-Br-hexadecanoic acid, creates a 2D chiral template that further effects resolution of the brominated molecules into chirally pure domains.

The ability to distinguish between and separate chiral conformers has become an important necessity in today's chemical world since a molecule's chirality plays a significant role in its function. Optical microscopy was the technique first used by Louis Pasteur 150 years ago to separate enantiomers of sodium tartrate. Methods such as NMR (1), circular dichroism (2), polarimetry (3), and x-ray crystallography (4), have since been added to the list of available technologies used to elucidate chiral configurations. The latest newcomer to this roster of techniques is the family of scanning probe microscopies, which includes scanning tunneling microscopy (STM) (5) and atomic force microscopy (AFM) (6). The advantage of STM and AFM over traditional analysis tools, with the exception of X-ray methods (4), is that these high-resolution imaging methods allow for a determination of absolute chirality while obviating the need for a standard sample or a compound with known chirality.

Several chiral systems have already been explored with STM and AFM. Separation of achiral molecules into lattices with chiral packing (7) and separation of chiral phases of chiral organic molecules in Langmuir-Blodgett films have been observed by AFM (8). Chiral liquid crystals have also been investigated with STM and found to form domains that exhibit two-dimensional chirality (9). Recently, supramolecular clusters of 1-nitronaphthalene on gold have been observed to aggregate in 2D domains that are mirror images of each other (10). Monolayers of enantiomerically pure isophthalic acid derivatives coadsorbed onto a graphite surface with the achiral solvent, 1-heptanol have also been studied (11). This work has shown that the orientation of the achiral solvent molecules expresses the chirality of the domains. Finally, there has also been a beautiful demonstration of the principles of chiral HPLC using chemical force microscopy (12). By attaching molecules with known chirality to the probe tips, two enantiomers of mandelic acid have been distinguished through differences in both adhesion and frictional forces. In all these cases, the submolecular or even atomic resolution necessary to determine the absolute chirality for individual enantiomers was not achieved. Assignment of absolute chirality has been recently accomplished in our work described below (13) and in studies of the configuration of chiral centers formed by the chemisorption of simple alkenes on silicon (14,15).

In this paper, we describe a series of works predicated on the ability to assign absolute chirality via the use of chemical marker groups for physisorbed molecules of racemic 2-Br-hexadecanoic acid on a graphite surface at the liquid-solid interface (13). We further illustrate the effect of a third element (a coadsorbate in the form of another solute) on the self-assembly process in a study of a mixture of racemic 2-Br-hexadecanoic acid with hexadecanoic acid physisorbed on graphite. These two solutes are found to coadsorb onto the surface, and the mixed monolayer film resolves into domains of hexadecanoic acid with either (R) or (S)-2-Br-hexadecanoic acid. Further analysis reveals that achiral hexadecanoic acid forms domains exhibiting non-superimposable mirror image morphology (enantiomorphous domains) on a graphite surface, similar to that observed for behenic acid (16), and these 2D chiral domains in fact effect enantiomeric resolution of 2-bromohexadecanoic acid in the self-assembled structure (17).

Racemic 2-Br-hexadecanoic acid

When a solution of racemic 2-bromohexadecanoic acid is applied to a graphite surface, different domains are formed ranging in width from 30 nm to more than 100 nm as measured from STM topographs. These domains appear to be composed of single enantiomers. Parts a and b of Fig. 1 represent typical STM images collected from two different domains. These are the only two structures that can be identified on the surface. Despite the difference in resolution between these two images, it is apparent that these topographs reflect mirror images of each other.

The images shown here were obtained in the constant current mode (STM current is held fixed and the tip height varies). Brighter regions in the images correspond to a retraction of the STM tip and darker regions to a movement of the tip toward the surface. Thus, formally, "bright" regions correspond to topographical protrusions while "dark" regions correspond to topographical depressions. Nevertheless, at any point on the surface, the tunneling probability is a convolution of factors that include electronic coupling to the surface and the distance from tip to surface ad-atom or ad-molecule and its orientation. The tunneling probability maps the density of states at the Fermi level of the surface, and atoms that physically protrude from the surface can actually appear as depressions if their effect is to decrease the number of surface states near the Fermi level. Thus, Xe physisorbed on Ni appears as an expected topographical protrusion (18) but He on Ni is predicted to appear as a depression (19). A further illustration of this concept is acetylene on copper, which is observed as a dumbbell-shaped depression (20). A final example lies in partially fluorinated long chain carbon compounds physisorbed at the liquid solid interface, for which a counterintuitive topographical depression is observed at the positions of the F atoms, despite the fact that they are closer to the tip than the H atoms, which in turn appear as topographical protrusions (21).

For the present study a key observation is that Br atoms in 2-bromohexadecanoic acid appear as bright, topographical protrusions in the STM images of the molecule. A great deal of both experimental (13, 22-25) and theoretical (25,26) evidence has now been assembled indicating that these Br atoms are in fact physically protruding above the H atoms of long chain carbon compounds physisorbed at the liquid-solid interface whenever the Br atoms are located at a nonterminal position along the hydrocarbon chain. This orientational assignment (Br's that appear in the STM images as bright, topologically raised features actually physically protrude above neighboring H atoms) has been confirmed in studies of enantiomerically pure 2-bromohexadecanoic acid. In the case of pure enantiomers, the topographical position of Br for an adsorbed species is known with certainty allowing an unambiguous correlation between the STM features and the orientation of the atom (27).

Fig. 1a is a 12x12 nm^2 STM image taken at the interface of the phenyloctane solution and graphite. A computer generated molecular model deduced from this image is shown in Fig. 1c. Based on the molecular structure of 2-bromohexadecanoic acid and previous STM studies (13,22-24), each black bar in the image is attributed to one molecule lying flat on the surface adopting an all trans conformation. The length of the molecule measured in the STM image is 2.38±0.06 nm, which matches the length of the molecule with an all

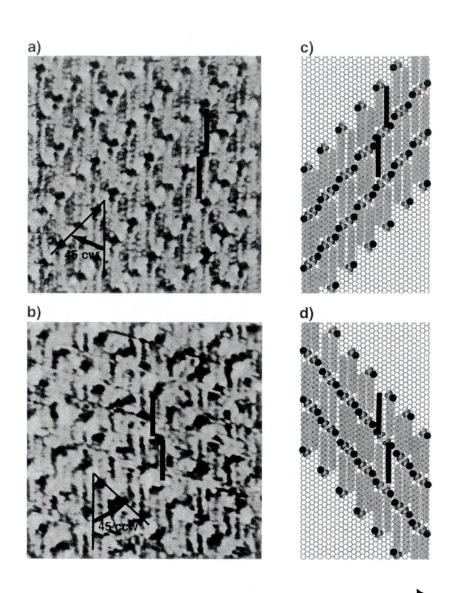

Figure 1. (a) A 12x12 nm² STM constant current image of (R)-2-bromohexadecanoic acid at the interface of a phenyloctane solution and the graphite basal plane. One molecular length is indicated by a thick black bar. Two thick black bars correspond to the chiral dimer formed by two molecules hydrogen bonded through carboxyl groups. The 45° angle "clockwise" is from the direction of the long axis of the molecule to the direction of the lamellar orientation. The imaging parameters are 1.4V (sample negative) and 300 pA. (b) A 12x12 nm² STM image of (S)-2-bromohexadecanoic acid at the interface of a phenyloctane solution and the graphite basal plane. One molecular length is indicated by a thick black bar. Two black bars correspond to the chiral dimer formed by two molecules hydrogen bonded through carboxyl groups. The 45° angle "counterclockwise" is from the direction of the long axis of the molecule to the direction of the lamellar orientation. The imaging parameters are 1.4V (sample negative) and 300 pA. (c) Top view of a model of (R)-2-bromohexadecanoic acid molecules on a graphite surface. White represents bromine atoms, gray represents carbon atoms, light gray represents hydrogen atoms, and dark gray represents oxygen atoms. Two black bars mark the chiral dimer formed by two molecules hydrogen bonded through the COOH acid groups. (d) Top view of a model of (S)-2-bromohexadecanoic acid molecules on a graphite surface. Two black bars mark the chiral dimer formed by two molecules hydrogen bonded through the COOH acid groups.

trans conformation (2.24 nm). Each molecular image is composed of three features: a large bright spot, a dark part adjacent to the bright spot, and a chain of a few small protrusions attached to the bright and dark features. This picture is much clearer in Figure 2, which is an enlargement of Figure 1a. Previous studies on long chain n-carboxylic acids have shown that these molecules physisorb on graphite with the carboxyl end groups oriented towards each other in order to facilitate hydrogen bonding (28,29). These studies also reveal that the carboxyl end group always appears with "darker" contrast compared to the rest of the hydrocarbon chain. The dark part in the image can, thus, be ascribed unambiguously to the carboxyl groups. We attribute the large "bright" spot to the location of the bromine atom in the "up" position (as noted above, protruding Br atoms appear bright in STM images of physisorbed molecules) at a non-terminal position along the hydrocarbon chain (13, 22-25). Given the relative position of the bromine atom and the carboxyl group, the orientation of the rest of the alkyl chain is determined, since the bromine atom is between the carboxyl group and the rest of the alkyl chain (as can be seen in the molecular structures depicted in Fig. 3a). The long alkyl tail of the molecule must lie down on the opposite side of the bromine atom from the carboxyl group. Thus, the chain of a few small protrusions, numbered 1-14 in Figure 2, can be attributed to the hydrogen atoms on the methylene carbons of 2-bromohexadecanoic acid. The spacings between these small protrusions match very well with those of hydrogen atoms on the methylene carbons of 2-bromohexadecanoic acid when it assumes an all trans conformation with the carbon backbone parallel to the surface (the spacings between the two closest protrusions along each row are 0.26 ± 0.01 nm and those between the protrusions in adjacent rows are 0.26 ± 0.02 nm as measured from the STM images; the spacings obtained from crystallographic data are 0.25 nm for both).

Based on the above assignment, the molecule can be easily identified as the R conformer since the relative positions of three of the four groups attached to the chiral carbon have been directly determined. The fourth atom on the chiral center (hydrogen) cannot be seen in the STM images because it is right beneath the bromine atom. Nevertheless, the position of this atom is firmly established by the topographical position of the Br atom as well as the positions of the labeled H atoms and their spacing, which establish the orientation of the carbon chain. The atomic resolution obtained in this STM image has enabled us to determine the absolute configuration of a single organic molecule directly, i.e., without the help of any compound of known chirality. Each molecule in Fig. 1a can be examined individually, and all are in the same configuration, (R)-2-bromohexadecanoic acid.

Fig. 1b is a 12x12 nm^2 STM image taken from another area of the surface. The resolution of this image is not as good as that attained in Fig. 1a, but the important features of the two dimensional structure can still be readily discerned. Again, each black bar in the image corresponds to one molecule lying flat on the surface in an all trans conformation. Each molecular image is composed of three parts: a large bright spot, Br, next to a dark region, the

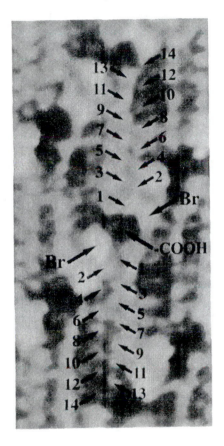

Figure 2. An enlarged area of Figure 1a. The detailed assignment of individual atoms on a chiral pair of molecules indicated by two black bars in Figure 1a is displayed here. The two Br designations mark the positions of the bromine atoms and -COOH points to the position of the carboxyl groups. Two sets of numbers 1 to 14 point to the positions of 14 hydrogen atoms on the 14 methylene carbons of the two molecules extending away from the -COOH groups.

212

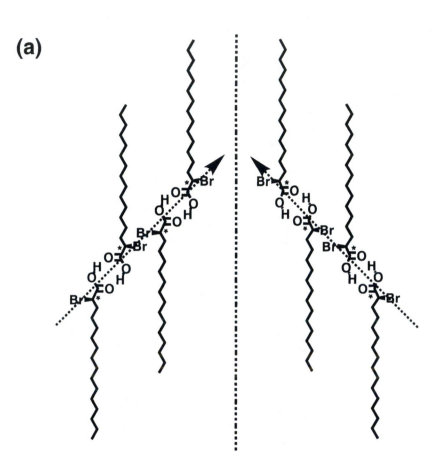

(a)

Figure 3. (a) Schematic of the molecular structures of (R)/(S)-2-bromohexadecanoic acid physisorbed on a graphite surface with an all trans conformation. The single chiral carbon atom on the molecules is indicated by an asterisk. Each molecule is hydrogen bonded to the next one through the COOH acid groups. The dashed lines with arrows are just a guide to show the orientation of the lamella. (b) Schematic of the self-assembly of hexadecanoic acid physisorbed on a graphite surface. The 2D organization is driven by hydrogen-bonding COOH groups and results in interdigitating acid head groups. In contrast to (a), here the lamellar axis is perpendicular to the molecular axis.

(b)

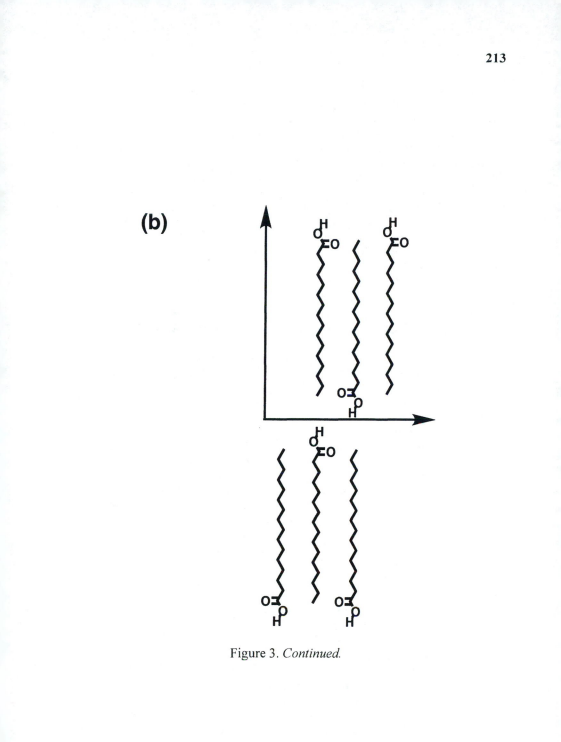

Figure 3. *Continued.*

carboxyl group, and a chain of a few small protrusions. Each molecule hydrogen bonds to another one in the next column and forms a dimer (marked as two black bars in the image), which is a chiral structure as well. Despite the fact that this image looks very much like the one in Fig. 1a, the topographs displayed in Fig. 1a and b are mirror images of each other. The dimers formed in Fig. 1b are mirror images of those formed in Fig. 1a. Following the assignment of the locations of the Br, -COOH, and alkyl chain in the same manner as outlined above, all the molecules in the image can be identified as (S)-2-bromohexadecanoic acid. A molecular model of this film is shown in Fig. 1d for comparison with the images formed by the R enantiomers. In the case of the S enantiomer, the quality of the STM images is not sufficient to assign the absolute chirality from a high resolution image such as that in Figure 2. Rather, the assignment of this domain is based on its overall mirror image symmetry with respect to the R enantiomeric domains. Furthermore, our work on chirally pure molecules confirms the unique self-assembly patterns formed by R-2-Br-hexadecanoic acid and S-2-Br-hexadecanoic acid (27). We find that R-2-Br-hexadecanoic acid only forms patterns such as the ones depicted in Figure 1a and S-2-Br-hexadecanoic acid self-assembles only in configurations similar to Figure 1b. The reason for the lower resolution of the S domains is not clear at this time, though it may simply represent changes in tip quality between the images shown in Fig 1a and Fig 1b.

R and S enantiomers of 2-bromohexadecanoic acid clearly form chiral structures that are mirror images of each other, as shown in the schematic of Fig. 3a and the topographs of Figure 1. The atomically resolved STM images have enabled us to "see" the absolute configuration of these molecules on the graphite surface directly, by taking advantage of the bright Br and dark -COOH STM "marker" groups. Moreover, the graphite lattice puts additional restrictions on the formation of a stable self-assembled layer structure. When the enantiomers of 2-bromohexadecanoic acid adsorb onto the graphite surface, an S enantiomer only hydrogen bonds to another S enantiomer to form a dimer; an R enantiomer only hydrogen bonds to another R enantiomer and forms another dimer, which is the mirror image of the one generated by the S enantiomers. This phenomenon observed in the STM images indicates that the interactions with the surface for dimers possessing the same chirality are stronger than for dimers having different chirality. If an R enantiomer hydrogen bonds to an S enantiomer, the bromine group on one of the stereoisomers must point down into the graphite surface as discussed below. When this occurs, the bulky bromine group lifts the molecule from the graphite surface thereby decreasing the interactions of the remaining atoms in the molecular chain with the surface. (Both the van der Waal's radius of Br (1.95 Å) and the C-Br bond length (1.937 Å) are much larger than the 1.2 Å H atom van der Waal's radius and 1.101 Å C-H bond length.) While this structure may form initially, especially at domain boundaries that are dynamic and poorly resolved, it will eventually be replaced by the more stable dimers in which R enantiomers bond to R enantiomers and S enantiomers bond to S enantiomers. It is possible, in principle, to further examine the chiral segregation observed in the STM images by comparing the bromine-bromine distances of a hydrogen-bonded pair consisting of an R conformer and another R enantiomer with a pair composed of an R conformer and an S enantiomer; however x-ray crystallography results suggest that the difference in Br-Br distances in these two cases would be only on the order of 0.5Å. Further, because of the templating effect of the graphite surface and the presence of solvent in these ambient STM images, the Br-Br distance for R-R versus R-S physisorbed dimers cannot be assumed to the same as found in the 3D crystal structure. As a result, the Br-Br distance does not provide a clear distinction between R-R and R-S dimer pairs.

The dimers formed by S enantiomers and those formed by R enantiomers thus segregate on the graphite surface and form different domains in which mirror image patterns can be easily detected by STM. By examining larger scale STM images, e.g. 20x20 nm^2, the chirality of the molecules can be simply determined from the overall long range packing order of the molecules.

Mixture of racemic 2-Br-hexadecanoic acid and hexadecanoic acid

Mixtures of racemic 2-Br-hexadecanoic and hexadecanoic acid exhibit long-range order when physisorbed onto a graphite substrate. Multiple domains are formed extending in width from 30nm to over 100nm. Figure 4a shows a representative 14nm x 14nm STM topograph of a 1:1 mixture of hexadecanoic acid with 2-Br-hexadecanoic acid imaged at the phenyloctane-graphite interface. This image also depicts well-ordered lamellae separated by dark troughs. The length of a given molecule, 2.25 ± 0.08 nm, corresponds to the expected 2.25 nm length of either a hexadecanoic acid or a 2-Br-hexadecanoic acid molecule configured in an all-trans conformation lying flat on the surface. Further, the lamellae form a 90° angle with the molecular axis in this two-dimensional array.

Troughs consisting of two different elements are evident in this constant-current topograph. The majority of each trough is comprised of dark, circular groups that are located at the end of each molecule, while a small portion of every trough has bright spots accompanying these dark regions. As described previously, the dark areas of lower image contrast are associated with the hydrogen-bonding carboxyl functional groups. The bright spots (areas of higher image contrast) correspond to bromine atoms, as observed for other bromines located at a non-terminal position along the hydrocarbon chain (13, 22-25). These chemical marker groups, therefore, allow for a straightforward differentiation between molecules of hexadecanoic acid and those of 2-Br-hexadecanoic acid present in this self-assembled, chemically mixed monolayer. A schematic of the self-assembly of pure hexadecanoic is depicted in Figure 3b. All of the carboxyl groups in Figure 4 form an interdigitating pattern similar to the one reported for pure hexadecanoic acid and shown in Figure 3b. As

216

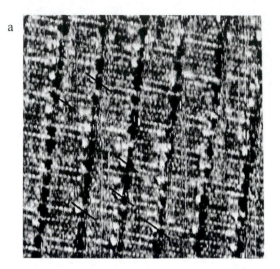

a

Figure 4. (a) and (b) show constant-current STM topographs of a 1:1 mixture by volume of 2-Br-hexadecanoic acid to hexadecanoic acid physisorbed onto a graphite surface, imaged at 1.5V (sample negative) and 300pA. The 14nm x 14 nm image in (a) displays hexadecanoic acid molecules and an occasional (R)-2- Br-hexadecanoic acid molecule, both of which are configured in the all-trans conformation. A trough reveals dark regions that are attributed to the locations of hydrogen-bonding COOH groups accompanied sporadically by bright groups attributed to bromine atoms. The lamellar axis is perpendicular to the molecular axis. There is a consistent orientation of bromine atoms relative to COOH groups marked by superimposed black lines; either the bromine lies above and to the left or below and to the right of the COOH group. These molecules are identified as (R)-2-Br-hexadecanoic acid. The 11nm x 11nm topograph in (b) exhibits a bromine/COOH orientation that is the mirror image of that presented in (a). Here the bromine atom lies above and to the right or below and to the left of the COOH group. These brominated molecules are identified as (S)-2-Br- hexadecanoic acid. A top view of a model of hexadecanoic acid interspersed by (R)-2-Br-hexadecanoic acid physisorbed onto the graphite surface is shown in (c). The black line superimposed on one of the bromine/COOH combinations mimics the arrangement found in the corresponding STM image (a). A top view of a model of hexadecanoic acid interspersed by (S)-2-Br-hexadecanoic acid on the graphite surface is shown in (d). The black line superimposed on a bromine/COOH combination shows the same pattern depicted in the corresponding STM image (b).

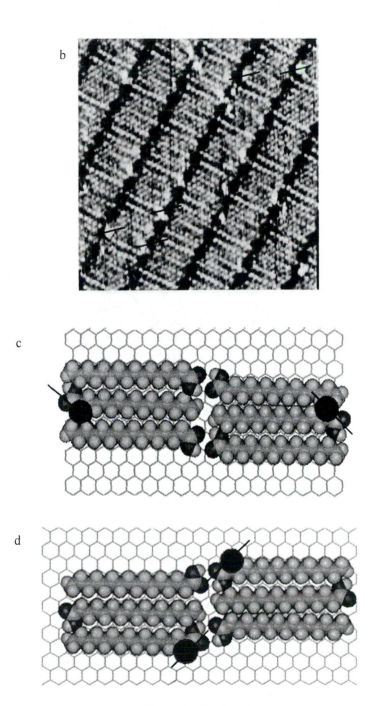

Figure 4. *Continued.*

demonstrated by a random array of bright spots visible in this topograph, the 2-Br-hexadacanoic acid molecules are sporadically located throughout the two-dimensional structure. Hexadecanoic acid molecules dominate the monolayer while some 2-Br-hexadecanoic acid molecules are able to squeeze into the dominant ordering structure of the non-brominated moiety.

The relative orientation of the bromine atoms to the carboxylic acid groups remains consistent throughout the STM images of a single domain. The assignment of a bromine atom to a particular carboxyl group is achieved by carefully comparing distances between the bromine atom and neighboring carboxyl groups within a given trough. This assignment is further facilitated by taking advantage of the interdigitation patterning of the hexadecanoic acid molecules that separates neighboring carboxyl groups. When this procedure is followed, over 90% of the time the bromine atom (bright spot) is found to be closest (directly adjacent) to only one carboxyl group (dark region). In Figure 4a, all of the bromine atoms lie either above and to the left of the carboxyl group or below and to the right of the carboxyl group. Black lines have been superimposed on some of the individual carboxyl/bromine combinations to denote this arrangement. A different marker group positioning is found in the 11nm x 11nm constant current topograph in Figure 4b where the bromine/carboxyl group orientation is the mirror image of that presented in Figure 4a. Each bromine in Figure 4b lies either above and to the right or below and to the left of its respective carboxyl group, as noted by the superimposed black lines.

Figures 4c and 4d provide molecular models based on a general portion of the self-assembled monolayer; these models are not meant to represent a specific region in the STM images previously described. The interdigitating pattern has been accurately depicted, and the orientation of carboxyl groups with their accompanying bromines (Figures 4a and 4b) are accordingly reflected in the respective models. A black line superimposed on a carboxylic/bromine combination in Figure 4c shows a bromine atom above and to the left of its carboxyl group, similar to the pattern found in the corresponding STM image in Figure 4a. A similar comparison arises in Figure 4d where the model's Br/COOH arrangement, highlighted by a black line, mimics the orientation found in the corresponding STM image (Figure 4b). Using the guidelines to directly assign chirality described in the previous section, the brominated molecules in Figure 4a and 4c are uniquely identified as (R)-2-Br-hexadecanoic acid whereas the brominated molecules in Figure 4b and 4d are exclusively (S)-2-Br-hexadecanoic acid.

Finally, note that unlike the 45° angle between the lamella direction and molecular axis present in pure 2-Br-hexadecanoic acid from the previous section and from Figure 1, these images display a 90° angle between these two axes for both hexadecanoic acid with (R)-2-Br-hexadecanoic acid and hexadecanoic acid with (S)-2-Br-hexadecanoic acid. Recall that pure hexadecanoic acid self-assembles with a 90° lamellar-molecular axis angle as shown in the schematic in Figure 3b. The 45° structure shown in Figure 3a is driven by favorable Br-Br attractive interactions, which are effectively shut down in mixtures where the probability that two Br-containing molecules will be found adjacent to each other is low.

Two phenomena clearly stand out in the STM images of the mixed monolayer. First, the ability to effect a specific self-assembly of 2-Br-hexadecanoic by introducing hexadecanoic acid into the solution has been demonstrated in Figure 4. The element of control for this feature is the "forced" opening of the angle between the molecular axis and the lamellar direction from 45° to 90° in the mixed self-assembly. Second, racemic 2-Br-hexadecanoic acid co-adsorbed with hexadecanoic acid segregates into separate chiral domains when the solution is applied to a graphite surface. This phenomenon is attributed to a curious effect of the two-dimensional graphite surface on one element in the mixture, the achiral hexadecanoic acid molecules. In fact, physisorption of hexadecanoic acid on a graphite surface causes it to separate into two distinct domains that exhibit non-superimposable mirror image morphology (enantiomorphous domains), a general feature observed for fatty acids possessing an even number of carbon atoms (16).

Hexadecanoic acid is an achiral molecule in three dimensions in that it is always superimposable with its mirror image. This property changes dramatically when hexadecanoic acid is adsorbed onto a graphite surface, for now the molecule is fixed in a plane instead of being able to rotate freely in three dimensions (16). Similarly, in the language of crystallography, the acid in three dimensions belongs to a non-chiral point group. Physisorption onto the surface removes the molecule's mirror plane and thus hexadecanoic acid adsorbed onto the surface now belongs to a chiral point group. Two hexadecanoic acid molecules will adsorb onto a graphite surface in a manner that allows the carboxylic acid functional groups to hydrogen bond together; Figure 5a depicts two such molecules whose conformations have been labeled "A". When this pair is reflected through a mirror plane demarcated by the black line, the resultant molecules on graphite adopt the configuration labeled "B", which (in the restricted two-dimensional world of the interface) are non-superimposable mirror images of the "A" molecules on the surface. The surface ultimately drives hexadecanoic acid to self-assemble in two distinct enantiomorphous domains that now possess 2D chirality, as defined in Reference 9. Achiral molecules forming enantiomorphous domains on a surface have been previously observed for discotic liquid crystals studied by STM (30), for a liquid crystal phase of achiral molecules investigated via optical microscopy (31), for calcium arachidate assembled on mica probed by AFM (7), and for adenine molecules adsorbed onto molybdenum disulfide observed via STM (32).

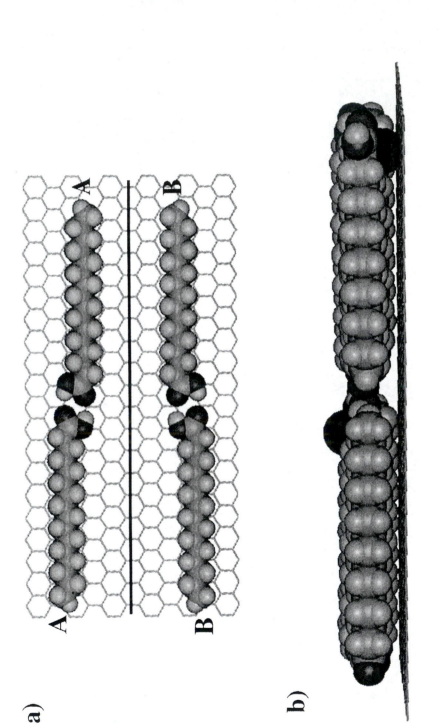

a)

b)

Figure 5. A series of models showing different self-assembly patterns for hexadecanoic acid on a graphite surface. The two-dimensional chiral nature of achiral hexadecanoic acid physisorbed onto a graphite surface is shown in (a). The pair of molecules on top have been arbitrarily labeled as "A" while the orientation on the bottom is "B". The "B" molecules on graphite are a mirror image of the "A" molecules on the surface, reflected through the mirror plane indicated by a black bar. A side view of a possible model for a domain consisting of both (R) and (S)-2-bromohexadecanoic acid mixed with hexadecanoic acid molecules physisorbed on a graphite surface is shown in (b). Hexadecanoic acid molecules are arranged in their classic interdigitating pattern with a sample brominated molecule on the left and on the right (the black ball signifies the position of the bromine). The conformer on the left-hand side of the figure is identified as R while the one on the right is the S conformer. The R enantiomer has a bromine protruding up and away from the graphite substrate, while the (S)-2-Br-hexadecanoic acid has the bromine pushing into the graphite. The Br down configuration is energetically unfavorable because it pushes the physisorbed molecule up off the graphite surface and is, therefore, also rejected as a model for the mixed monolayer on a thermodynamic time scale.

Resolution of (R)/(S)-2-Br-hexadecanoic acid within the mixture

The 2D chiral assembly of hexadecanoic acid molecular pairs on a graphite surface described in Figure 5a drives the mixture of 2-Br-hexadecanoic acid and hexadecanoic acid to segregate into enantiomerically pure domains on the graphite surface. Figure 4a shows an image of hexadecanoic acid with exclusively (R)-2-Br-hexadecanoic acid while Figure 4b reveals a monolayer of hexadecanoic acid with uniquely (S)-2-Br-hexadecanoic acid. The self-assembly of hexadecanoic acid into two distinct domains exhibiting 2D chirality constrains only one of the two chiral isomers of 2-Br-hexadecanoic acid to co-exist with the unsubstituted fatty acid.

A possible way of combining the R and S enantiomers of 2-Br-hexadecanoic acid within a hexadecanoic acid domain is depicted in Figure 5b with a side-view model. Eight hexadecanoic acid molecules are arranged on the surface in the classic interdigitating pattern with one molecule on each side of the central trough being substituted with a bromine at the alpha position. The brominated molecule on the left represents an R enantiomer whose bromine atom (colored in black) protrudes up and away from the graphite surface. The substituted hexadecanoic acid on the right is an S enantiomer with the bromine (colored in black) pointing down towards the graphite surface. This configuration is not energetically favorable because it forces the physisorbed molecule up off the surface (17). The more energetically favorable self-assemblies where chiral segregation takes place and registry with the surface can be maintained are more stable and are those "seen" by the STM.

Conclusion

A series of studies have been undertaken involving the chiral molecule 2-Br-hexadecanoic acid and the achiral molecule hexadecanoic acid. It has been clearly demonstrated that STM can be used to determine the chirality of organic molecules directly. The chiral molecules, (R)/(S)-2-bromohexadecanoic acid, adsorbed on a graphite surface, have been imaged by STM with atomic resolution. These two enantiomers segregate on a graphite surface and form different domains that exhibit mirror image packing patterns on the graphite surface. The self-assembly of a mixture of hexadecanoic acid with racemic 2-Br-hexadecanoic acid has also been investigated. Hexadecanoic acid has the ability to control the two-dimensional structure formed at the interface by creating a template that sets the angle between the molecular and lamellar axes in the mixed monolayer. Further, inherently achiral hexadecanoic acid self-assembles on the graphite surface into two distinct, although almost indistinguishable domains that exhibit mirror image morphology, similar to that observed for other even-numbered fatty acids (16). This property leads to resolution of (R)/(S)- 2-Br-hexadecanoic acid into enantiomerically pure (R and S) domains when physisorbed onto a graphite surface from the mixture.

Acknowledgments

The authors would like to thank Professor Gerard Parkin and Martin Leivers for helpful discussions. This work was supported by grants from the National Science Foundation (CHE-97-27205), jointly by the National Science Foundation and the U.S. Department of Energy through a grant to the Environmental Molecular Sciences Institute at Columbia University (NSF CHE-98-10367), and in part by the Columbia Materials Research Science and Engineering Center (DMR-98-09687).

References

1. Dale, J. A.; Mosher, H. S. *J. Am. Chem. Soc.* **1973**, *95*, 512.
2. Harada, N.; Nakanishi, K. *Circular Dichroic Spectroscopy - Exciton Coupling in Organic Stereochemistry*; University Science Books: Mill Valley, CA, 1983.
3. Lyle, G. G.; Lyle, R. E. In *Asymmetric Synthesis*; Morrison, J. D., Ed.; Academic Press: New York, 1983; Vol. 1, p 13.
4. Lipscomb, W. N.; Jacobon, R. A. In *Physical Methods of Chemistry*; Rossiter, B. W., Hamilton, J. F., Eds.; Wiley: New York, 1986; Vol. V, p 1.
5. Binnig, G.; Rohrer, H.; Gerber, C.; Weibel, E. *Phys. Rev. Lett.* **1982**, *49*, 57-61.
6. Binnig, G.; Quate, C. F.; Gerber, C. *Phys. Rev. Lett.* **1986**, *56*, 930-933.
7. Viswanathan, R.; Zasadzinski, J. A.; Schwartz, D. K. *Nature* **1994**, *368*, 440-443.
8. Eckhardt, C. J.; Peachey, N. M.; Swanson, D. R.; Takacs, J. M.; Khan, M. A.; Gong, X.; Kim, J.-H.; Wang, J.; Uphaus, R. A. *Nature* **1993**, *362*, 614-616.
9. Walba, D. M.; Stevens, F.; Clark, N. A.; Parks, D. C. *Accts. of Chem. Res.* **1996**, *29*, 591-597.
10. Bohringer, M.; Morgenstern, K.; Schneider, W.; Berndt, R. *Angew. Chem. Int. Ed.* **1999**, *38*, 821-823.
11. DeFeyter, S.; Grim, P. C. M.; Rucker, M.; Vanoppen, P.; Meiners, C.; Sieffert, M.; Valiyaveettil, S.; Mullen, K.; DeSchryver, F. C. *Angew. Chem. Int. Ed.* **1998**, *37*, 1223-1226.
12. McKendry, R.; Theoclitou, M. E.; Rayment, T.; Abell, C. *Nature* **1998**, *391*, 566.

13. Fang, H.; Giancarlo, L. C.; Flynn, G. W. *J. Phys. Chem. B* **1998**, *102*, 7311-7315.
14. Lopinski, G. P.; Moffatt, D. J.; Wayner, D. D. M.; Wolkow, R. A. *Nature* **1998**, *392*, 909-911.
15. Lopinski, G. P.; Moffatt, D. J.; Wayner, D. D. M.; Zgierski, M. Z.; Wolkow, R. A. *J. Am. Chem. Soc.* **1999**, *121*, 4532-4533.
16. Hibino, M. S., A.; Tsuchiya, H.; Hatta, I. *J. Phys. Chem. B* **1998**, *102*, 4544-4547.
17. Yablon, D. G.; Giancarlo, L. C.; Flynn, G. W. *J. Phys. Chem. B,* in press.
18. Eigler D.M.; Weiss P.S.; Schweizer E.K.; Lang N.D. *Phys. Rev. Lett.* **1991**, *66*, 1189-1191.
19. Lang N.D. *Phys. Rev. Lett.* **1986**, *56*, 1164.
20. Stipe, B.C.; Rezaei, M.A.; Ho, W. *Science* **1998**, *280*, 1732-1735.
21. Stabel A.; Dasaradhi L.; O'Hagan, D. Rabe, J.P. *Langmuir* **1995**, *11*, 1427.
22. Cyr, D. M.; Venkataraman, B.; Flynn, G. W.; Black, A.; Whitesides, G. M. *J. Phys. Chem.* **1996**, *100*, 13747-13759.
23. Fang, H.; Giancarlo, L. C.; Flynn, G. W. *J. Phys. Chem. B* **1998**, *102*, 7421-7424.
24. Fang, H.; Giancarlo, L. C.; Flynn, G. W. *J. Phys. Chem. B* **1999**, *103*, 5712-5715.
25. Claypool, C. L.; Faglioni, F.; Matzger, A. J.; W.A., G.; N.S., L. *J. Phys. Chem. B.* **1999**, *103*, 9690-9699.
26. Crystal, J.; Zhang L.; Friesner R.; Flynn G.W. "Computational Modeling for Scanning Tunneling Microscopy of Physisorbed Molecules via Ab Initio Quantum Chemistry", submitted.
27. Yablon, D.G.; Guo, J.; Knapp D; Fang H.; Flynn, G.W. "Obtaining unambiguous 3D images: STM investigations of chirally pure molecules at the liquid-solid interface", in preparation.
28. Hibino, M.; Sumi, A.; Hatta, I. *Jpn. J. Appl. Phys.* **1995**, *34*, 3354-3359.
29. Hibino, M.; Sumi, A.; Hatta, I. *Jpn. J. Appl. Phys.* **1995**, *34*, 610-614.
30. Charra, F.; Cousty, J. *Phys. Rev. Lett.* **1998**, *80*, 1682-1685.
31. Link, D. R.; Natale, G.; Shao, R.; Maclennan, J. E.; Clark, N. A.; Korblova, E.; Walba, D. M. *Science* **1997**, *278*, 1924-1927.
32. Sowerby, S. J.; Heckl, W. M.; Petersen, G. B. *J. Molec. Evol* **1996**, *43*, 419-424.

Chapter 16

A Spectroscopic Study of Chiral Discrimination in Jet-Cooled van der Waals Complexes

K. Le Barbu, F. Lahmani, and A. Zehnacker

Laboratoire de Photophysique Moléculaire du CNRS. Bât. 213,
Université de Paris, XI 91405 Orsay Cédex, France

We have combined the use of supersonic jet techniques with laser-induced fluorescence to study weakly bound complexes formed between chiral molecules. By using a chiral naphthalene derivative as a fluorescent selectand, we have been able to discriminate between the enantiomers of a chiral secondary alcohol, 2-butanol, and of chiral aminoalcohols, on the basis of different excitation spectra. Non chiral systems have also been studied for the sake of comparison. Ground-state depletion spectroscopy (hole-burning) experiments have shown that, in most cases, two isomers of the same diastereoisomer coexist in the jet. We discuss these results which shed light on the role of hydrogen bonding *vs* dispersion in the observed spectral discrimination.

Introduction

Chiral discrimination is of the widest importance in life chemistry and takes place through stereoselective interactions with an optically active selecting agent in diastereoisomeric contact pairs implying short-range forces. Since the molecular interactions responsible for chiral discrimination are often weak, the diastereoisomeric forms at play are mostly transient in nature and thus difficult to investigate in solution at room temperature. We have thus undertaken a spectroscopic study of van der Waals complexes of chiral molecules by means of the supersonic-jet technique combined with laser-induced fluorescence: by stabilising weakly bound diastereoisomers in the gas phase, this method provides us with a powerful tool for studying at the molecular level the short-range molecular interactions involved in chiral recognition. The experiments rest on the formation of a complex between the fluorescent chiral chromophore 2-Naphtyl-1-ethanol, denoted NapOH hereafter, acting as a selectand, and the chiral solvating agent. The spectroscopic properties of the selectand are modified by complexation with the chiral solvating agent and one expects this modification to be different for the **RR** and the **SR** pairs. On the basis of this methodology, we have demonstrated for the first time that chiral discrimination could be obtained in jet-cooled complexes on the basis of different fluorescence excitation spectra [1-4] and a different efficiency of the deactivation processes of the excited state [5]. The binding energy of the **RR** and **SR** associations has also been directly obtained by Giardini-Guidoni *et al.* [6-7], and our group [19], by measuring the dissociation threshold of diastereoisomeric complexes by resonance-enhanced-multiphoton ionisation. Complexes of NapOH with chiral aliphatic alcohols have been studied by combining double-resonance experiments (hole-burning) with a theoretical approach. In the case of the simplest chiral alcohol 2-butanol [4] or its non chiral model, 1-propanol, the results have been explained in terms of two different conformers (*anti* and *gauche*), which coexist under supersonic-jet conditions and are efficiently discriminated by the NapOH selectand. In this paper, we explore more in detail the role of conformational isomerism in chiral discrimination. The properties of linear alkyl alcohols will be first recalled. They are compared to those of aminoalcohols, whose geometry is influenced by the formation of an internal H bond. We discuss how the strength of the H bond influences the spectroscopic properties of the system by comparing two similar chiral systems (2-amino1-propanol and 1-amino-2-propanol) for which the H-bond strength is different. For the sake of clarity, the assignment of the 0_0^0 transition of the different complexes under study is summarised in table I.

Experimental

The experiment rests on the laser-induced fluorescence of van der Waals complexes formed in a continuous supersonic expansion of Helium (2-3 atm) and has been described previously[4]. The molecules are excited in the cold region of the jet by means of a frequency-doubled dye laser (DCM) pumped by the second harmonic of a YAG laser (BM Industrie or Quantel). The fluorescence is observed at right angle trough a WG355 cut-off filter by a Hamamatsu R2059 photomultiplier. The signal is monitored by a Lecroy 9400 oscilloscope connected to a PC computer.

Hole-Burning experiments in a supersonic jet were first reported by Lipert and Colson[8] and are used to identify, among the numerous bands of the excitation spectrum, those spectral bands which belong to the same species. The principle of the experiment involves a pump-probe excitation scheme, with two counterpropagating lasers delayed in time. The wavelength of the first intense laser (the pump) is scanned through the region of interest: when its energy matches that of a transition arising from a given ground-state species it induces the depopulation of the corresponding species. The wavelength of the second laser (the probe) is fixed on a transition arising from a given ground state and the induced fluorescence is a measure of the (de)population of the selected ground state. When both lasers excite transitions which arise from the same ground-state species, the pump-beam-induced depopulation manifests itself by a decrease in the intensity of the fluorescence induced by the probe (spectral hole). If both lasers excite different ground-state species, such a decrease is not observed. It is therefore possible to record separately the absorption spectra of different species which absorb in the same energy range.

Experimental results:

The laser-induced fluorescence spectrum of NapOH has already been reported[1]. Its 0^0_0 transition located at 31738.4 cm^{-1} is followed by two low-frequency features at 39 and 76 cm^{-1} assigned to the torsion of $CH(CH_3)OH$.

1-Complexes with alcohols.

Figure 1 shows the fluorescence excitation spectrum of the S-NapOH complexes with alcohols, in the region of the $S_0 \rightarrow S_1$ origin of the bare chromophore, together with the hole-burning spectra.

a - NapOH / 1-Propanol (1-PrOH) complexes

The formation of the complex manifests itself by the appearance of four main features located at -138, -112, -60 and -55 cm^{-1} respectively from the 0_0^0 transition of NapOH. Hole-burning experiments have been performed with the probe laser tuned on the features at -138, -60 and -55 cm^{-1} from the origin. These hole-burning spectra show unambiguously that these features correspond to the 0_0^0 transitions of three isomers denoted hereafter **a**, **g$_1$**, and **g$_2$**. The hole-burning spectra of **g$_1$** as well as **g$_2$** display a single intense feature. These bands are separated from each other by only 5 cm^{-1} and appear as a doublet in the fluorescence excitation spectrum. The hole-burning spectrum of **a** shows a more red-shifted 0_0^0 transition located at -138 cm^{-1} and two weaker bands which involve a progression built on a 26 cm^{-1} mode.

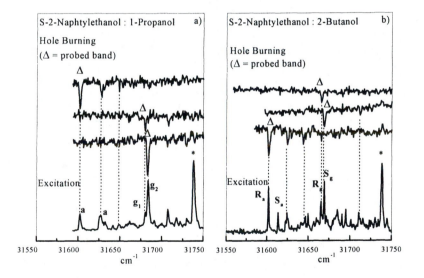

*Figure 1: Laser-induced fluorescence spectrum (bottom) and hole-burning spectra (top) of the S-NapOH complexes with a) 1-propanol. The bands attributed to the complexes with the anti form (respectively gauche) of 1-propanol are denoted by **a** (respectively **g**) b)2-butanol. The bands attributed to the complexes with the **R** enantiomer (respectively **S**) of 2-butanol are denoted by **R** (respectively **S**). **a** and **g** have the same meaning as for 1-propanol. The band denoted by * is the 0_0^0 transition of the bare chromophore.*

b - NapOH / 2-Butanol (2-BuOH) complexes

The formation of the complex manifests itself by the appearance of five main bands in the fluorescence excitation spectrum. The excitation spectrum recorded with pure enantiomers allows one to assign the bands located at -136, -114 and -73 cm^{-1} from the origin of NapOH to the **SR** diastereoisomer, and those located at -125 and -69 cm^{-1} from the origin to the **SS** complex[2-4]. Hole burning experiments show that four different isomers coexist in the jet, two of them correspond to the **SR** complex and the two others to the **SS** complex. The 0^0_0 transitions of the **SR** isomers are located at -136 and -73 cm^{-1} from the 0^0_0 transition of the bare molecule: these isomers will be denoted respectively **SR$_a$** and **SR$_g$** hereafter. The hole burning spectrum of the **SR$_a$** isomer shows (as **a** in the case of the complex with 1-PrOH) a vibrationnal progression built on a 22 cm^{-1} vibration (bands at -114 cm^{-1} and -92 cm^{-1}). The 0^0_0 transition of the **SS** isomers are located at -125 and -69 cm^{-1}: these isomers will be denoted **SS$_a$** and **SS$_g$** respectively. Notice that **SR$_g$** and **SS$_g$** origins are only separated by 4 cm^{-1}, and that no vibrational feature appears for these isomers, as was the case for **g$_1$** and **g$_2$**.

2-Complexes with aminoalcohols.

a - S-NapOH/1-amino-2-propanol (1am2pro) complexes.

The fluorescence excitation spectra of the S-NapOH/1am2pro complex with each enantiomer of 1am2pro in the region of the $S_0 \rightarrow S_1$ origin of the bare chromophore are presented in Figure 2, together with the hole burning spectra. It can be deduced from these spectra that the features located at -145 and -90 cm^{-1} from the origin of the bare chromophore correspond to two different isomers of the **SR** complex, and those located at -110 and -49 cm^{-1} to two different isomers of the the **SS** complex. The hole-burning spectrum obtained when probing the 0^0_0 transition of the **SR** complex located at –145 cm^{-1} shows, besides the origin, an additional band due to the excitation of a low-frequency mode (20 cm^{-1}). The hole-burning spectrum of the **SR** complex whose 0^0_0 transition is located at –78 cm^{-1} displays a single band. As already observed for other secondary alcohols, the 0^0_0 transition of **SR** is more shifted than that of **SS**.

Table I. Assignment of the 0^0_0 Band of the different Ground State Isomers of the Complexes under Study.

Complexing alcohol	Shift (cm^{-1})	Isomer
1-propanol	-138	a
	-60	g_1
	-55	g_2
2-Butanol	-147	SRa
	-125	SSa
	-73	SRg
	-69	SSg
2-amino-1-propanol	-96	SS
	-60	SR
1-amino-2-propanol	-145	SR
	-110	SS
	-90	SR
	-49	SS

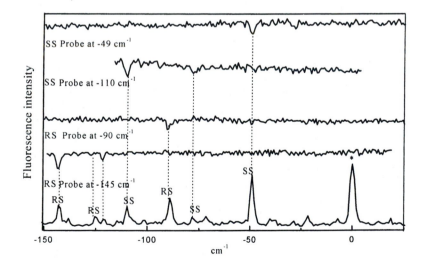

Fig 2. : Fluorescence excitation spectra (bottom) and hole-burning spectra of the 1-amino-2-propanol/ S-NapOH complex. The bands due to the bare molecule are denoted by*

b - S-NapOH/2-amino-1-propanol (2am1pro) complexes.

The fluorescence excitation spectra of S-NapOH complexed with racemic, R- and S-2am1pro in the region of the $S_0{\rightarrow}S_1$ origin of the bare chromophore are presented in Figure 3. It is to be noted that the **SR** complex as well as the **SS** complex display a single feature, located at -60 cm^{-1} from the origin of the bare chromophore for **SR** and at -96 cm^{-1} for **SS**. As already observed [1,3] for primary alcohols, the electronic transition of **SS** is more shifted to the red than that of **SR**.

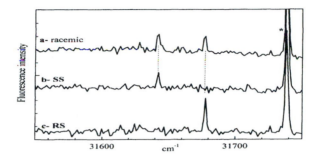

*Figure 3. : Fluorescence excitation spectra of the complexes between S-NapOH and a) racemic mixture of 2-amino-1-propanol, b) S-2-amino-1-propanol, c) R-2-amino-1-propanol. The band denoted by * is the $S_0{\rightarrow}S_1$ origin of the bare molecule.*

Discussion:

1-Description of the systems under study:

a-Chromophore : 2-naphtylethanol (NapOH).

Both hole-burning experiments and theoretical results [4] show that only one conformation of the chromophore exists under supersonic-jet conditions, in which the O atom lies in the naphthalene plane and the hydroxyl H atom is out of the plane. This geometry is different from that of benzylalcohol [9] for which an

232

out-of-plane configuration of the OH group has been proposed. This difference probably results from the stronger steric hindrance introduced by the CH_3 substituant on the side-chain carbon.

b-Aliphatic alcohols : bare molecules

1-PrOH : Two *gauche* and one *anti* conformers of 1-PrOH have been calculated and are denoted g_1, g_2 and **a** respectively.

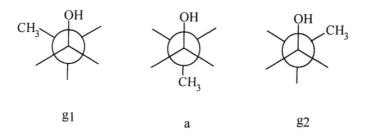

g1   a   g2

The *gauche* conformer is more stable by 0.4 kcal/mol than the *anti* one, in agreement with the microwave spectrum of 1-PrOH which shows an energy difference of (0.29 ± 0.15) kcal/mol [10]. The attraction between the hydroxyl and methyl group is responsible for this difference in stability. As there is a barrier for rotation of about 2.3 kcal/mol, the ground state conformers do not interconvert in the supersonic expansion under our experimental conditions. Their relative abundance reflects the Boltzmann distribution at room temperature: the relative populations are 40% of each *gauche* conformer and 20% of the *anti*. It is here important to notice that under these conditions, the rotation around the C_1-C_2 axis is blocked and g_1 and g_2 must be considered as distinct conformers which are axial enantiomers. At the low temperature achieved in the jet, an axial chirality may therefore be revealed by complexation with a chiral partner for molecules which are not chiral at room temperature.

2-BuOH : The difference between 1-PrOH and the chiral secondary alcohol 2-BuOH is the substitution in the 1-position of a hydrogen atom by a methyl group. The MP2-optimisation of the geometry shows that two different conformers of *R*- (respectively *S*-)2-BuOH exist under jet-cooled conditions, namely **Rg**, **Ra** (respectively **Sg**, **Sa**). In this notation, the index denotes the *anti* or *gauche* position of the OH relative to a CH_3 group. The population of the other possible conformations is negligible. The **a** and **g** conformations of the solvent are similar to *anti* and *gauche* propanol respectively, following this scheme:

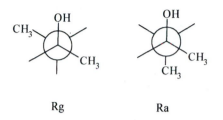

Rg   Ra

According to a Boltzmann distribution at room temperature, the relative populations of **g** and **a** are 0.79 and 0.21. The **Rg** and **Sg** conformers of 2-BuOH are built from the g_1 and g_2 conformers of 1-PrOH respectively whereas the **Sa** or **Ra** forms of 2-BuOH correspond to the **a** conformer of 1-PrOH. As the different conformers of the solvating agent coexist in the jet, they can all be complexed by the NapOH chromophore. These different complexes have been calculated for 1-PrOH as well as 2-BuOH.

c-Complexes with alcohols:

We have calculated the geometry and stabilisation energy of the complexes between NapOH and the three possible conformers of 1-PrOH. The calculations, which rest on the exchange perturbation method, have been described in an other publication[4]. In this method, both subunits of the complex are frozen in the most stable geometry (ies) cooled down in the jet. The results we have obtained are depicted in figure 4. All the calculated complexes involve an Hydrogen-bond between the chromophore acting as a donor and the alcohol solvent. The most stable structure of the complex with a *gauche* forms of 1-propanol has an extended geometry, with the alkyl chain of 1-PrOH far from the aromatic ring. The contribution of the dispersion energy to the total binding energy (–6.5 kcal/mol) is -5.3 kcal/mol. On the other hand, the most stable structure of the complex with the *anti* form of 1-PrOH is folded, which means that the alkyl chain of the solvating agent is bent over the aromatic ring of the chromophore. This results in a more important contribution of the dispersion energy (-6.0 kcal/mol) to the total interaction energy (-6.9 kcal/mol).

Similarly to 1-propanol, the Rg and Sg forms of 2-butanol form "extended complexes" with S-NapOH, whose geometry is not very different from that of the complexes with 1-propanol *gauche*. In other words, the addition of a methyl group doesn't introduce some steric effect which could modify the geometry and the interaction energy of the system. Owing to the large distance between the aromatic ring and the alkyl chain of the solvent, there is no strong modification of the molecular interaction when the shape of this chain is slightly modified. In contrast to that, the diastereoisomeric complexes with the *anti* forms strongly differ from each other (see figure 4). One of them, the **SR$_a$** one, has almost the same geometry as the complex with the 1-propanol *anti* conformer. The **SS$_a$** one displays a "less folded" geometry, owing to the steric hindrance introduced by the additive methyl substituent. This results in a smaller contribution of the dispersive interaction when compared to the heterochiral complex (-6.3 vs –6.0 kcal./mol. for a total interaction energy of –7.0kcal./mol.).

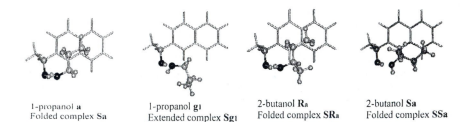

| 1-propanol **a**
 Folded complex **Sa** | 1-propanol **g**₁
 Extended complex **Sg**₁ | 2-butanol **R**$_a$
 Folded complex **SR**$_a$ | 2-butanol **S**$_a$
 Folded complex **SSa** |

Fig 4. : Examples of the calculated geometry for the 1-propanol/NapOH and 2-butanol/NapOH complexes.

d-Complexes with amino alcohols:

The theoretical description of the potential energy surface of aminoalcohols has been the subject of numerous studies [10-15-20]. All these studies agree on the fact that, in the gas phase, the most stable conformer of amino-substituted secondary alcohols shows an intramolecular hydrogen bond directed from the hydroxyl group to the nitrogen atom. Experimental data [14-16-20] show that an other form coexist with the OH…N form, which could be a NH…O form, since infrared results [15-20] suggest that the *gauche* NH…O form is more stable than the *anti* (NH$_2$ and OH in *anti* position). However, the relative abundance of the OH…N and NH…O is different in the gas phase and in the liquid phase[20], because of the competition between the formation of intramolecuar and intermolecular H-bonds. Similarly in the complex, the intramolecular H-bond will compete with the formation of an intermolecular H-bond with the chromophore. This competition strongly depends on the strength of the intramolecular H bond, which seems to be different for the two aminoalcohols studied here, as evidenced by their infrared spectra [16]. On the basis of its infrared spectrum in the gas phase, the intramolecular H-bond of 1-amino-2-propanol seems to be stronger (observation of an intense "bonded" OH stretching mode, a weak "free" OH stretching mode, and a NH_2 stretching mode). This system will be referred to as "strong intramolecular H bond". 2-amino-1-propanol probably displays weaker H-bond (observation of an intense "free" OH stretching mode, a weak "bonded" OH stretching mode, no NH_2 stretching mode). This system will be referred to as "weak intramolecular H bond".

A theoretical method such as the one applied for aliphatic alcohols cannot be applied in these systems, because the geometry of amino-alcohols can be modified upon complexation (disrupting of the internal H bond). In what follows, we shall qualitatively use the structural arguments presented above to discuss the experimental results. Calculations at the MP2 level are currently in progress to confirm these hypotheses.

Assignment of the isomers experimentally observed: alcohols

We now discuss a tentative assignment of the different isomers observed in the hole-burning experiments. In what follows we call "spectral discrimination" the difference in the 0_0^0 transition energies of the **SR** and **SS** complexes.

For 1-propanol and 2-butanol, this assignment rests on the comparison between theoretical results and hole-burning spectra [5]. It is based on the following hypotheses:

i) The spectral characteristics between the complexes of NapOH with 1-PrOH and 2-BuOH (similar red-shift and vibrational progression) must correspond to similarities between the calculated structures. Thus, the comparison between the two systems can be used to assess the proposed attribution.

ii) The dispersion forces are responsible for an increase in the binding energy of the excited state of the complex relative to its ground state. By comparing the spectroscopic properties of complexes of NapOH with several chiral and non-chiral alcohols [3], we have shown that the difference in the dispersion term was responsible for the variation in the red shift observed for the different complexes under study: the higher the dispersion in the ground state, the larger the red-shift of the electronic transition.

a - The most red-shifted isomers : ***a*** *compared with* ***SR****$_a$ and* ***SS****$_a$*

Under the reasonable assumption that the complex which exhibits the most important dispersion term corresponds to the most red-shifted isomer (hypothesis ii), we may associate the **a** isomer of 1-PrOH complex (located at -138 cm^{-1} from the origin) with the complex involving the *anti* conformer of 1-PrOH. This hypothesis is reinforced by the fluorescence excitation spectra of complexes of NapOH with short chain alcohols such as methanol or ethanol [3], which do not display any feature in the range of -130 to -140 cm^{-1} from the bare molecule origin and exhibit mainly an intense single band at around -70 cm^{-1}. This latter band may be assigned [3] to the shift of the electronic transition induced by the H-bond, with only a very weak participation of the dispersive interactions between the alkyl chain of the alcohol and the aromatic ring.

We may apply the same consideration to the R-2-BuOH/S-NapOH complexes and assign the most stable **SR**$_a$ calculated complex to the most red-shifted isomer observed in the hole-burning experiment. The **SR**$_a$ (for R-2-BuOH) and **Sa** (for 1-PropOH) complexes have a very similar geometry, which explains the similarity between their spectroscopic properties (see figure 1 and 4). The ethyl chain is in the same position in both complexes, and interacts strongly with the ring. In the case of R-2-BuOH, the H atom on the chiral centre is in the same position as the H atom of propanol and shows a repulsive interaction with the ring. The added methyl group is away from the naphthalene ring: its presence modifies neither the complex geometry nor its interaction energy, in particular the repulsive part.

The most red-shifted **SS** diastereoisomer can also be assigned to the most stable **SS**$_a$ calculated complex. However, its geometry cannot be simply deduced from that of the **S**$_a$ isomer of 1-PrOH/NapOH. The position of the CH$_3$ group and the H atom of the chiral centre are inverted in **Sa** 2-BuOH relative to the **Ra** enantiomer. The repulsion part of the interaction energy is thus increased when H is replaced by CH$_3$, which results in a slipping of the ethyl group on the edge of the ring and in a decrease in dispersive interactions. This distorted geometry could explain the smaller red-shift and the absence of vibrational progression.

This example clearly shows that in this case, the chiral discrimination is achieved by means of the repulsion-dispersion part of the interaction energy.

b -The less red-shifted «doublet» : g_1, g_2 *compared with* ***SRg*** *and* ***SSg***

As mentioned previously the excitation spectrum of both complexes with 1-PrOH and 2-BuOH exhibits a slightly red shifted ($\approx$ -60 cm^{-1}) doublet with a small separation ($\approx$ 5 cm^{-1}) which has been shown to arise from different isomers. In the case of 2-BuOH, these isomers are different diastererisomers.

As discussed previously, the **Rg** and **Sg** enantiomers of 2-BuOH are built from the g_1 and g_2 conformers of 1-PrOH. Moreover, there is a strong resemblance between the calculated structures of the NapOH complexes with g_1 and **Rg**, and g_2 and **Sg** respectively. Because of this correlation between the spectral characteristics and the calculated structures, the doublet has been asigned to the complexes with g_1 and g_2 of 1-propanol for which there is no folding of the alkyl chain on the naphthalene ring, which explains the small red-shift of the electronic transition. It has to be noticed here that despite the fact that the alkyl chain lies away from the naphthalene ring, which means that the ring "does not see" whether the solvent is in the g_1 or g_2 conformation, a spectral discrimination, even weak (5 cm^{-1}) is observed.

It is interesting to stress here that the chiral discrimination obtained with the stable optical isomers of 2-BuOH reveals the spectral behaviour of the complexes involving 1-PrOH by providing an explanation of the origin of the **a** and g_1 discrimination as being due to the presence of rotational enantiomers of 1-PrOH which cannot be separated at room temperature.

Assignment of the isomers experimentally observed: aminoalcohols

a-Primary alcohols: 2-amino-1-propanol.

As mentioned previously, we can distinguish between the "weak" and "strong" intramolecular H bonds. In weakly H-bonded species, namely 2-amino-1-propanol, the OH...N bond is easy to disrupt. The disrupting of an internal H bond has been recently observed by IR/UV double resonance experiments on several systems: for example, we have shown that the internal H bond of o-cyanophenol was disrupted by the addition of a single water or methanol molecule [17]. The same phenomenon has been evidenced in water complexes of a noradrenaline analogue, 2-amino-1-phenylethanol, by Mons and co-workers [18], who have shown that the water molecule could insert between the OH group and the N atom to form a cyclic structure. This latter molecule is a primary alcohol bearing an amino group in 2-position, as the 2-amino1-propanol studied here. As an alcohol, NapOH has a larger gas-phase acidity than water and is thus expected to disrupt more easily the internal H bond of an aminoalcohol. We can thus safely assume that the internal H bond of 2-amino1-propanol is disrupted by complexation. The simplicity of the excitation spectrum of the complex between S-NapOH and 2-amino-1-propanol is therefore easy to understand: one isomer only of each diastereoisomer exists in the jet, in which the OH group of NapOH is supposed to insert between the OH and the NH2 group of the 2-amino-1-

propanol, as schematically shown on figure 5. The presence of two H bonds result in a quite rigid structure, which accounts for the lack of vibrational progression observed in the fluorescence excitation spectrum.

We can notice that the spectral discrimination, that is the difference in the transition energies between the homo and heterochiral complexes is larger than the one observed for aliphatic alcohols such as 2-butanol (35 cm^{-1}). This can be accounted for if one considers that its lack of flexibility prevents the complex from modifying its structure to accommodate steric constraint and optimise the binding energy.

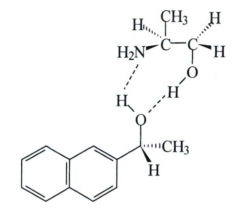

Figure 5: Schematic view of the proposed geometry of the NapOH/2-amino1-propanol complex. Note that the internal H bond is disrupted by complexation.

b-Secondary alcohols: 1-amino-2-propanol.

The first clue for the understanding of the complexity of the spectrum is given by the examination of the infrared spectra of the 1-amino-2propanol. As mentioned before, the large intensity of the band assigned to the H-bonded OH stretching mode shows that in this secondary alcohols, the H bond is much stronger than it was in 2-amino-1-propanol. It is therefore more difficult to disrupt: this leads to the formation of complexes between NapOH and both the H-bonded form and the free OH form of 2-amino-1-propanol for each diastereoisomer. On the basis of the similarity between their spectroscopic fingerprints and those of the 2-amino-1-propanol complex (shift and lack of vibrational structure), we assign the **SS** and **SR** diastereoisomers located at –49 and -90 cm^{-1} respectively to a rigid insertion complex in which the intramolecular H bond is disrupted, similar to that depicted figure 5.

The bands located at -145 cm^{-1} (**SR**) and -110 cm^{-1} (**SS**) can be assigned to a complex in which an intramolecular H bond still exists in the aminoalcohol molecule. Two structures can be proposed: In the first one, there is an intramolecular OH...N bond as observed in aminoethanol in the gas phase [20]. As the lone pair of the nitrogen atom is already involved in a H bond, the oxygen atom acts as an alternative binding site for the formation of the intramolecular H bond. The second possibility is that 1-amino-2-propanol shows an intramolecular NH...O hydrogen bond when complexed to NapOH, as observed for aminoethanol in solution[20]. The binding site for the intermolecular Hydrogen bond between NapOH and 1-amino-2-propanol would thus be the nitrogen atom. In both cases, the resulting structure is much more flexible than the inclusion complex, because there is only one intermolecular H bond. The transition origin of these latter complexes is much more shifted to the red than the transition of the insertion ones. According to the hypotheses mentioned above, this means that the dispersion term is larger for the "floppy complexes" which can present a folded geometry. Such a folded geometry would also explain the appearance of a low-frequency intermolecular mode.

Conclusion

We have proposed that conformational isomerism plays an important role in chiral discrimination: the *anti* conformers of chiral aliphatic secondary alcohols are much better discriminated than the *gauche* ones, because the complex adopts a folded geometry which optimises the dispersion forces. We have also suggested that the formation of a complex between NapOH and aminoalcohols results in the disrupting of the intramolecular H bond. For those of them with a weak intramolecular H bond (2-amino-1-propanol), this results in the formation of only one **SR** and only one **SS** complex, which have been assigned to "insertion" complexes in which the NapOH molecule is inserted within the aminoalcohol molecule. These complexes which involve two H bonds are quite rigid and show a very good chiral discrimination. Aminoalcohols with a stronger intramolecular H bond (1-amino-2-propanol) display more complicated spectra. An insertion complex, similar to that proposed above, is therefore thought to coexist with a "floppy" species in which an intramolecular bond still exists. We are currently doing *ab initio* calculations to confirm this interpretation. Since the frequency of the OH stretching mode is very sensitive to the formation of an hydrogen bond, experiments resting on a IR/UV double resonance technique are currently in progress to confirm the site of interaction with aminoalcohols.

References:

1. A. R. Al-Raaba, E. Bréhéret, F. Lahmani, A. Zehnacker. *Chem. Phys. Letters.* **1995,** *237* 480.
2. A. R. Al Rabaa, K. Le Barbu F. Lahmani, A. Zehnacker-Rentien. *J. Photochem. Photobiol. Part A Chemistry.* **1997,** *105,* 277
3. Al Rabaa, K. Le Barbu F. Lahmani, A. Zehnacker-Rentien. *J. Phys. Chem.* **1997,** *101,* 3273
4. K. Le Barbu , V. Brenner, Ph. Millié, F. Lahmani, A. Zehnacker-Rentien *J. Phys. Chem. A* **1998,** *102,* 128-137
5. F. Lahmani, K. Le Barbu, A. Zehnacker-Rentien *J. Phys. Chem. A* **1999,** *103,* 1991
6. A. Latini, M. Satta, A. Giardini Guidoni, S. Piccirilo, M. Speranza. *Chem. Eur. J.* **2000,** *6,* 1 and ref. therein
7. A. Giardini Guidoni, S. Piccirillo *Isr. J. Chem.* **1997,** *37,* 439
8. R.J. Lippert , S.D. Colson. *J . Phys. Chem.* **1989,** *93,* 2093.
9. Im, H.S.; Bernstein, E.R.; Secor, H.V. ; Seeman, J.I. *J. Am. Chem. Soc.* **1993,** *113,* 4422.
10. L. Radom, W. A. Lathan, W. Hehre, J.A. *Pople J. Am. Chem. Soc.* **1973,** *95,* 693
11. Y.P. Chang, , T.M. Su, T.W. Li, I. Chao. *J. Phys. Chem.* **1997,** *101,* 6107
12. G. Buemi Int. *J. of Quantum. Chemistry* **1996,** *59,* 227
13. a) R.E. Penn, R.F. Curl *J. Chem. Phys.* **1971,**55 , 651
    b) R.E. Penn, R.J. Olsen *J. Mol. Spectrosc.* **1976,** *62,* 423
14. M. Räsänen, A. Aspiala, L. Homanen, J. Murto. *J. Mol. Structure* **1982,** *96,* 81
15. P.J. Krueger, H.D. Mettee *Canadian Journal of Chemistry* **1965,** *43,* 2970
16. NIST Chemistry webbook (http://webbook.nist.gov/chemistry)
17. M. Broquier, F. Lahmani, A. Zehnacker. to be published
18. R. J. Graham, R.T. HKroemer, M. Mons, E. G. Robertson, L. C. Snoeck, J.P. Simons *J. Phys. Chem. A* **1999,** *103,* 98706
19. M. Mons, F. Piuzzi, I. Dimicoli, F. Lahmani, A. Zehnacker. Submitted for publication in *CPPC.*
20. a) C. Cacela, M.L. Duarte, R. Fausto *Spectrochemica acta PartA: Molecular spectroscopy* **2000,** *56,* 1051
    b) Silva C.F. Duarte M.L. Fausto R. *J. of Mol. Structure* **1999,** *483,* 591.

Surface Chirality and Chiral Nanostructures

Chapter 17

Spontaneous Generation of Chirality via Chemistry in Two Dimensions

I. Weissbuch, L. Leiserowitz, and M. Lahav

Department of Materials and Interfaces, The Weizmann Institute of Science, Rehovot 76100, Israel

An experimental model for the spontaneous separation of enantiomers of racemic α-amino acids, between crystals of glycine grown at the air-aqueous solution interface and the solution itself, is presented. This process involves several steps. A small enantiomeric excess of chiral α-amino acids is first achieved by oriented growth of few glycine crystals at the solution surface. The growing glycine crystals occlude only one of the α-amino acid enantiomers through the appropriate enantiotopic {010} face exposed to the solution, yielding an enrichment in the solution of the other enantiomer. This chiral bias created in the solution is preserved and amplified by virtue of two effects. First is the formation, at the air-solution interface, of monolayer clusters of partially resolved hydrophobic α-amino acids arranged in a way akin to the glycine layer structure. Such monolayer clusters serve as templates for an oriented crystallization of fresh glycine crystals. Second, the water-soluble hydrophobic and hydrophilic α-amino acids enantiomerically enriched in the aqueous solution inhibit nucleation of "wrongly" oriented glycine crystals. Recent grazing incidence diffraction studies have provided direct information on the structure and dynamics of the monolayer clusters at the solution surface. Of particular importance are the observations that certain crystallites of racemic amphiphilic α-amino acids undergo spontaneous resolution in two dimensions. This effect opens new opportunities for the generation of homochiral oligopeptides from hydrophobic α-amino acids by their polymerization at interfaces.

Introduction

Theories for the origin of a single chirality in the biological world fall into two major categories, biotic and abiotic. The abiotic scenario implies that chiral resolved materials had been formed prior to the biopolymers. Such asymmetry could have emerged provided a small fluctuation from the racemic state can be amplified to a state useful for biotic evolution. Mathematical models proposed by Franck (*1*) and later by Selig (*2*) and Decker (*3-6*) have suggested that an efficient amplification of a small chance fluctuation from racemic mixtures is feasible. In Franck's model enantiomeric excess would be efficiently amplified provided the system is designed such that one of the enantiomers acts as a catalyst for its own formation and an inhibitor for the formation of the second enantiomer. In order to materialize such a model we exploited the possibility of transferring chiral information across an interface.

Some years ago we applied cooperative crystallization processes for the spontaneous separation of racemic mixtures of α-amino acids rich with glycine into optically pure enantiomers.(*7*) The experiment involves slow evaporation of aqueous solution of a racemic mixture of α-amino acids rich with glycine. The oriented glycine crystals that are spontaneously formed at the air-solution interface contain enantioselectively occluded only one enantiomer of the α-amino acids, leaving the solution enriched with the other enantiomer. Here we review our experimental approach leading to a model for the spontaneous generation and amplification of optical activity for α-amino acids.

The overall process comprises the following steps:
I. Spontaneous generation of chirality via enantioselective occlusion of one of the enantiomers of the racemic α-amino acids within crystals of glycine grown at the air-water or glass-water-interface;
II. Several amplification processes:

(a) self-aggregation of hydrophobic α-amino acids into chiral clusters that operate as templates for an oriented crystallization of fresh crystals of glycine;

(b) spontaneous segregation of the racemates of some of the hydrophobic α-amino acids into two-dimensional chiral domains at the air–solution interface;

(c) enantioselective inhibition of embryonic clusters of glycine generated at the air-solution interface by enantiomerically pure water-soluble α-amino acids.

I. Spontaneous generation of chirality

Enantioselective Occlusion of Amino acids Within Glycine Crystals

The interactions of growing or dissolving crystals with molecules of the environment are done through the surfaces that delineate them. The arrangement of molecules at crystal surfaces is only a subset of the overall structure of these molecules in the bulk. This characteristic can be expressed in terms of the two-dimensional (2-D) symmetry relating molecules at a certain surface, which is generally lower than the three-dimensional (3-D) symmetry within the bulk. For example, a surface structure cannot be centrosymmetric nor have a glide plane or rotation axis parallel to the surface plane. Thus molecules that are related in the bulk of the crystal via a center of symmetry are not so at the crystal surface.

Chiral molecules may be enantioselectively adsorbed at a subset of the bulk symmetry-related sites. This principle is illustrated here for the crystals of glycine. Glycine is trimorphic. Its stable monoclinic centrosymmetric α–form is grown from water. The crystal faces relevant to the present discussion are of the type {010}, as shown in Fig.1. Of the four symmetry-related molecules (labeled *1, 2, 3* and *4* in Fig 1) *1* and *2* are homochiral, since they are related to one another via a two-fold screw symmetry. Both molecules are oriented such that the vector of their C-H*re* bonds points in the +*b* direction and so emerge from the (010) face. By crystal symmetry, the vector C-H*si* of molecules *3* and *4*, that are related to molecules *1* and *2* by center of inversion, points toward –*b* direction and so emerge from the enantiopic (0$\bar{1}$0) face. Replacement of the C-H*re* and C-H*si* groups by α–amino acids bearing hydrocarbon side chains would lead to molecules of (R*)* and (S) configuration, respectively. Thus during growth of the glycine crystals from a solution containing a mixture of racemic amino acids only (R)–amino acid additives can substitute glycine molecules at sites *1* and *2* on face (010) and only (S)-amino acids can be adsorbed at sites *3* and *4* on face (0$\bar{1}$0). This results in the formation of platelike crystals of glycine containing ~ 0.02 to 0.2% wt/wt of the racemic amino acids.(7) As expected by the above mechanism, the occluded racemic amino acids are spontaneously segregated within two different sectors at the two poles at the +*b* and –*b* sides of the crystal, according to the HPLC enantiomeric analysis (Fig. 2).

When the crystal is grown at an interface, one of the {010} faces is exposed to the solution whereas the opposite enantiotopic face is blocked. Consequently, such crystals pick up from the solution, through the exposed face, food-stock molecules of glycine together with only one of the enantiomers of the racemic amino acids. The absolute configuration of the occluded additive depends upon the face of the crystal that is exposed to the solution. Consequently, the centrosymmetric host is transformed into a chiral mixed crystal upon occlusion of the additive, as shown in Scheme 1. A large number of crystallization experiments at the air-aqueous solution interface have shown, as expected, that the (R)-amino acids were incorporated only into those floating glycine crystals whose (010) faces were exposed to the solution, and the (S)-amino acids only into crystals with the exposed (0$\bar{1}$0) faces, as determined by HPLC enantiomeric analysis (Fig. 3).(7) Platelike crystals of glycine grown at the glass-solution interface showed similar resolution, albeit with a lower enantiomeric excess.

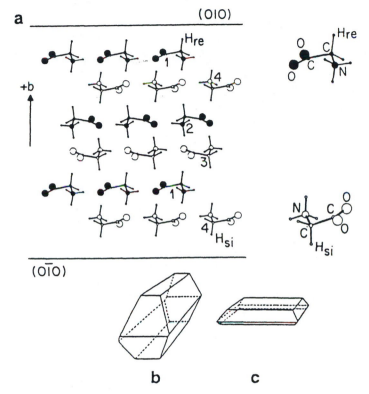

Figure 1. (a) Packing arrangement of α-form of glycine. (b-c) Crystal morphologies: (b) pure α-glycine and (c) grown in the presence of (R,S)-amino acid additives.

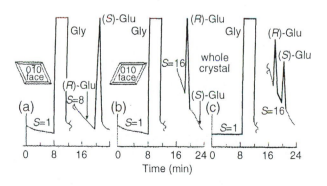

Figure 2. HPLC enantiomeric distribution of occluded (R,S)-glutamic acid: (a-c) material taken from (010) crystal face, (0$\bar{1}$0) face and remaining whole crystal.

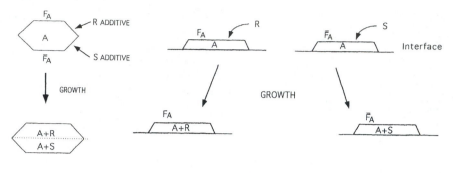

Scheme 1

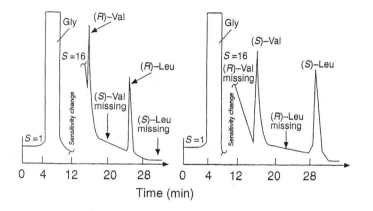

Figure 3. HPLC enantiomeric analysis of single crystal specimens floating at the air-solution interface when grown in the presence of (R,S)-leucine and valine and exposing either their (010) or (0$\bar{1}$0) face to air.

II. Amplification of chirality

(a) Self-aggregation of Hydrophobic α-Amino Acids at the Air-Solution Interface

In the absence of an amplification mechanism, no systematic net resolution of the α-amino acids is anticipated from a large number of crystals grown in the above described way at the air-solution interface in a thermodynamically closed system. If, however, the α-amino acids that were partially resolved in the first random crystallization are able to induce correct orientation in the subsequently grown glycine crystals floating at the air-solution interface, the initial asymmetry may be both preserved and amplified.

The 3-D packing arrangements of some racemic, hydrophobic α-amino acids show that such molecules are interlinked by hydrogen bonds forming centrosymmetric bilayers akin to those of α-glycine. Within each layer the molecules are related by translation symmetry and so layers are homochiral (Scheme 2). In the event that hydrophobic α-amino acids, as additives to glycine aqueous solutions, form at the air-solution interface, ordered hydrogen-bonded, 2-D chiral clusters with layer arrangement akin to that observed in their own 3-D crystals, such aggregates can furthermore form pseudo-centrosymmetric hydrogen-bonded bilayers with glycine solute molecules. Following this line of thought, a hetero-bilayer at the air-solution interface composed of hydrophobic R–amino acids bound to glycine solute molecules would resemble the top (010) bilayer of an α–glycine crystal face and so enantioselectively trigger its nucleation.(7) Hydrophobic α-amino acids such as valine, leucine, norleucine, phenylalanine, phenylglycine and α-amino butyric acid accumulate at their air-aqueous solution interface,(8) as indicated by surface tension measurements and X-ray reflectivity experiments.(9) The open question to be answered was whether these amino acids self-organize themselves in the form of structured clusters serving as templates for an oriented growth of the glycine crystals.

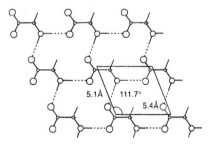

Scheme 2

Systematic studies demonstrated that the presence of the above-mentioned hydrophobic amino acids as co-solutes to glycine aqueous solutions indeed induced the expected oriented crystallization of the glycine crystals floating at the air-solution interface. The presence of the (R)-amino acids in a concentration as low as 1%wt/wt to glycine induce complete orientation of the glycine crystals with their (010) faces directed towards the air (Fig. 4). Based on

248

the above findings, a novel induced resolution of α-amino acids via occlusion inside glycine crystals was performed by oriented crystallization of glycine in the presence of, say, (S)-leucine and mixtures of other (R,S)-amino acids such as p-hydroxy-phenylglycine, glutamic acid, methionine, etc. The floating glycine crystals were formed with their (0$\bar{1}$0) faces poiting upwards, according to the X-ray diffraction measurements, and occluded exclusively the (R)-amino acids from the solution through the exposed growing (010) face, according to HPLC enantiomeric analysis (Fig. 5). Furthermore, even under conditions where polycrystalline crusts of glycine were formed, the occluded α-amino acids had an enantiomeric excess higher than 70% in the examples studied.

Figure 4. Photographs of α-glycine crystals grown at the air-solution interface and oriented with their (010) face exposed to solution (left) or to air (right).

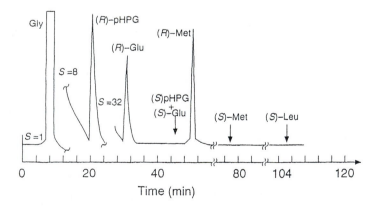

Figure 5. HPLC enantiomeric analysis of α-glycine crystals grown in the presence of 1% of (S)-leucine and other racemic amino acids.

(b) Spontaneous Resolution of Racemates of some Hydrophobic Amino Acid Amphiphiles into Structured clusters at the Air–Aqueous Solution Interface

More direct information on the structure of the nucleating domains was obtained by grazing incidence X-ray diffraction (GIXD) studies using synchrotron radiation. The GIXD measurements were performed on self-assembled clusters of enantiomerically pure N^{ε}–alkanoyl-R-lysine, $C_nH_{2n+1}CONH(CH_2)_4CH(NH_3^+)CO_2^-$ n =15,17,21, monolayers in the uncompressed state at the air-water interface.(10)

The amphiphilic molecules were designed such that the hydrophilic head groups at the air-water interface would form a hydrogen-bonded layer arrangement similar to that of glycine and thus creating a template akin to a layer of the to-be-grown glycine crystal (Scheme 3). Indeed, fast nucleation of floating glycine crystals was obtained by spreading the amphiphile molecules on saturated glycine solutions. The floating glycine crystals appeared with either their (010) or (0$\bar{1}$0) face attached to the monolayer, depending upon the absolute configuration of the monolayer molecules. Monolayers comprising molecules of R configuration induced the formation of glycine crystals oriented with their (010) face attached to the monolayer, whereas a monolayer of S configuration induced the formation of oppositely oriented glycine crystals.(11)

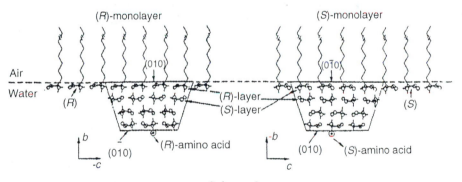

Scheme 3

The GIXD data analysis demonstrated that indeed the packing arrangement of the amino acid head groups of the monolayer is almost identical to that of layers of glycine molecules at the (010) crystal face.(9) Furthermore, recent GIXD comparative studies on 2-D clusters self-assembled from the racemic and enantiomerically pure α-amino-acid amphiphiles demonstrated that they display very similar diffraction patterns, Fig. 6.(12) These observations imply that at least the crystalline fraction of the racemic amphiphiles undergoes spontaneous separation into homochiral 2-D clusters. The packing arrangement of the enantiomorphous crystals containing R-molecules is shown in Fig. 7 as viewed perpendicular and parallel to the water surface.

More recent studies demonstrated that other systems such as salts of amphiphilic racemic mandelic acid and phenethyl amine derivatives that yield stable diastereoisomers undergo spontaneous resolution at the interface.(*13,14*) This segregation effect is currently being extended for racemates of activated amino acids, such as N-carboxy anhydrides and active esters of α-amino acid amphiphiles, that undergo efficient polymerization at the air-water interface to yield oligopeptides. Indeed, GIXD experiments on some of these systems

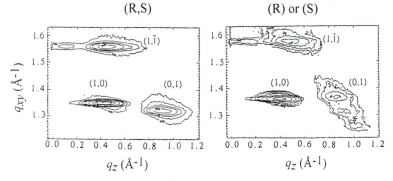

Figure 6. GIXD patterns, presented as two-dimensional intensity contour plots $I(q_{xy}, q_z)$, from the monolayer of: (left) racemic(R,S) and (right) enantiomeric (R) or (S) α-amino acid amphiphiles on water.

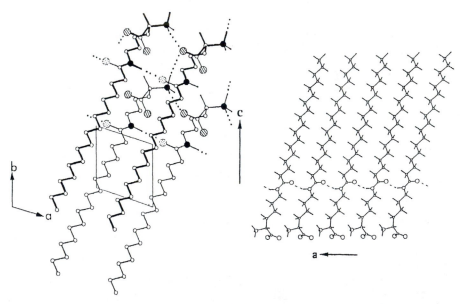

Figure 7. The packing arrangement of the 2-D crystals containing (R)-molecules as viewed perpendicular and parallel to the water surface.

indicated that such systems also undergo spontaneous segregation at the air-liquid interface.(*15*) These findings open the possibility of finding new experimental ways for the spontaneous generation of homochiral oligopeptides from racemates in prebiotic conditions.

(c) Enantioselective Inhibition of Glycine Crystal Nucleation at the Air-Water Interface

Another effect that helps to orient the freshly grown crystals of glycine at the air-solution interface is the enantioselective inhibition of crystal nucleation and growth of the "wrongly" oriented crystals as induced by the water-soluble hydrophilic amino acids. In a series of kinetic experiments, the oriented crystallization of glycine was performed in the presence of a mixture of racemic hydrophobic R,S-leucine (1% wt/wt of glycine) and optically pure α-amino acids such as S-alanine, S-serine and S-histidine (1-4% wt/wt). The racemic hydrophobic additive induced the formation of glycine crystals floating on the solution surface. When small amounts of the hydrophilic additives were introduced, two types of floating {010} glycine plates were formed, thick crystals with their $(0\bar{1}0)$ face oriented towards air and thin crystals oriented with their $(0\bar{1}0)$ face pointing to the aqueous solution. The number of the thick plates was larger and increased with the amount of the hydrophilic additive in solution. When 4% wt/wt hydrophilic S-amino acid was present in solution, only the thick plates were observed the growth of the "wrongly" oriented crystals being completely inhibited.(*7*)

Moreover, crystallization of glycine in the presence of partially enriched mixtures of R,S-leucine in various concentrations showed complete orientation of the glycine crystals at the air-solution interface. Such a complete oriented crystallization was achieved when leucine "tailor-made" additive was present at an enantiomeric excess as low as 6% (i.e. R/S ratio of 53:47) in the glycine solution and its total concentration was of 2.4 wt/wt of glycine. Such a concentration of the leucine was required for inducing a single orientation of the glycine crystals floating with their $(0\bar{1}0)$ face pointing to the solution (Fig. 8).

More recently, the inhibition of crystallite formation in the presence of "tailor-made" inhibitors has been demonstrated directly by grazing incidence diffraction measurements at the air–solution interface. (*16*)

Overall model for the spontaneous generation and amplification of optical activity

The possibility of inducing a complete orientation of the glycine crystals floating at the air-solution interface provides evidence for the feasibility of an overall model for the spontaneous generation and amplification of optical activity under prebiotic conditions.

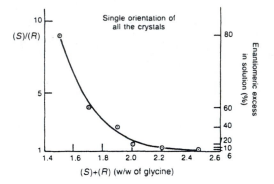

Figure 8. Corelation between the initial leucine enantiomeric ratio and its total concentration required for inducing a single orientation of the floating crystals.

Let us assume that an aqueous solution of glycine containing a variety of α-amino acids is slowly evaporating. Spontaneous fluctuations of concentration can promote the appearance of the first crystal or a small number of crystals with the same {010} orientation. The enantioselective occlusion of the hydrophobic α-amino acids into these crystals will generate in solution a small enantiomeric excess of molecules of opposite handedness. This excess might be sufficient to trigger amplification such that subsequent glycine crystals appearing at the solution surface will be oriented in the same way as the first crystals. The newly formed crystals oclude only one enantiomer, the solution contains both enmantiomers of the α-amino acids but the enantiomeric excess in solution is increased. Repeated crystallization cycles could, in principle, eventually lead to a pronounced separation of chiral teritories.

Conclusions

The several processes described above operate in unison for the complete resolution of a racemic mixture of α–amino acids with the assistance of glycine crystals grown at the air-aqueous solution interface.

The experiments provide us with a simple model for both generation and amplification of the optically resolved amino acids under pre-biotic conditions. Furthermore, reactivity of non-chiral or racemic mixtures at liquid or solid interfaces might yield macromolecules of a chiral conformation induced by the interface. With the advent of modern analytical tools (16) it has become possible not only to design such experiments but also to characterize the system. Studies along these lines are under current investigations.

Acknowledgments. We thank the Israel Science Foundation for financial support.

References

1.   Frank, F. C. *Biochem. Biophys. Acta*, **1953**, *11*, 459.
2.   Selig, F. F.; *J. Theor. Biol.*, **1971**, *31*, 355.
3.   Decker, P. *J. Mol. Evol.*, **1973**, *2*, 137.
4.   Decker, P. *Nature*, **1973**, *241*, 72.
5.   Decker, P. *J. Mol. Evol.*, **1974**, *4*, 49.
6.   Decker, P. *Origins of Optical Activity in Nature*, Elsevier, New York, 1977; p 109.
7.   Weissbuch, I.; Popovitz-Biro, R.; Leiserowitz, L.; Lahav, M. In *The Lock-and-Key Principle The State of the art – 100 Years on*; Behr, J.-P., Ed.; John Wiley & Sons Ltd: New York, 1994; Vol. 1, p 173.
8.   Weissbuch, I.; Frolow, F.; Addadi, L.; Lahav, M.; Leiserowitz, L. *J. Am. Chem. Soc.*, **1990**, *112*, 7718.
9.   Weissbuch, I.; Leiserowitz, L.; Lahav, M. unpublished results.
10.   Jacquemain, D.; Grayer Wolf, S.; Leveiller, F.; Deutsch, M.; Kjaer, K.; Als-Nielsen, J.; Lahav, M.; Leiserowitz, L. *Angew. Chem., Int. Ed. Engl.*, **1992**, *31*, 130.
11.   Landau, E. M.; Grayer Wolf, S.; Levanon, M.; Leiserowitz, L.; Lahav, M.; Sagiv, J. *J. Am. Chem. Soc.*, **1989**, *111*, 1436.
12.   Weissbuch, I.; Berfeld, M; Bouwman, W.; Kjaer, K.; Als-Nielsen, J.; Lahav, M.; Leiserowitz, *J. Am. Chem. Soc.*, **1997**, *119*, 933.
13.   Kuzmenko, I.; Buller, R.; Bouwman, W. G.; Kjaer, K.; Als-Nielsen, J.; Lahav, M.; Leiserowitz, L. *Science*, **1996**, *274*, 2046.
14.   Kuzmenko, I.; Kjaer, K.; Als-Nielsen, J.; Lahav, M.; Leiserowitz, L. *J. Am. Chem. Soc.*, **1999**, *121*, 2657.
15.   Weissbuch, I.; Bolbach, G.; Leiserowitz, L.; Lahav, M. *in preparation*, **2000**.
16.   Rapaport, H; Kuzmenko, I.; Berfeld, M.; Kjaer, K.; Als-Nielsen, J.; Popovitz-Biro, R.; Weissbuch, I.; Lahav, M.; Leiserowitz, L. *J. Phys. Chem. B*, **2000**, *104*, 1399-1428.

Chapter 18

Chirality at Well-Defined Metal Surfaces

Gary A. Attard[1], Jean Clavilier[2], and Juan M. Feliu[3]

[1]Department of Chemistry, University of Wales, Cardiff, P.O. Box 912,
Cardiff CF10 3TB, United Kingdom
[2]CNRS, 1, Place A. Briand, 92195 Meudon Cedex, France
[3]Departament de Química Física, Universitad d'Alacant, Apartat 99,
E–03080, Alacant Spain

The surface geometry necessary to impart chirality in clean metal surfaces is described. The importance of kink sites and the symmetry of their microstructure is emphasized together with a discussion of "non-ideal" kink structures formed as a consequence of kink coalescence and compound-step reconstruction. Experimental data verifying the intrinsic chirality of kink sites is also presented in relation to the electro-oxidation of glucose. The possible implications of the work for enantioselective heterogeneous catalytic reactions are briefly outlined.

Introduction

Our present understanding of "handedness" or chirality (from the Greek word "chir" meaning "hand") in molecular systems stems from a series of key experiments performed in the nineteenth century by, among others, Biot, Fresnel, Pasteur and Faraday *(1)*. During the twentieth century, this concept has spawned major advances in chemistry, physics and biology and continues to play an important role including the quest for more effective drugs and pharmaceuticals. Although attempts have been made in the past to confer handedness on metals deposited upon chiral "templates" such as quartz and silk *(2,3)*, these experiments lacked precision in relation to the microstructuring of the metal surfaces generated by such procedures and suffered from irreproducibility in the catalytic response that they engendered.

In this short review article, an analysis of surface chirality in terms of kink geometry will be outlined together with experimental verification of this analysis. Although work on single crystal surfaces is still at a relatively early stage, some pointers as to future trends and areas of investigation may even now be discerned.

EXPERIMENTAL- CRYSTAL PREPARATION.

All platinum single crystal samples used were prepared using the method of Clavilier (4). Very briefly, the end of a 0.5mm diameter platinum wire is melted and cooled carefully, without vibration, to produce a platinum single crystal bead. The method is based on a melting zone technique, which concentrates impurities in the final part of the crystal to solidify. Hence it may be used to eliminate impurities contained in the base metal. This may be achieved by both etching the bead in aqua regia followed by re-melting of the bead or removal of the "contaminated" region by grinding and polishing (see later). The melting/etching procedure may be repeated several times. The presence of impurities may be suspected when difficulties appear in the crystallisation of the metal (irregular shape of the final region of crystallisation from the melt).

When placed at the centre of a goniometer cradle situated at one end of a 2m long optical bench, light from a He-Ne laser (placed at the opposite end of the optical bench) incident on the spherical crystal is diffracted by the edges of {111} and {100} facets arranged symmetrically around the crystal surface. Utilising these {111} and {100} diffracted beams, the crystal may readily be oriented, taking advantage of this visually "direct" imaging of the crystal axis positions in order to determine the mirror symmetry planes from which the chirality of the kinked surface will be subsequently determined. The crystal is then embedded in epoxy resin and a grinding wheel employed to remove approximately half of the crystal. After grinding to produce the hemispherical crystal (the flat face of which will become the active, orientated surface) finer and finer grades of diamond paste are used to polish the crystal surface until a mirror finish is produced, the reflection at the He-Ne laser from the polished surface being used to judge the quality of the finish and the precision of the cut (usually ± 3 minutes of arc). After removal from the goniometer, the epoxy resin is dissolved in dichloromethane and the crystal is annealed in a Bunsen flame for one hour at ~ 1000°C to remove surface impurities and restore any surface damage produced by the grinding and polishing procedure (5). Careful cooling of the crystal to room temperature in one atmosphere of hydrogen[1] (DANGER, DO NOT TRY THIS UNLESS YOU KNOW WHAT ARE YOU DOING!) and subsequent protection of the surface with a droplet of ultra-pure water allows transfer of the clean, well-ordered electrode surface either to an electrochemical cell or to a UHV system for surface analysis. Following flame annealing and cooling, low energy electron diffraction (LEED) analysis always gives rise to sharp, (1x1) diffraction patterns and negligible impurity levels according to Auger electron spectroscopy (AES). Details of the electrochemical and surface science analytical procedures may be found in references *(6)* and *(7)*.

THE ORIGIN OF CHIRALITY AT SOLID SURFACES

Gellman and co-workers *(8)* originally postulated that certain kinked metal surfaces should be chiral. Subsequent theoretical work by Sholl *(9)* confirmed that chiral discrimination between kink site and asymmetric molecular centres should give rise to an enantiomeric response. These workers proposed that so long as the step lengths comprising the kink site of a single crystal surface were of unequal magnitude, reflection through a plane normal to such a surface produces a new surface which itself cannot be superimposed on the original, that is, such kink sites should be chiral (figure 1). In other words, chirality should be absent when the surfaces contain steps of equal length on each side of the kink.

However, in two subsequent papers *(10,11)* it was proposed that this interpretation was an approximation since it did not take into account the surface geometry of the individual Miller index planes of the two steps whose junction forms the kink site. In this alternative analysis, all kink sites should be considered as being chiral, even when the two steps have equal lengths. This point is illustrated using figure 2 which depicts schematically the stereographic projection, centred on the (010) pole, of crystal planes in face-centred and body-centred cubic crystals. A point in the figure corresponds to a particular Miller index plane. The x- and y- axes represent rotations of the crystal in the directions from (010) to the enantiomorphic (111) and (1 $\bar{1}$1) planes and (1 $\bar{1}$ $\bar{1}$) and (11 $\bar{1}$) planes, respectively. In the present context we shall denote angles of rotation along the x- and y- axes as "α". Similarly, the outer circumference of the circle represents a rotation of the crystal in a plane orthogonal to α in the direction from {111}→{110} planes, that is a rotation of the crystal around the axis perpendicular to the {100} plane. Rotation of the crystal in this direction shall be denoted as β. Hence, any point in the stereographic projection, starting from (010) may be reached by applying two angles of rotation. The first, α, the second β in the plane orthogonal to α. The segment of the circle generated by

---

[1] It is recommended that dilute hydrogen (20% by volume) in an argon stream is used rather than pure hydrogen.

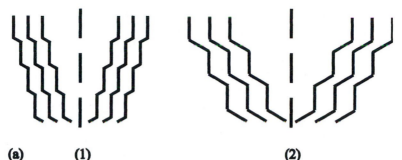

(a) **(1)** **(2)**

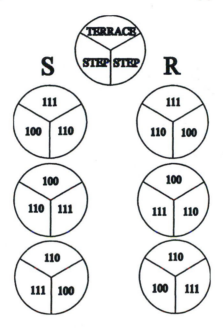

(b)

Figure 1. (a) Representation of surface chirality based on reference 8. In (1), the length of the steps forming the kink are different. Hence reflection of the kinked surface in a plane perpendicular to the surface generates a non-superimposable enantiomer. In (2), because the step lengths comprising the kinks are equivalent, no enantiomeric surface is generated after reflection in a plane perpendicular to the surface.

(b) Surface chirality based on the microstructure of the individual sites forming the kink. The handedness of the kink is decided by viewing the surface from the vacuum / electrolyte side. If the sequence of sites {111} – {100} – {110} is clockwise, the kink is denoted R-, if counter-clockwise it is denoted S-. The possible configuration of step and terrace sites forming a chiral kink are also depicted schematically as the junction between three components, terrace, step and step.

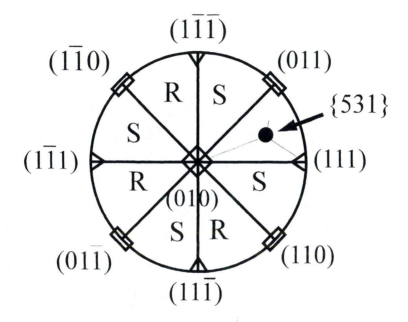

Figure 2. Family of stereographic triangles representing the various R- and S- zones surrounding the (010) pole. In the figure, one of the stereographic triangles has been modified to show the {531} pole at its centre and the corresponding "turning lines" (see text below).

any particular set of {111}, {100} and {110} planes is more commonly referred to as a "stereographic triangle". The stereographic triangle may be thought of as being analogous to a three-component phase diagram in which the poles of the triangle represent flat {111}, {100} and {110} terraces (one component), the edges of the triangle to stepped surfaces (n(111) x (100), n(100) x (111), n(100) x (110), n(110) x (100), n(110) x (111) and n(111) x (110) in microfacet notation *(12)*, i.e., consisting of two components, terraces and linear steps) and all points inside the triangle denoting kinked surfaces (three components: terraces, steps and kinks, in microfacet notation p(111) x q(100) x r(110), were p, q and r are integers ≥ 1). If one arbitrarily assigns a "priority" to each of the planes in terms of surface packing density (hence {111}> {100}> {110} for fcc crystals), then it is found that two types of stereographic triangle may be distinguished. Using an analogy with Cahn-Ingold-Prelog sequence rules found in introductory text books in organic chemistry *(13)*, it is observed that one set of triangles may be designated "R-" (from the Latin *rectus*) since the sequence {111}→ {100}→ {110} around the triangle runs <u>clockwise</u>. The other, alternate set of stereographic triangles give rise to a counter-clockwise sequence {111}→ {100}→ {110} and hence may be designated "S-" (from the Latin *sinister*). Each "couple" (R- and S-) of adjacent triangles is symmetric about a mirror plane defined by the common edge of both triangles. Because the sequence {111}→ {100}→ {110} defines the handedness of kink sites in the <u>real</u> crystal surface associated with R- or S- stereographic triangles, it is relatively straightforward to produce a kinked surface consisting entirely of R- and S- kinks.

If the single crystal is aligned for example along the {100} plane, as stated earlier, in order to produce a kinked surface, the angles of rotation α and β are required. However, although an R- or S- surface may be generated using a singular value of α, for the final rotation, a choice of "+β" or "-β" is possible, i.e. a rotation of "+β" could generate an R- surface whereas a rotation "-β" would generate the S-surface- an equivalent, but non-superimposable crystal plane. An alternative way of obtaining R- and S- kinked surfaces would be to purchase a standard "disc" single crystal of the required kinked crystal plane from a commercial manufacturer. If both crystal faces are cut parallel to one another, as noted in reference *(8)*, one face of the disc would correspond to the "R-" surface and the opposite face to the "S-".

It is worth mentioning that in the case of an "ideal" kinked surface (we shall define "non-ideal" kinked surfaces later), the condition for chirality outlined in reference 8 is generally fulfilled because, of the two steps forming the kink, the first is always monoatomic long in the surface plane, whereas the other is multiatomic. The only exceptions to the Gellman interpretation *(8)* of chirality occur when the surfaces considered belong to one of the three "turning lines" joining an "apex" of the stereographic triangle to the {531} pole (see figure 2). These turning lines represent crystal surfaces of constant kink geometry but varying terrace width. All other lines from the {531} pole to the sides of the triangle represent varying kink geometry at constant terrace width. For kinked surfaces situated on the turning line, the kink is made up of two steps, both of which are monoatomic in length in the surface plane, i.e. both steps are of equal length and should not be chiral according to reference 8. However, the difference in step site symmetry, as outlined in figure 2 ensures in fact that such surfaces should be chiral (see later for Pt{321}) and, depending on the type of stereographic triangle in which the plane is situated, the R- and S- symmetry of the plane should be preserved.

Deviations from the "ideal kink model" outlined above may be envisaged as a consequence of kink coalescence. In this case the two steps forming the kink are both multiatomic long in the crystal plane. An analogy can be drawn here with surface facetting *(14)*. For many stepped surfaces, instead of a regular array of monoatomic high steps separated by terraces, facetting may occur such that multiatomic high steps form. This means that although the nominal crystal plane direction is preserved, the facets giving rise to the surface do not reflect the microstructure expected from a simple truncation of the bulk crystal. In a similar way, kink reconstruction/ coalescence may lead to a nominal "compound-step" direction, but facetting of the compound-step into varying, multiatomic step lengths may lead to the formation of "non-ideal" kinks. Nonetheless, <u>on average</u>, even for reconstructed/facetted/non-ideal kinked surfaces, the R- or S- symmetry of the nominal Miller index plane should be preserved. Hence, all crystal planes located inside the stereographic triangle should preserve their R- or S- symmetry. Equivalent points inside adjacent stereographic triangles having a common side should form a pair of chiral surfaces.

It is interesting to note that non-chiral kinks may also be identified since the {110} site is actually composed of the intersection of two {111} planes or two {100} planes, i.e., (110) ≡ {111} x {111} or (110) ≡ {100} x {100}. Hence, stepped surfaces along the {100} → {110} direction actually correspond to three intersecting planes (although two of the planes are of the same character) and have therefore sometimes in the past been referred to as "kinked". Such surfaces should not, according to the previous analysis, exhibit an enantiospecific response. An example of an achiral "kinked" surface would be Pt{210} which may be written in microfacet notation as 2(100) x (110) ≡ 2(100) x (111) x (111). When written as 2(100) x (110), the surface is stepped (two intersecting planes) whereas when written as 2(100) x (111) x (111) it is kinked (three interacting planes).

EXPERIMENTAL VERIFICATION OF THE INTRINSIC CHIRALITY OF KINK SITES

In order to detect any chiral response one first requires a chiral probe. A suitable electrochemical probe molecule which also contains a number of chiral carbon centres would be D-glucose. Its principal advantages are that it is cheap, it can be obtained in a very pure form and it gives rise to a strongly surface-sensitive electrochemical oxidation current at platinum single crystal electrodes *(15)*. Structure sensitive reactions involve adsorption steps in the reaction mechanism. This means that if the initial step in the electro-oxidation of D-glucose is (presumably) chemisorption onto the electrode surface, the energetics of this adsorption step may be influenced by the "handedness" of chiral kink sites and this should be detectable by following changes in the electro-oxidation current. In addition, the L- isomer may be obtained from commercial suppliers (although it is somewhat more expensive than the D- form). Since glucose undergoes mutarotation in aqueous solutions *(13)*, one should always prepare such solutions at least 12 hours prior to any electrochemical measurement to allow time for equilibration to be reached.

The disadvantages of the use of glucose as a probe molecule are linked to the complexity of its electro-oxidation reaction. Glucose oxidation leads to the production of gluconolactone, gluconic and glucaric acids *(16)*, but CO is also formed at the very beginning of the electro-oxidation of the molecule. The problem with carbon monoxide is that its strong adsorption on platinum blocks surface sites and hence precludes glucose reactions in the potential range in which glucose is normally electro-oxidised. Moreover, as stated previously, the rates of all these reactions depend on the surface structure, and also on anion adsorption, to a different extent for each type of surface site (or ensemble) *(15)*. These disadvantages, linked to the complexity of the molecule are compensated by the fact that both glucose stereoisomers can be obtained with a reasonable level of purity. In Surface Electrochemistry the main experimental difficulty is always surface contamination.

Figure 3 shows the electro-oxidation of D-glucose using Pt{111} and Pt{211} electrodes. Pt{111} is a hexagonal close-packed surface, whereas Pt{211} consists of an arrangement of 3- atom wide {111} terraces followed by a {100} linear step. It is closely similar to the structure of the Ag{643} surface used by Gellman and co-workers to detect an enantiomeric response in the thermal desorption of chiral alcohols *(8)* with the same 3-atom wide {111} terrace sites. However, it consists of linear steps, not compound-steps (zig-zag) as in the Gellman studies, and so should not be chiral. The cyclic voltammograms (CV) show how the rate of glucose electro-oxidation (as measured on the y- axis by the current density) varies as a function of potential.

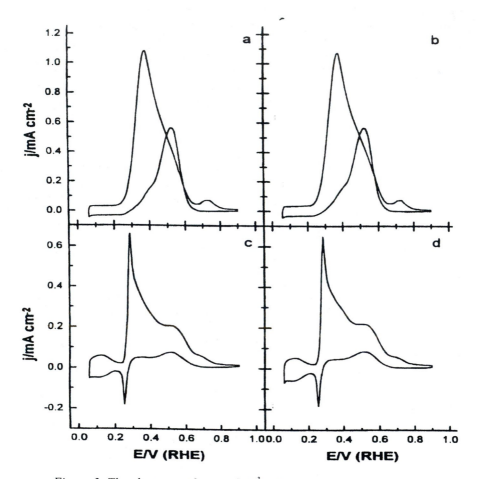

Figure 3. The electro-oxidation of $10^{-3}M$ D-glucose using (a) Pt{111} and (c) Pt{211} in 0.1M H_2SO_4 , the electro-oxidation of $10^{-3}M$ L-glucose in 0.1M H_2SO_4 using (b) Pt{111} and (d) Pt{211}. Sweep rate = 50mV/s. So far as these achiral solutions are concerned, both solutions are identical.

The form of the CV is similar to that found in previous studies *(15)*. The important point to realise, however, is that so far as the achiral Pt{111} and Pt{211} surfaces are concerned, the D- and L- glucose solutions are indistinguishable although there are differences between Pt{111} and Pt{211} demonstrating the structure-sensitivity of the reaction.

When this experiment is repeated, but using the chiral Pt{643}^{R} and Pt{643}^{S} electrodes, the glucose solutions are no longer equivalent (figure 4). In fact, the combination of R- crystal with D- glucose gives the same result as an S-crystal with L- glucose. The R- crystal with L- glucose and S- crystal with D-glucose also give the same result, but nonetheless a result different to the previous one. Inspection of figure 4 shows a classical diastereomeric result involving two stereogenic reactive centres. Hence, the electro-oxidation of glucose at kinked single crystal platinum surfaces corresponds to the first experimental proof of the intrinsic chirality of surface kinks. That the length of the two steps constituting the kink site is not the determining characteristic of the "handedness" of the kink, as originally proposed in reference *(8)* may be shown using a Pt{321} electrode. All step sites in this surface are two atoms long, yet it too gives rise to an enantioselective response when reacted with glucose (figure 5). In fact the magnitude of the enantioselective response (as measured by the difference in electric current density for R- and S- crystals) close to 0.3 V (Pd/H) is found to be greater than for Pt{643}. For Pt{531} (a surface composed entirely of kink sites and hence lacking terraces or linear steps), the enantioselectivity is found to be still greater *(10)*. Hence, at least for the electro-oxidation of glucose, the enantioselectivity of the reaction at platinum surfaces appears to be strongly correlated with the surface density of kink sites.

FUTURE DIRECTIONS

The notion of being able to control the inherent chirality of a two-dimensional surface, particularly a metal which is catalytically active, is an attractive one. In principle, it would enable the experimentalist to investigate chirality in a new and fundamental manner, not only in terms of the differences in reactivity of asymmetric centres on molecules, but also to control the activation of these centres via their interaction with particular surface sites of a preferred handedness. The huge potential of, for example, synthesising nanostructured materials with a definite handedness and to probe their interaction with electromagnetic radiation, molecules and other "handed" materials will probably ensure intensive research efforts well into the twenty first century. One particular area of application could be the elucidation of the mechanism of a heterogeneous catalytic reaction, first reported by Orito and co-workers in 1979 involving the hydrogenation of pro-chiral bonds in α-ketoesters *(17)*. In the presence of a supported Pt catalyst, at high pressures of hydrogen, the following reaction will occur:

264

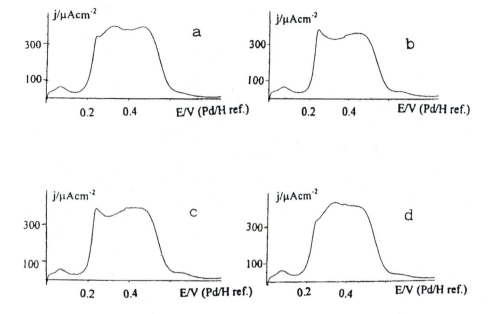

Figure 4. The electro-oxidation of 10^{-3} D-glucose using (a) Pt{643}^{S} and (c) Pt{643}^{R} in 0.1M H$_2$SO$_4$. The electro-oxidation of 10^{-3} L-glucose using (b) Pt{643}^{S} and (d) Pt{643}^{R} in 0.1M H$_2$SO$_4$. Note that in this case there is a clear enantiomeric response. Sweep rate = 50mV/s.

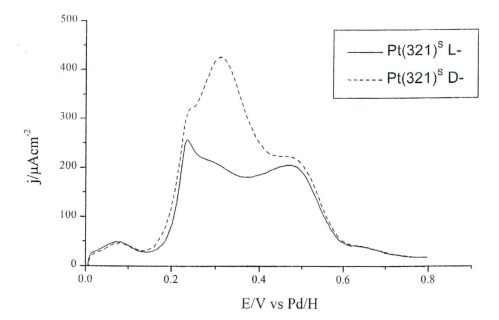

Figure 5. The electro-oxidation of D- and L- glucose in 0.1M H$_2$SO$_4$ using Pt{321}^{S}. Sweep rate = 50mV/s. Although predicted to be achiral in reference 8, the surface clearly displays a significant eneantiomeric response at 0.3V (Pd/H).

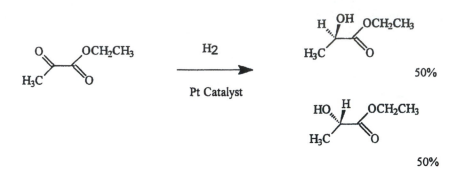

50%

50%

i.e., statistically equal amounts of the two optical isomers of ethyl lactate will be produced. However, in the presence of the alkaloid cinchonidine, an enantiomeric excess of >90% in the R- isomer may be achieved.

CINCHONIDINE

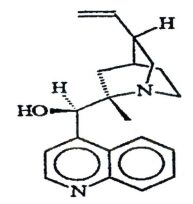

Not only this, but a huge <u>increase</u> in the absolute rate of hydrogenation is observed in the presence of the chiral auxiliary. Interestingly, when two of the stereogenic centres on the alkaloid are reversed (to give cinchonine), enantiomeric excesses >70% are produced in favour of S- ethyl lactate. There is an extensive literature covering this fascinating reaction and the reader is referred to references (18-20) for greater detail. However, in all of the models proposed to explain this reaction and the role of the modifier (for example complexation in solution of alkaloid with reactant followed by hydrogenation of the complex at the surface, generation of chiral "pockets" adjacent to the chiral modifier at the platinum surface followed by hydrogenation of reactant at these special sites or combinations of both models), none has highlighted the fact that the platinum catalyst particles are <u>racemates</u> containing, on average, equal numbers of left- and right- handed kinks. An alternative mechanism as a consequence of the present analysis would be preferential reaction of the chiral auxiliary at R- or S- kinks, leaving an "excess" of the opposite handedness kink to act as a hydrogenation centre for the reactant. Interestingly, this model would also lead to selective blocking of sites which is known from electrochemical studies of organic reactions at surfaces to enhance reaction rate by preventing any "poisoning" reactions running in parallel, due to the requirements of a different ensemble of surface sites for "poisoning" and "reactive" pathways. Adsorption studies of chiral auxiliaries at well-defined platinum kinked surfaces are already underway *(21)* to detect enhancements in the rate of adsorption of the alkaloids as a function of kink microstructure and chirality.

ACKNOWLEDGEMENTS

J.M.F. acknowledges support from DGES through project PB97-0409.
G.A.A acknowledges the support of the EPSRC Grant Number M65724

REFERENCES

1. Barron L.D. in "*New Developments in Molecular Chirality*"; Mezey, P.G., Ed.; Kluwer Academic, The Netherlands, 1991.
2. Schwab G.M. and Rudolph L. *Naturwiss.* **1932**, *20*, 362.
3. Akabori S., Sakurai S., Izumi Y. and Fujii Y. *Nature* **1956**, *178*, 323.
4. Clavilier J. in "*Interfacial Electrochemistry- theory, experiment and applications*", Wieckowski A., Ed.; Marcel Dekker, N. Y., 1999.
5. Clavilier J., El Achi K., Petit M., Rodes A. and Zamakhchari M.A., *J. Electroanal. Chem.* **1990**, *295*, 333.
6. Attard G.A. and Bannister A., *J. Electroanal. Chem.* **1991**, 300, 467.
7. Attard G., Price R., Al-Akl A., *Electrochim. Acta* **1994**, *39*, 1525.
8. McFadden C.F., Cremer P.S., Gellman A.J., *Langmuir* **1996**, *12*, 2483.
9. Sholl D.S., *Langmuir* **1998**, *14*, 862.
10. Ahmadi A., Attard G., Feliu J. and Rodes A., *Langmuir* **1999**, *15*, 2420.
11. Attard G.A., Ahmadi A., Feliu J., Rodes A., Herrero E., Blais S. and Jerkiewicz G., *J. Phys. Chem. B* **1999**, *103*, 1381.
12. Van Hove M.A. and Somorjai G.A., *Surf. Sci.* **1980**, *92*, 489.
13. Morrison R.T. and Boyd R.N., *Organic Chemistry, 3rd. Ed.*, Allyn and Bacon, Boston, 1973.
14. Blakeley D.W. and Somorjai G.A., *Surf. Sci.* **1977**, *65*, 419.
15. Rodes A., Llorca M.J., Feliu J.M. and Clavilier J., *An. Quím. Int. Ed.* **1996**, *92*, 118.

16. Popovic, K.D., Tripkovic, A.V. and Adzic, R.R., J. Electroanal. Chem., 339, 227 (1992).
17. Orito, Y., Imai, S. and Nina, S., J. Chem. Soc. JPN. 8, 1118 (1979).
18. Baiker A., *Molec J.. Catal. A* **1997**, *115*, 473.
19. Wells P.B. and Wilkinson A.G., *Topics in Catalysis* **1998**, *5*, 39.
20. Blaser H.-U., Jalett H.P., Muller M. and Studer M.S., *Catal. Today* **1997**, *37*, 441.
21. Attard G.A., Harris C. and Feliu J.M., in preparation.

Chapter 19

Enantiospecific Properties of Chiral Single-Crystal Surfaces

Joshua D. Horvath, Andrew J. Gellman[*],
David S. Sholl, and Timothy D. Power

Department of Chemical Engineering, Carnegie Mellon University,
Pittsburgh, PA 15213

Single crystalline surfaces with structures having kinked steps are inherently chiral. As such these surfaces have enantiospecific properties that have been explored both theoretically and experimentally. Experimental studies of adsorption have been performed using two chiral adsorbates ((S)-1-chloro-2-methylbutane and (R)-3-methylcyclohexanone) on chiral Cu(643)^R and Cu(643)^S surfaces. Preliminary work has shown desorption kinetics for (R)-3-methylcyclohexanone adsorption that are influenced by the handedness of the Cu(643) surface. Theoretical simulation of the adsorption of small chiral molecules on chiral Pt surfaces has been used to attempt to understand the role of surface and adsorbate characteristics in determining enantiospecificity.

Introduction

A type of chiral surface is generated when crystalline surfaces are terminated in structures with kinked steps. McFadden *et al.* first described kinked chiral surfaces such as the (643) and $(\overline{643})$ surfaces of a face-centered cubic crystal (fcc) that are nonsuperimposable mirror images of each other. They proposed a naming convention for chiral surfaces that is analogous to the Cahn-Ingold-Prelog rules for naming chiral organic molecules.[1] The (643) surface was designated $(643)^S$ and the $(\overline{643})$ surface was designated $(643)^R$. To determine if enantiospecific adsorption of molecules on chiral single crystal surfaces could be observed, McFadden et al. examined the adsorption of (R)- and (S)-butan-2-ol on Ag(643)^R and Ag(643)^S surfaces.[1] Two experiments designed to detect enantiospecific interactions between the chiral Ag(643) surfaces and these chiral alcohols were conducted. The first experiment measured the desorption energies of (R)- and (S)-butan-2-ol from the Ag(643)^R and Ag(643)^S surfaces. The second experiment measured the activation barriers to β-hydride elimination for (R)- and (S)-butan-2-oxide on the Ag(643)^R and Ag(643)^S surfaces. No enantiospecific energy differences greater than ~0.1 kcal/mol were found for either of these systems. If the adsorption of these molecules was influenced by the chirality of the surface, the differences were too small to measure.

A later theoretical paper by Sholl describes the adsorption of chiral hydrocarbons on the chiral Pt(643) surface.[2] Binding energies of several chiral hydrocarbons adsorbed on Pt(643) were calculated using Monte Carlo simulations. The chiral molecules limonene, 1,2-dimethylcyclopropane, and 1,2-dimethylcyclobutane have predicted binding energies that differ significantly between the two enantiomers of the same chiral compound. These theoretical results suggest that for several chiral hydrocarbons on Pt(643), the enantiospecific desorption energy differences are large enough to measure experimentally.

More recently, experiments by Attard et al. investigated the electro-oxidation of the chiral molecule glucose at chiral platinum electrodes.[3,4] They also proposed a more general nomenclature for the chirality of high Miller index fcc surfaces based on the orientations of the planes that form the steps and kinks. As expected, the achiral Pt(211) and Pt(332) electrodes did not demonstrate enantiospecific differences between the rates of oxidation of D- and L-glucose. On the other hand, the chiral Pt(643) and Pt(531) electrodes did show enantiospecific oxidation of the two enantiomers of glucose. This was the first experimental demonstration of the enantiospecific adsorption properties of chiral single-crystal surfaces.

To determine if enantiospecific adsorption of chiral organic molecules can be detected on the Cu(643)^R and Cu(643)^S surfaces, we have investigated the desorption of (S)-1-chloro-2-methylbutane and (R)-3-methylcyclohexanone from the chiral Cu(643) surfaces using temperature programmed desorption (TPD).

Experimental Section

All experiments were conducted in a stainless steel chamber built to achieve ultrahigh vacuum conditions. A titanium sublimation pump and a cryopump produce a base pressure of less than 10^{-10} Torr. The chamber is equipped with a quadrupole mass spectrometer, used to perform TPD and other analyses of the gas composition in the chamber. Two leak valves are installed to introduce gases and vapors into the chamber. One leak valve has a ½ inch diameter dosing tube terminating two inches from the center of the chamber. This allows the crystal to be positioned directly in front of the vapor stream flowing into the chamber through the leak valve. The other leak valve does not have a dosing tube and is used for background (non-line-of-sight) exposures. Exposures are reported in Langmuirs (1 L = 1.0 x 10^{-6} Torr-sec) and are not corrected for ion gauge sensitivity to different gas species. A sputter ion gun is installed to perform inert gas ion etching (cleaning) of surfaces. A four-grid retarding field analyzer is used to perform low energy electron diffraction (LEED) experiments and Auger electron spectroscopy (AES).

A single crystal Cu disk 12.5 mm in diameter and 2 mm thick was purchased from Monocrystals Company. One side of the sample exposes the (643) surface, and the other side of the sample exposes the $(\overline{643})$ surface. The crystal was spot-welded between two tantalum wires attached to a sample holder, at the bottom of a manipulator. The manipulator allows 360-degree rotation of the crystal about the chamber axis and movement in the x, y, and z directions. The crystal could be cooled with liquid nitrogen to less than 90 K and heated resistively to over 1000 K. The crystal temperature is measured by a chromel/alumel thermocouple junction spot-welded to the top edge of the crystal. During heating the sample temperature is computer controlled to ensure a linear heating profile.

Both sides of the crystal were initially cleaned by cycles of 1.0 keV argon ion bombardment and annealing to 950 K until no contaminants were detected by AES. Each of the chiral adsorbates investigated in this paper were found to contaminate the surface with carbon after cumulative exposures greater than 1 L. To remove carbon contamination after each experiment, the surface was cleaned by several 15-minute cycles of 1.0 keV argon ion sputtering and annealing to 950K. The surface cleanliness was verified by AES and the (643) structure was verified by sharp LEED patterns.

The chiral adsorbates we have used are liquid at room temperature and have relatively high vapor pressures, allowing them to be introduced to the vacuum chamber through leak valves. (S)-1-chloro-2-methylbutane and a racemic mixture of (R/S)-1-chloro-2-methylbutane were obtained from TCI American at 99% purity. (R)-3-methylcyclohexanone (99%) and a racemic mixture of (R/S)-3-methylcyclohexanone (97%) were obtained from Aldrich Chemical. The chiral compounds were each transferred to glass vials and subjected to several cycles of freezing, pumping, and thawing to remove dissolved air or other high pressure impurities. The purity of each sample was verified by mass spectrometry.

Temperature programmed desorption experiments were performed by cooling the Cu(643) surface with liquid nitrogen to less than 90 K. The clean Cu(643) surface was then exposed to vapors admitted to the chamber through a leak valve. Exposures of (S)-1-chloro-2-methylbutane and its racemic mixture were performed with the Cu crystal positioned 1 inch in front of the dosing tube. Exposures of (R)-3-methylcyclohexanone were done using background (non-line-of-sight) exposures. The quadrupole mass spectrometer used for desorption measurements is enclosed in a stainless steel tube, terminating in a circular aperture approximately 0.75 cm in diameter. Desorption measurements were made with the sample positioned 3-4 mm from the aperture. The aperture is smaller than the diameter of the crystal and meant to minimize the measurement of species desorbing from surfaces other than the crystal surface facing the mass spectrometer. Desorption measurements were performed by heating the sample at a constant rate while the mass spectrometer monitored the species desorbing from the surface.

Results

Low Energy Electron Diffraction (LEED)

LEED patterns for stepped and kinked surfaces arise from diffraction off the terraces and spot splitting due the steps in the surface.[5] The terrace widths of stepped surfaces can be calculated from the magnitude of the spot splitting.[5,6] The angle between the primary spots from the terraces and the split spots from the steps provides information about the orientation of the kinks with respect to the lattice vectors of the terraces in real space. The real space structures and the LEED patterns for the Cu(643)^R and Cu(643)^S surfaces are shown in Figure 1. Both LEED patterns were taken with a primary electron beam energy of 183.0 eV while the crystal temperature was 85 K.

The lattice parameter of a bulk face-centered cubic Cu crystal is 3.61 Å.[7] The terrace width was calculated to be 11.4 Å, following the kinematic method of Henzler.[6] This corresponds to three atom wide terraces. A vector connecting the split spots on the (643)^R diffraction pattern are at an angle of approximately -23° degrees with respect to the vertical direction and a vector connecting the split spots on the (643)^S diffraction pattern is at an angle of approximately +23° with respect to the vertical direction. The direction of the spot splitting is reversed between the (643)^R and (643)^S surfaces and indicates the "handed" relationship between the two surfaces. The handedness of the real space

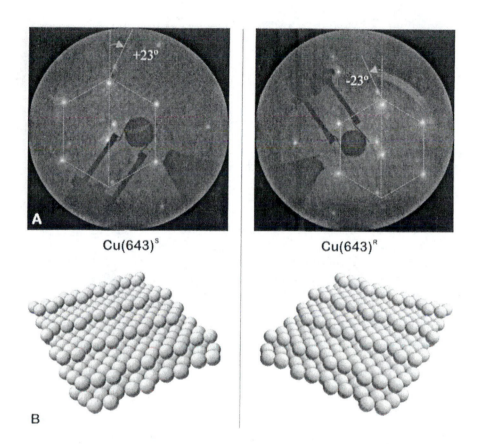

Figure 1. (A) Photographs of the LEED patterns from the Cu(643)^S and Cu(643)^R surfaces. The sample temperature is 85 K and the electron beam energy is 183.0 eV. (B) Three dimensional ball models of the Cu(643)^S and Cu(643)^R surfaces. The handedness of the real space structures is evident in the handedness of the diffraction patterns.

structure is evident in the handedness of the diffraction patterns. The LEED patterns observed for the $Cu(643)^R$ and $Cu(643)^S$ surfaces are in good agreement with previously published LEED patterns for Ag(643) and Pt(643) surfaces.[1,4]

Temperature-Programmed Desorption (TPD)

1-Chloro-2-methylbutane

TPD spectra were acquired for (S)-1-chloro-2-methylbutane and a racemic mixture of (R/S)-1-chloro-2-methylbutane on $Cu(643)^R$. If enantiospecific desorption occurs, the TPD spectra for these two samples should be different. Specifically, differences should exist between peak desorption temperatures for (S)-1-chloro-2-methylbutane and (R/S)-1-chloro-2-methylbutane desorbing from the same chiral surface. Peak temperature differences as small as 1 K are detectable with the equipment currently in use.

TPD spectra were acquired for (S)-1-chloro-2-methylbutane on $Cu(643)^R$ by direct exposure to the clean surface. After exposure to the chiral molecule, the surface was heated at a constant rate of 2 K/sec while positioned in front of the mass spectrometer detector. The fragments at $m/q=29$ and 41 were monitored. The background mass spectrum analysis of (S)-1-chloro-2-methylbutane showed that the signal from the fragment at $m/q=29$ was 5% more intense than the signal at $m/q=41$. However, when measuring the species desorbing from the Cu(643) surface during TPD experiments, the fragment at $m/q=41$ was 10% more intense than the signal at $m/q=29$. We believe that the desorption of (S)-1-chloro-2-methylbutane from the Cu(643) surface is molecular, but a small fraction decomposed on the surface to give alkyl groups which desorbed as olefins. Figure 2 shows a series of TPD spectra taken after increasing exposures from 0.05 L to 0.40 L. At the lowest exposure (0.05L), a single peak was observed at 260.5 K, corresponding to desorption from the monolayer. Due to some noise in the TPD spectra, peak temperatures were determined by Gaussian curve fitting on an interval ±10 K from the temperature at which the maximum desorption rate was recorded. Following an exposure of 0.20 L, the monolayer is saturated and the peak temperature has shifted slightly to 252.2 K. At an exposure of 0.40 L, a zero order peak corresponding to multilayer desorption is visible at 143.0 K. TPD spectra for (R/S)-1-chloro-2-methylbutane on $Cu(643)^R$ were obtained under the same conditions. Figure 3 shows the monolayer desorption peak temperatures for (S)-1-chloro-2-methylbutane and (R/S)-1-chloro-2-methylbutane on $Cu(643)^R$ as a function of increasing coverage. Coverages are given in terms of monolayers (ML), where 1 ML is the exposure at which the multilayer desorption peak becomes visible in the TPD spectrum. Coverages were determined by scaling the area under each TPD curve by the area under the saturated monolayer peak. As Figure 3 illustrates, the peak desorption temperatures for the pure enantiomer and the racemic mixture of 1-chloro-2-

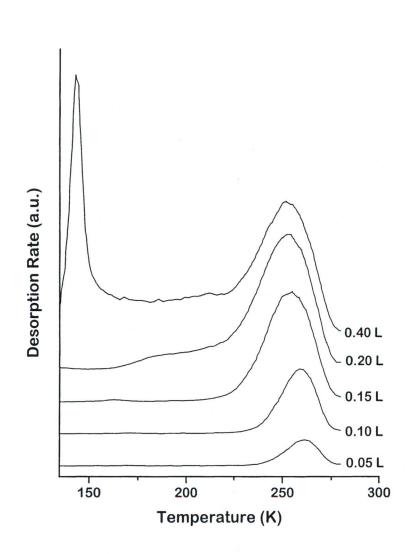

Figure 2. A series of TPD Spectra taken following increasing exposures of (S)-1-chloro-2-methylbutane to the $Cu(643)^R$ surface, while monitoring the fragment at m/q = 41. The high temperature peak (~250 K) corresponds to monolayer desorption of molecules adsorbed directly to the $Cu(643)^R$ surface, while the low temperature peak (143 K) corresponds to multilayer desorption. Curves are offset for clarity.

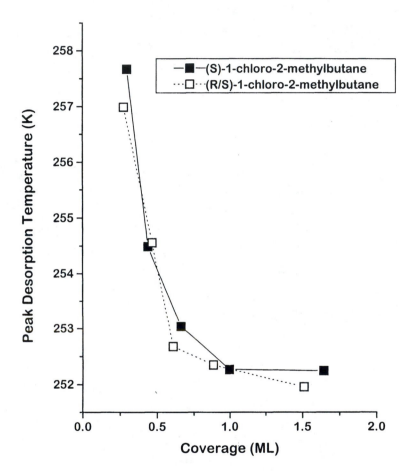

Figure 3. Peak desorption temperatures for (S)-1-chloro-2-methylbutane and (R/S)-1-chloro-2-methylbutane on Cu(643)^R plotted as a function of surface coverage. No enantiospecific differences are visible between the desorption characteristics of (S)-1-chloro-2-methylbutane and (R/S)-1-chloro-2-methylbutane on the Cu(643)^R surface.

methylbutane on the Cu(643)^R surface are nearly identical. Thus, any enantiospecific effects are too small to measure for this particular system.

3-Methylcyclohexanone

TPD spectra were acquired for (R)-3-methylcyclohexanone ((R)-3MCHO) and a racemic mixture of (R/S)-3-methylcyclohexanone ((R/S)-3MCHO) on the Cu(643)^R and Cu(643)^S surfaces. These TPD experiments were conducted with background exposures to the clean Cu(643) surface at 90 K. After exposure to the chiral molecule, the surface was heated at a constant rate of 1 K/sec while positioned in front of the mass spectrometer. Figure 4 shows TPD spectra for (R)-3-MCHO on Cu(643)^S following exposures from 0.05 L to 0.45 L. (R)-3-MCHO shows several peaks in its TPD spectrum. At the lowest exposure (0.05 L) two overlapping peaks are visible. Again the positions of the maxima have been determined by fitting a Gaussian function through the points lying ±10K from the maximum. A more intense peak is present at 386.4 K and a less intense peak is present at 349.1 K. The origin of the two peaks is not known at this time, however they must correspond to different adsorption states on the Cu(643)^S surface. It is possible that one peak corresponds to desorption of molecules adsorbed in the kink sites of the surface and the other peak corresponds to molecules adsorbed to the flat (111) terraces of the surface. At the highest exposure (0.45 L), a zero-order desorption peak corresponding to multilayer desorption appears at 171.0 K. TPD spectra for (R)-3-MCHO on Cu(643)^R and (R/S)-3-MCHO on Cu(643)^R and Cu(643)^S were also obtained and the results were qualitatively similar to the spectra in Figure 4. Initially several ionization fragments were monitored in the mass spectrometer in order to determine if 3-methylcyclohexanone was reacting on the surface. The relative intensities of the fragments monitored in the TPD experiments corresponded to the relative intensities of the fragments monitored during a background analysis of the vapor and there was not evidence of decomposition.

Careful examination of the TPD spectra described above reveals that the peak temperatures depend on the chirality of the adsorbed species and of the surface. Figure 5 shows the peak desorption temperatures for (R)-3-MCHO and (R/S)-3-MCHO on Cu(643)^R and Cu(643)^S as a function of surface coverage. Figure 5 only considers the highest temperature peak in the TPD spectra for 3-methylcyclohexanone on the Cu(643) surfaces. Surface coverage was determined by scaling the area under each TPD curve by the area under the two highest temperature peaks, once saturated. Figure 5 shows that the racemic mixture of (R/S)-3-MCHO has peak desorption temperatures on the Cu(643)^R and Cu(643)^S surfaces that are within 1.0 K for similar exposures. As the racemic mixture contains an equal amount of both enantiomers of 3-methylcyclohexanone, enantiospecific behavior is not expected and is not

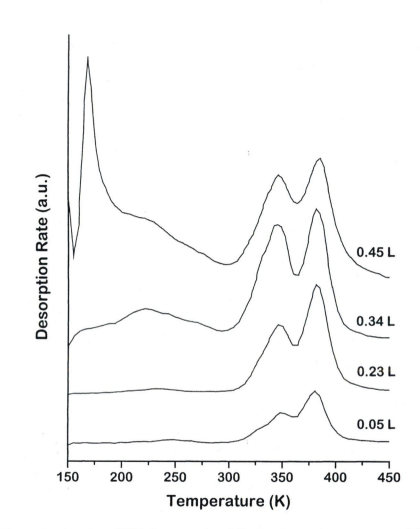

Figure 4.   A series of TPD Spectra taken following increasing exposures of (R)-3-methylcyclohexanone to the Cu(643)^S surface, while monitoring the fragment at m/q = 39.   The two highest temperature peaks (~350K, ~385 K) correspond to different adsorption states for (R)-3-MCHO on the Cu(643)S surface, while the low temperature peak (171 K) corresponds to multilayer desorption.   Curves are offset for clarity.

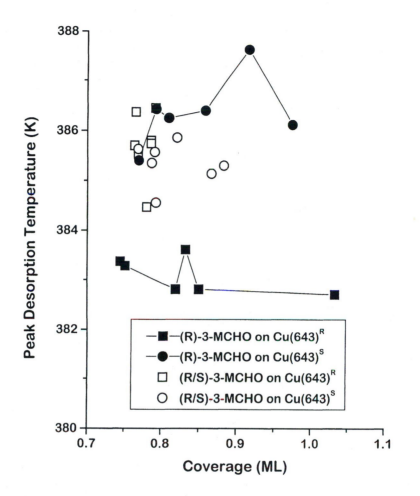

Figure 5. Peak desorption temperatures for (R)-3-methylcyclohexanone and (R/S)-3-methylcyclohexanone on Cu(643)^R and Cu(643)^S as a function of surface coverage. The racemic mixture of (R/S)-3-MCHO has similar peak desorption temperatures on both Cu(643)^R and Cu(643)^S surfaces. Enantiospecific effects are evident as the peak desorption temperatures for the pure enantiomer of (R)-3-MCHO differ by ~3.5 K between the Cu(643)^R and Cu(643)^S surfaces. This corresponds to a difference in desorption energies of 0.2 kcal/mol for (R)-3-MCHO on the Cu(643)^R and Cu(643)^S surfaces.

observed. (R)-3-MCHO on the Cu(643)^S surface has peak desorption temperatures that are approximately 1-2 K higher than the racemic (R/S)-3-MCHO on the same surface. In contrast, (R)-3-MCHO on the Cu(643)^R surface has peak desorption temperatures that are 2-3 K lower than the racemic (R/S)-3-MCHO on the same surface. These results clearly reveal enantiospecific adsorption is occurring on the chiral Cu(643) surface. Table I shows the average peak desorption temperature (T_p) and the average desorption energy (E_{des}) for each of the systems described above. Average peak desorption temperatures were calculated by averaging all data points collected for each adsorbate/surface combination. Average desorption energies were calculated using the standard relation for first-order desorption kinetics ($T_p^2=(\beta E_{des}/R\nu_1)\exp(E_{des}/RT_p)$) with an assumed preexponential factor (ν_1) of 10^{13} s^{-1}. In this equation, T_p is the peak desorption temperature, β is the heating rate, and E_{des} is the desorption energy for an unactivated process. The difference in E_{des} between (R)-3-MCHO on the Cu(643)^R and Cu(643)^S surfaces is ~0.2 kcal/mol.

Table I

System	Avg. T_p (K)	Avg. E_{des} (kcal/mol)
(R/S)-3-MCHO on Cu(643)^R	385.72 ± 0.65	24.83 ± 0.04
(R/S)-3-MCHO on Cu(643)^S	385.34 ± 0.42	24.80 ± 0.03
(R)-3-MCHO on Cu(643)^R	383.10 ± 0.37	24.65 ± 0.02
(R)-3-MCHO on Cu(643)^S	386.37 ± 0.72	24.87 ± 0.05

Note: Error ranges given are ± 1 standard deviation from the calculated values.

Theoretical Results

To complement the experiments described above, we have performed a series of theoretical calculations examining the adsorption of chiral hydrocarbons on chiral Pt surfaces. These calculations extend our previous studies of these systems.[2,8] For brevity, we will only describe our calculations involving the Pt(643) surface. We have also simulated a range of other Pt surfaces and will report on these results elsewhere.[9] In our simulations, the Pt surface is held rigid in the relaxed structure predicted for Pt(643) using density functional theory in the local density approximation.[10] Hydrocarbon molecules are modeled using the united atom approximation with bond bending and torsion angles described using the parameterization of Grest et al.[11] A bond stretching potential was fitted to spectroscopic data[8] and an anti-inversion potential[11] was included to avoid the unphysical interconversion of enantiomers that is possible with united atom models. We have confirmed that gas phase simulations of these molecules reveal the correct distribution of molecular conformations for the molecules discussed below.[12] The interaction potential between hydrocarbon molecules and the Pt surface was defined using a potential fitted by Stinnett *et al.* to molecular beam scattering data.[13] The possible configurations for isolated adsorbates were explored using hybrid Monte Carlo trajectories including $> 10^7$ sample configurations per adsorbate. Energy minimization was applied to these sample configurations to determine adsorption geometries that are local minima on the potential energy surface.

The methods outlined above were applied to four chiral hydrocarbons, namely, trans-1,2-dimethyl-cyclo-X, where X = propane, butane, pentane, and hexane. We will abbreviate these molecules by DMCPr, DMCB, DMCPe, and DMCH. For each molecule, the stable adsorption geometries for both enantiomers were computed on the Pt(643) surface. Although every molecule exhibits multiple adsorption geometries, here we will only discuss the most stable (i.e., global energy minimum). For DMCPr and DMCB, the R enantiomer is more stable than the S enantiomer by 0.27 and 0.31 kcal/mol, respectively. The R enantiomer is also more stable for DMCPe, but only by 0.054 kcal/mol. By contrast, the S enantiomer's energy minimum lies 0.11 kcal/mol below the R enantiomer for DMCH. Qualitatively, the same trends are observed when thermal averages over all accessible minimum energy states are considered.[9] Three of the four molecules simulated yield enantiospecific differences in energy similar to those reported above for (R)-3-MCHO on Cu(643) and for glucose on Pt(643).[4] However, for one molecule, the enantiospecific energy difference is small enough that it would not be detected in our TPD experiments, even though this quantity is nonzero. This type of situation may account for our experimental results with (S)-1-chloro-2-methylbutane.

Conclusions

Several demonstrations of the enantiospecific properties of chiral single-crystalline surfaces have been observed. Through the observations of LEED patterns that are nonsuperimposable mirror images of each other, the Cu(643)^R and Cu(643)^S are shown to be enantiomers of each other. Similar LEED results have been observed elsewhere for Ag(643) and Pt(643) surfaces.[1,3,4] Theoretical calculations have shown that certain hydrocarbons on chiral Pt(643) surfaces exhibit enantiospecific adsorption characteristics. Experimental results from TPD show that the desorption energies of (R)-3-methylcyclohexanone on Cu(643)^R and Cu(643)^S differ by approximately 0.2 kcal/mol. Other work has shown enantiospecific adsorption of glucose enantiomers on chiral Pt electrodes through electro-oxidation experiments.[3,4]

Acknowledgement.

This work was supported by NSF grant CTS-9813937. DSS and TDP acknowledge access to the CLAN cluster managed by the Pittsburgh Supercomputer Center.

References

1. McFadden, C.F.; Cremer, P.S.; Gellman, A.J. *Langmuir.* **1996**, 12, 2483-2484.
2. Sholl, D.S. *Langmuir.* **1998**, 14, 862-867.
3. Ahmadi, A.; Attard, G.; Feliu, J.; Rodes, A. *Langmuir.* **1999**, 15, 2420-2424.
4. Attard, G.A.; Ahmadi, A.; Feliu, J.; Rodes, A.; Herrero, E.; Blais, S.; Jerkiewicz, G. *Journal of Physical Chemistry B.* **1999**, 103, 1381-1385.
5. Ertl, G.; Küppers, J. *Low Energy electron and Surface Chemistry*, 2nd ed.; VCH: Weinheim, 1985; pp 246-2488.
6. Henzler, M. *Surface Science.* **1970**, 19, 159-171.
7. Kittel, C. *Introduction to Solid State Physics*, 7th ed.; John Wiley and Sons: NY, 1996.
8. Power, T.D.; Sholl, D.S. *Journal of Vacuum Science Technology A.* **1999**, 17, 1700-1704.
9. Power, T.D.; Sholl, D.S. submitted to *Topics in Catalysis*.
10. Feibelman, P. J. *personal communication*.
11. Mondello, M.; Grest, G.S. *J. Chem. Phys.* **1995**, 103, 7156.
12. Eliel, E.L.; Wilen, S.H. *Stereochemistry of Organic Compounds*; John Wiley and Sons: New York, 1994.
13. Stinnett, J.A.; Madix, R.J.; Tully, J.C. *J. Chem. Phys.* **1996**, 104, 3134.

Chapter 20

Single Molecule Absolute Chirality Determination, Isomerization, and Asymmetric Induction at a Silicon Surface

D. J. Moffatt, G. P. Lopinski, D. D. M. Wayner, and R. A. Wolkow*

Steacie Institute for Molecular Sciences, National Research Council, 100 Sussex Drive, Ottawa, Ontario K1A 0R6, Canada

STM is used to identify and discriminate between adsorbed molecules differing only in the geometric configuration of methyl groups. The absolute configuration (either R or S) for each newly formed asymmetric centre is identified. The degree of stereoselectivity of the reaction of 2-butene with the Si(100) surface is also been studied. A small degree of isomerization is observed, indicating that the reaction, although 98% stereoselective, is not stereospecific. This implies that the [2+2] cycloaddition reactions of alkenes on this surface proceed in a step-wise manner with the time scale between formation of the two Si-C bonds on the order of a few picoseconds. We show the first example of an enantiospecific reaction with a silicon surface. 1S(+)-3-carene adsorption on the Si(100) surface results in the formation of a chiral surface.

Introduction

There is increasing interest in the modification of semiconductor surfaces via the covalent attachment of organic molecules for applications ranging from optoelectronics to molecular sensors (1). Fabrication of hybrid devices integrating the wide range of functionality of organic molecules with existing microelectronics technology will require the ability to form highly ordered continuous films (2), as well as nanostructured patterns of these molecules (3,4). The chemical and physical properties of these films are highly dependent upon molecular structure and conformation (5,6). Moreover, molecular recognition events, particularly chiral recognition, will be dependent on the stereochemistry of the adsorbed molecules. In order to control and monitor the stereochemical outcome of a reaction extraordinary tools capable of probing subtle structural details are required. The scanning tunneling microscope (STM), with its ability to image surfaces at the atomic scale, is uniquely important in this regard (7). Our interest in the stereochemistry of adsorbates on semiconductors has led us to assess the capability of STM to identify and discriminate between adsorbed molecules differing only in the geometric configuration of methyl groups. Through examination of propylene, trans-2-butene and cis-2-butene on the Si(100) surface we find that individual methyl groups can be readily detected allowing the geometric configuration (cis or trans) to be determined. We illustrate that STM can be used to measure the degree of stereoselectivity of such reactions with an accuracy limited only by the number of molecules inspected. Furthermore, because the position and orientation of the methyl groups can be seen, it is possible to determine the absolute configuration (either R or S) for each newly formed asymmetric centre. The ability of STM to probe the structure and absolute stereochemistry of molecules on surfaces is crucial for the design of enantioselective processes at these interfaces. These processes are expected to lead to the development of highly ordered chiral interfaces capable of recognition of complex organic or bioorganic molecules.

We have studied the (100) crystal face of silicon, the surface generally used in silicon-based microelectronic devices. The Si(100) surface has been extensively investigated by STM and other techniques (8). The salient feature of this surface is the silicon dimer. Each surface atom has two back bonds to the substrate below, shares in one dimer bond, and has one unsatisfied or dangling bond. The dimers form rows that appear as bars in STM images. Alkenes adsorb molecularly on the silicon dimers forming two Si-C bonds with the dangling bonds on each dimer (9-15). Our measurements were performed in an ultra high vacuum chamber (with a base pressure of 5×10^{-11} Torr) which allows the crystal to be kept largely free of contaminants. The n-type (As doped, 0.005 Ohm.cm) Si(100) wafers were resistively heated to 680 °C for 12hrs. to degas the sample and holder, followed by heating to 1250 °C for several seconds to obtain a clean surface. Adsorption of the subject molecules was achieved by controllably leaking the gaseous molecules into the vacuum chamber.

Determination of the Absolute Chirality of Individual Adsorbed Molecules Using the Scanning Tunneling Microscope

Figures 1 a), b) and c) show respectively STM images of propylene, trans-2-butene and cis-2-butene on Si(100). In each case, small protrusions are observed that we associate with the adsorbed molecules. While in the case of propylene the protrusions are randomly distributed the butenes exhibit what are clearly paired protrusions. These bright features can be associated with the methyl groups within each of these molecules. It is also apparent from Figure 1 that the cis- and trans- images are different. The methyl groups of the cis isomer define a line that is at a right angle to the dimer rows. In contrast, the methyl units of the trans isomer define a line that is angled approximately 30° to the dimer row. We note that in the particular imaging mode employed here, the dimer rows give rise to bars centred between the actual dimer rows. These images probe unoccupied electronic levels revealing antibonding states which have a node at the dimer centre. The images are therefore consistent with the adsorbed molecule being bound to a single Si dimer.

Assignment of the bright protrusions to methyl groups is further justified by consideration of the bonding geometries of these alkenes on the Si(100) surface, illustrated in the insets of Figure 1. Calculated adsorption geometries were obtained using the AM-1 (16) semi-empirical method and a two-dimer, 31 Si atom cluster to model the surface (17). Upon adsorption, the alkenes become alkane-like as each carbon atom rehybridizes from sp2 to sp3 forming a bond to one silicon atom (of the same dimer) thereby satisfying the dimer dangling bonds. The C-C bond, originally 1.33Å in the free molecule is calculated to increase to 1.50 Å upon adsorption, consistent with a change from a double to a single bond in agreement with previous theoretical studies of ethylene/Si(100)(13,14). The positions of the methyl groups in the images are in quantitative agreement with the calculated bonding geometries. By registering the methyl protrusions of adsorbed propylene with the underlying Si dimers, the methyl group is seen to be displaced approximately 1.1 Å from the centre of these units, in good agreement with the inset of Figure 1a. Likewise the distance between the paired protrusions of the trans-2-butene is measured to be 3.4±0.2 Å, as compared with 3.6 Å obtained from calculation.

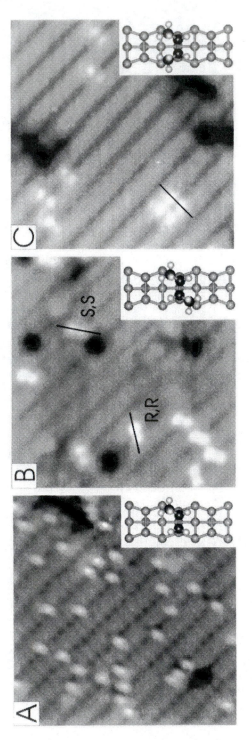

Figure 1. STM images of a) propylene, b) trans-2-butene and c) cis-2-butene on Si(100). The bar-like structures are the dimer rows of the silicon substrate. (The dark regions are due to pre-existing defects.) The white protrusions are due to the methyl groups of the adsorbed molecules. In b) and c) lines have been drawn through the methyl groups of selected molecules to show the orientation of these with respect to the underlying lattice. In b), an example of each enantiomer (RR or SS) of adsorbed trans-2-butene is labeled. The insets show the configuration of each of the alkenes studied on the Si(100) surface. (The dimer rows of the images are rotated ~450 relative to the insets.) Image area 75Å x 75Å, sample bias +2V, current 100 pA.

The observation of methyl groups as maxima in the STM images is somewhat surprising. Most adsorbates on silicon surfaces appear dark (topographic depressions) because surface-adsorbate bond formation eliminates the dangling bonds that give rise to image features associated with surface silicon atoms (18). In order for a feature to appear bright in an STM image it is generally not sufficient for it to be a topographic protrusion on the surface, it must also have electronic states that are energetically accessible for tunneling to (or from) the STM tip. Images of hydrogen atom-terminated dimers, for example, reveal a 1.5 Å depression on the surface relative to unreacted dimers (19). Adsorbed ethylene molecules also appear depressed (0.1-0.2 Å) relative to unreacted dimers (20). The methyl groups of the adsorbed alkenes, being fully saturated, might also be expected to appear as depressions, however, ab initio calculations using a Si cluster to model the surface indicate that the adsorbed molecules are indeed expected to be visible in STM, in accord with our observations (17).

In addition to testing the capacity of STM to determine the geometric configuration of the adsorbed molecule, the alkenes of this study provide an opportunity to monitor the degree of stereoselectivity of the reaction. Previously, cis and trans 2-butene have been shown to exhibit different desorption energies from Si(100) demonstrating that the isomers retain their configuration upon adsorption (15). Stereoselectivity in this class of reactions has also been suggested by an infrared study of dideuteroethylene/Si(100), relying on matching very small calculated frequency shifts with experiment to determine the dominant adsorbed isomer (21). These methods provide an average measurement from a macroscopic sample. However, the current drive to develop nanoscale devices will require the capability to determine the stereochemistry of extraordinarily small amounts of material with a high degree of precision. STM, with its molecule-by-molecule inspection capability allows us to determine the degree to which a reaction is stereoselective to arbitrary accuracy (simply determined by the number of molecules inspected). In the present case, inspection of approximately 500 adsorbed molecules revealed 2.1±0.7 % of the cis isomer and 1.2±0.5 % trans isomer in the in the trans and cis samples, respectively. Since these fractions are consistent with the specified isomeric impurity level in the gases used it is evident that no measurable isomerization occurs upon adsorption.

The observation that butene adsorption is highly stereoselective places important constraints on the reaction mechanisms. The retention of the stereochemistry of the reactants during addition to the surface implies either a concerted reaction or a stepwise process in which the time between the formation of the two Si-C bonds is insufficient to allow a C-C bond rotation. A concerted suprafacial reaction between ethylene and Si(100) has been suggested to be symmetry forbidden due to the Woodward-Hoffmann rules (21). However, we note that the Si dimers are in fact tilted (one end up, the other down (22)) at room temperature, which is likely sufficient to lift this restriction. This tilting, however, is also likely to favour a stepwise mechanism.

There is a further and unique benefit to direct observation of the geometric isomers of single adsorbed molecules, as illustrated by the present case. Knowledge of the spatial relationship between methyl groups of a single reacted alkene provides an opportunity to determine the absolute configuration (R or S) for each newly formed asymmetric (chiral) centre. As shown in Figure 2, an unsaturated bond such as that in trans-2-butene has two reactive faces, associated with two distinct products. These are called the Re and the Si faces; the Re face leading to a product with R chirality (pro-R) and the Si face leading to a product with S chirality (pro-S). Thus the reaction of trans-2-butene leads to two chiral products (either RR or SS) while the reaction of cis-2-butene results in a non-chiral product (RS or SR). Since the Re and Si faces of trans-2-butene react with equal probability an equal number of RR and SS enantiomers are formed. Nevertheless, it is clear from Figure 1 that RR is distinguishable from SS. Similarly, by registering the methyl groups with the underlying lattice the chiralities of adsorbed propylene (R or S) and cis-2-butene (RS or SR) can be determined. This is, to our knowledge, the first direct determination of the absolute chirality of individual adsorbed molecules. Although the Re and Si faces of trans-2-butene are identical, the introduction of a single chiral center adjacent to the double bond removes this degeneracy. It is expected that steric inhibition at one of the faces will result in a highly enantioselective process (asymmetric induction). This work demonstrates that it will be possible to determine quantitatively the enantioselectivities of these reactions.

How stereoselective are alkene addition reactions on Si(100)?

Reactions in which the molecules retain their geometric configurations upon reaction with the surface are particularly desirable. The addition of alkenes to the dimers of the reconstructed Si(100) surface have been suggested as an example of such a stereospecific reaction (*15,21*). These reactions are facile, resulting in the formation of two Si-C bonds as in a formal [2+2] cycloaddition (*23,24,20,13,14*) and have been exploited to controllably attach a number of different molecules to the Si(100) surface (*25-27*).

Previous studies of the stereochemistry of alkene addition, using macroscopic techniques, have concluded that the alkenes retain their stereochemistry upon adsorption (*15,21*). However, the limited sensitivity of the techniques used did not allow small amounts of isomerization (< 5%) to be detected. Previously we have demonstrated that scanning tunneling microscopy (STM) is a uniquely powerful

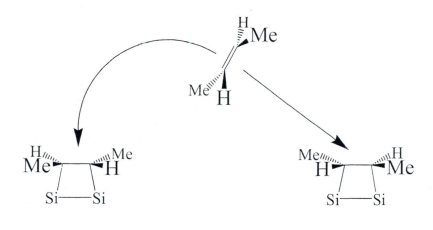

Figure 2. The planar trans-2-butene molecule has two distinct faces. Reaction at the Si face leads to the S,S structure shown at the left, while reaction at the Re face leads to the R,R structure at the right. While indistinguishable using optical and other spectroscopic techniques, the individual enantiomers are clearly identified by STM.

probe of the degree of stereoselectivity of surface reactions, with the sensitivity simply determined by the number of molecules examined (6). Here we report that, applying this technique to monitor the addition of cis and trans-2-butene to the Si(100) surface, a small but measurable degree of isomerization is observed indicating that this reaction, while highly stereoselective, is not stereospecific.

The STM studies were carried out in an ultrahigh vacuum (UHV) chamber with a base pressure below 5×10^{-11} Torr. This pressure is sufficiently low to maintain a clean surface for a day or more. A clean Si(100) surface was obtained by flashing a well-degassed crystal (Virginia Semiconductor, n-type, 0.005Ω cm) above 1200°C. The samples were exposed to 2-butene (purity 99+%) by controllably leaking the gaseous molecules into the vacuum chamber. The gases were analyzed by both NMR and gas chromatography and the isomeric purity was determined to be 99.8±0.1% for both cis and trans samples.

Figure 3 shows an unoccupied state STM image of a Si(100) surface exposed to a flux of trans-2-butene molecules. The bar-like structures running diagonally across the images are the dimer rows of the clean Si(100) surface. The dimers are 3.84 Å apart and the rows are separated by twice that distance. The various dark features are defects. The butene molecules are imaged as paired protrusions, centered along a dimer row. Comparison with the calculated bonding geometry (see Figure 4c) reveals these protrusions to be due to the two methyl groups on either end of the molecule (6). For the large majority of the adsorbed molecules, the two protrusions define a line that is inclined by ~30° with respect to the dimer row direction - consistent with an adsorbed molecule that has retained its configuration upon adsorption. However, a small fraction of molecules, an example is indicated in Figure 3, exhibit protrusions which define a line perpendicular to the dimer row direction, just as observed when the cis isomer has been adsorbed on the surface. By examining several hundred molecules, the fraction of molecules in the cis configuration is determined to be 2.1±0.7%. This is measurably higher than the level of cis contamination of the trans-2-butene gas sample, determined to be 0.2±0.1%. A similar analysis of images after exposure to cis-2-butene, indicates that 2.6±0.6% of the adsorbed molecule are in the trans geometry. These observations indicate that while the adsorption of the 2-butenes is highly stereoselective it is not stereospecific, with approximately 2% of the molecules undergoing isomerization upon adsorption. The STM images also show that isomerization occurs at dimers of the clean surface and is not induced by defects, steps or interactions with other adsorbates.

The current observations are not inconsistent with previous studies which concluded that alkene addition on Si(100) is stereoselective. Kiskinova and Yates showed that samples exposed to cis and trans-2-butene exhibit distinctly different desorption energies, demonstrating that most of the molecules retain their configuration upon adsorption (15). In a related study Liu and Hamers compared

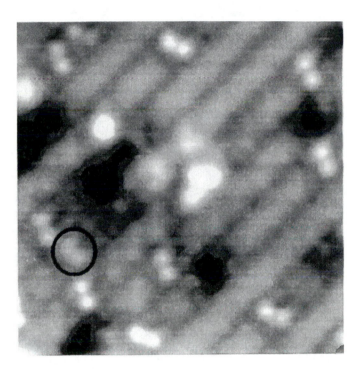

Figure 3. Unoccupied state STM image (75Åx75Å, 2V, 40pA) of a Si(100) surface exposed to trans-2-butene. While most of the molecules retain their configuration upon adsorption, the circled molecule has undergone isomerization to the cis form.

292

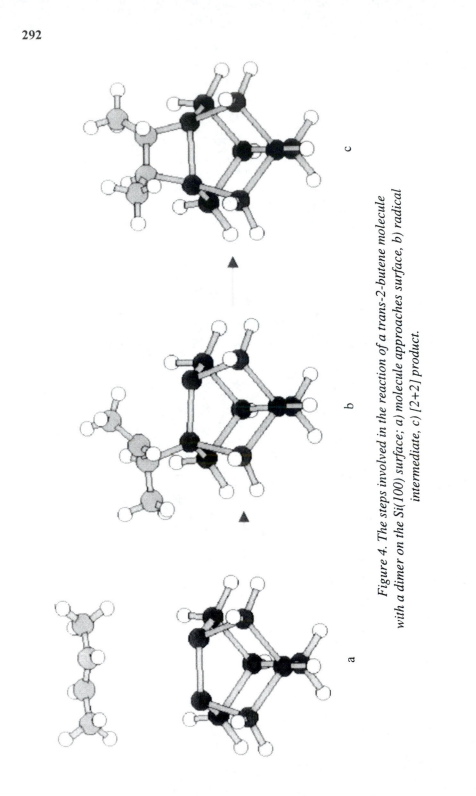

Figure 4. The steps involved in the reaction of a trans-2-butene molecule with a dimer on the Si(100) surface; a) molecule approaches surface, b) radical intermediate, c) [2+2] product.

measured and calculated splitting of the C-H stretch modes to conclude that dideuteroethylene adsorption was also stereoselective (*21*). Both these studies, however, employed macroscopic techniques not sufficiently sensitive to detect the small levels of isomerization described here.

The desorption study of *cis* and *trans*-2-butene revealed that the adsorbed *trans* isomer is 4 kcal/mol more stable than the *cis* species (*15*). It is notable then, that *if* the molecules could equilibrate, we would expect almost complete isomerization of the *cis* molecules while virtually all the *trans* molecules would retain their configuration. As the degree of isomerization reported here is similar for both the *trans* and *cis* isomers, we may conclude that the process is kinetically rather than thermodynamically controlled.

Our observation that a small but statistically significant fraction of the adsorbing 2-butene molecules isomerize has implications for the mechanism of alkene/Si(100) [2+2] addition reactions. Before discussing those, we note the possibility that isomerization occurred *before* adsorption on the silicon surface has been considered and found not to be supported by our measurements. Our reasoning is straight forward. The stainless steel gas handling vacuum systems used in preparation for NMR and gas chromatographic as well as for STM experiments were equivalent. Since the STM measurement alone showed relatively high isomer contamination, we conclude that the factor unique to that measurement, the silicon surface, was responsible for the observed isomerization.

The first conclusion to be drawn from the observation of isomerization upon adsorption is that a completely concerted addition along a symmetric pathway is ruled out, as that would require strict retention of the stereochemistry. Such a concerted [2+2] cycloaddition reaction of a symmetric dimer with an alkene is formally forbidden by Woodward-Hoffmann rules. However, the dimers are in fact buckled, with the "up" end of the dimer becoming electron rich and the "down" atom electron deficient (*28,22*). While this buckling breaks the symmetry and lifts this restriction, it also suggests that the reaction is likely to occur via an asymmetric path (*29,30*) and be step-wise rather than concerted, allowing for the possibility of isomerization upon adsorption. It is likely that adsorption proceeds via a single Si-C bonded radical species, as shown in Figure 4. A step-wise reaction mechanism allows for rotation about the central C-C bond. Within this model, the amount of isomerization will depend on the relative time-scales for rotation about the C-C bond versus formation of the second Si-C bond, closing the four-membered ring. If the probability of undergoing a rotation before the second bond is formed is high, a large degree of isomerization is expected, leading to an equal number of the two configurations on the surface. Observation of only a small degree of isomerization indicates that the second bond is formed relatively rapidly, allowing only a small probability of crossing the C-C bond rotational barrier before the ring is closed. The measured isomerization rate indicates that ~2% of the molecules execute a rotation before the second Si-C bond is formed. Assuming a barrier of 4 kcal/mol (typical for rotation about a single C-C bond) and a prefactor of 10^{13} s^{-1}, the time between the formation of the two bonds can be estimated to be ~2ps.

In the absence of a barrier to formation of the second Si-C bond, the degree of isomerization should increase rapidly as the adsorption temperature is raised, increasing the probability of surmounting the rotation barrier before the ring closes. For a rotational barrier of 4kcal/mol the degree of isomerization is predicted to increase to 5.7% at 350K (assuming the prefactor remains constant). In contrast, the level of isomerization of the *cis*-2-butene is measured to be 2.1±0.5% for adsorption at 350K. This is within the error bars of the 2.6±0.6% isomerization probability for room temperature adsorption, indicating a rather weak temperature dependence. The weaker than expected temperature dependence points to a deficiency in our simple model. It is most likely that a barrier, comparable to that for rotation about the C-C bond, exists for closing of the ring - this will include contributions related to rotation about the Si-C bond and to formation of the second Si-C bond once alignment is achieved. If ring closing is activated, this will compensate for more facile rotation about the C-C bond at elevated temperature. This scheme can account for either a weak increase or decrease in isomerization with increasing temperature (both consistent with the current data), depending on the relative heights of the two barriers. The present data yields an upper limit on the difference in barrier heights, with the barrier for ring closing less than 1 kcal/mol smaller or 2.6 kcal/mol larger than the rotational barrier. More detailed studies are required to determine the direction and magnitude of the temperature dependence. Theoretical calculations can also provide useful insight into the relative heights of these barriers. Although there are, as yet, no available calculations of the reaction pathway for 2-butene addition to Si(100), calculations for acetylene (*29*) and ethylene (*31*) addition via a radical intermediate state yield barriers to ring closure of 4.6 kcal/mol and 2 kcal/mol, respectively. These studies did not calculate rotational barriers in the radical state.

Before closing this section, we note that an alternate scenario can also account for the observed weak temperature dependence of isomerization. As energy relaxation on semiconductor surfaces is often inefficient, it is possible that the energy released upon reaction is converted solely into internal energy of the adsorbate. Forming the single Si-C bonded species is calculated to liberate approximately 10 kcal/mol (5000 K) of energy, some of which can be converted into the relevant rotational mode which drives isomerization. If this rotational temperature is higher than the substrate temperature, the isomerization rate would be temperature independent. This would also reduce our estimate of the time-scale for formation of the second Si-C bond. Although equipartition of the adsorption energy amongst all the modes of the adsorbate would suggest the temperature of the relevant rotational mode would be less than 166K, it is possible that individual molecules could have a rotational temperature considerably higher than this.

We have used scanning tunneling microscopy to study the degree of stereoselectivity of the reaction of 2-butene with the Si(100) surface. A small degree of isomerization is observed, indicating that the reaction, although 98% stereoselective, is not stereospecific. This implies that the [2+2] cycloaddition reactions of alkenes on this surface proceed in a step-wise manner with the time scale between formation of the two Si-C bonds on the order of a few picoseconds. These results also have implications regarding the extent to which molecular conformation can be controlled in organically 'modified surfaces made via this approach. More generally, this study demonstrates the usefulness of STM in monitoring the stereochemistry of surface reactions at the single molecule level.

Asymmetric induction at a silicon surface

Organically modified surfaces may enable the development of sensors capable of complex molecular recognition tasks such as the ability to discriminate between different enantiomers of the same compound. Progress in this area will require the design of enantiospecific reactions with the silicon surface. In this section we describe the first example of such a reaction. We use scanning tunneling microscopy (STM) to demonstrate that the adsorption of 1S(+)-3-carene on the Si(100) surface is enantiospecific, resulting in the formation of a chiral surface.

Previous studies have shown that STM is uniquely capable of directly determining the absolute chirality of individual adsorbed molecules (6,32). Trans-2-butene was observed to undergo a [2+2] cycloaddition to the Si(100) surface leading to the formation of (R,R) and (S,S) enantiomers that are clearly distinguished (6). However, in that case, reaction at both the Re and Si faces of the molecule was equally likely so that equal numbers of the two possible enantiomers were formed, resulting in a macroscopically achiral surface. One approach to achieving enantioselectivity in these reactions is to introduce steric hindrance at one of the faces. To test this hypothesis we studied the reaction of 1S(+)-3-carene(3,7,7-trimethyl bicyclo[4.1.0]hept-3-ene), a naturally occurring bicyclic alkene belonging to the terpene family, with the clean Si(100) surface. In this molecule a gem-dimethyl group significantly restricts access to one face of the double bond (Figure 5a).

The Si(100) surface consists of rows of Si dimers, with two dangling bonds per dimer. Simple alkenes have been shown to react readily with the dimers of this surface in a formal [2+2] cycloaddition, forming two Si-C bonds (23,24,20,13,14). Carene is expected to react with the surface in a similar manner, its double bond reacting with a single Si dimer as in Figure 5b. Adsorption introduces two additional chiral centers at positions C3 and C4. Addition at the face of the

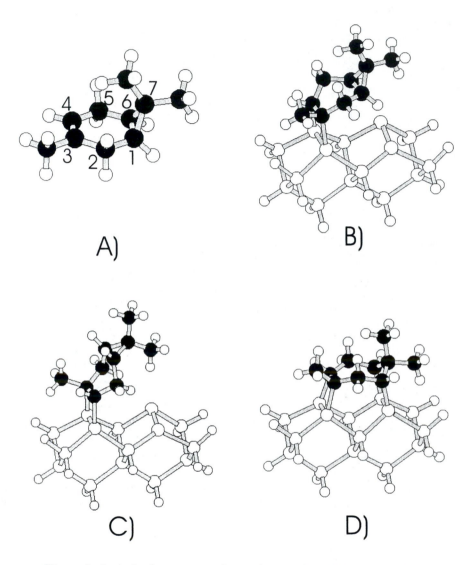

Figure 5. Optimized structures obtained using the B3P86 density functional and the 6-31G* basis set. The isolated 1S(+) 3-carene molecule (A) is shown along with three possible bonding geometries on the 15 Si atom cluster model of the Si(100) surface; B) unbridged, C) reverse unbridged, D) bridged.

molecule that is partially blocked by the *gem*-dimethyl group results in the formation of a different adsorbed enantiomer (Fig 1c). For the single dimer configuration shown in Figure 1b, it is possible for the molecule to undergo further reaction, opening the cyclopropyl ring on the molecule and forming two additional bonds to the silicon surface (Figure 5d).

Optimized structures and adsorption energies for the configurations of Figure 5 were obtained using gradient corrected density functional methods (B3P86 with the 6-31G* basis set) (*33*), employing a 15 Si atom cluster to model the Si(100) surface. Aside from the two dimers, all other edge Si atoms were terminated with hydrogen atoms, whose positions were fixed along bulk Si bond directions in order to prevent excessive relaxation of the cluster. Similar approaches have been used previously to successfully treat adsorption on Si(100) surfaces (*34,35,30,17*). Adsorption energies are extracted from the total energy of the molecule/cluster system by subtracting the energy of the separated molecule and cluster. The two unbridged structures (bound to a single dimer) have similar adsorption energies, 2.25eV (5b) and 2.07eV (5c), while the bridged configuration is predicted to be substantially more stable with a calculated adsorption energy of 3.96eV.

The STM studies were carried out in an ultrahigh vacuum (UHV) chamber with a base pressure below 5×10^{-11} Torr. This pressure is sufficiently low to maintain a clean surface for a day or more. A clean Si(100) surface was obtained by flashing a well-degassed crystal (Virginia Semiconductor, n-type, 0.005Ω cm) above 1200°C. Carene (Aldrich, 99%) was degassed by performing several freeze-pump-thaw cycles.

Figure 6a shows an STM image of the Si(100) surface after exposure to a low dose of 1S(+)-3-carene molecules. The bar-like structures running diagonally across the images are the silicon dimer rows. The six bright protrusions are due to adsorbed carene molecules. The protrusions all have a characteristic appearance, consisting of a bright round maximum with a smaller tail asymmetrically positioned off to one side. Four of the molecules are of the same orientation while the other two exhibit the same features but rotated by 180°. The observation of these two different orientations ensures that the tail is not a tip shape artifact. By comparison with the calculated geometries in Figure 5, the larger protrusion can be assigned as primarily due to the *gem*-dimethyl group while the smaller protrusion is due to the C3 methyl. While two orientations of the molecule are observed in the image, these are identical structures related by a rotation and hence with the same chirality. Inspection of several hundred molecules indicates that reaction at one face of the molecule is indeed blocked, confirming that the reaction is enantiospecific. As the observed chirality of the adsorbed molecules is inconsistent with the bonding geometry of Figure 5c we can conclude that this product is not obtained. Recalling that the calculated adsorption energies for the two unbridged structures are nearly equal, it is evident that asymmetric induction

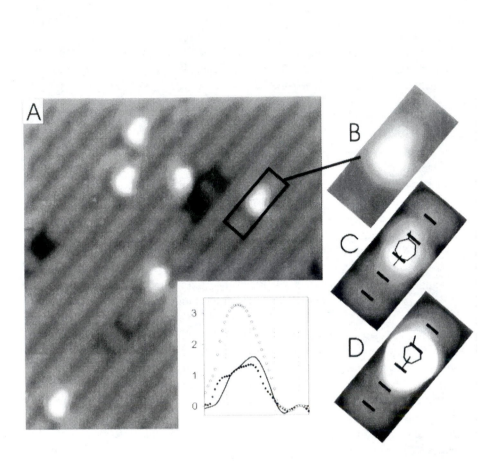

Figure 6. A) STM image (100Å², -2V, 100pA) of 1S(+) 3-carene molecules on Si(100). The inset shows the measured height profile (solid line) through a carene molecule along with calculated profiles for bridged (filled circles) and unbridged (unfilled circles). B) Enlarged view of a single adsorbed carene molecule. C) Simulated image for the bridging geometry of Figure 5d. D) Simulated image for the unbridged geometry of Figure 5b. The gray scale in the two simulated images is the same. The saturation (to white) of the unbridged simulated image reflects the much higher maximum predicted for that structure.

has been achieved as a consequence of kinetic rather than thermodynamic factors.

In order to distinguish between the bridged (5d) and unbridged (5b) bonding configurations the observed molecular features were compared to simulated images for these two geometries. The simulated images are obtained from constant electron density contours based on the highest occupied molecular orbitals. Within the Tersoff-Hamman approximation (36), these electron density contours correspond to a constant current STM image. We have previously demonstrated the success of this approach in simulating images of adsorbed benzene on Si(100) (17). While both of the simulated images exhibit the correct chirality, the calculated image for the bridged geometry (Figure 6c) is seen to give better agreement with the observed features. In the simulated image for the unbridged configuration (Figure 6d) the C3 methyl protrusion which gives the image its asymmetric appearance is relatively less distinct with the image being dominated by the protrusion due to the C7 dimethyl group. As seen by the height profiles shown in the inset in Figure 6, the largest distinction between the two simulated images is in the predicted maximum height with respect to a clean dimer; 1.3 Å and 3.3 Å for the bridged and unbridged configurations, respectively. The average maximum height of the molecules measured in the STM images varies somewhat with bias voltage, 1.71±0.12 Å at -2V and 1.37±0.10 Å at -1.5V. The good agreement between the measured and calculated profiles suggests that the molecule adopts the bridging geometry. We note that comparison of infrared spectra in the C-H stretching region with calculations for both geometries did not allow the bonding configuration to be definitively assigned.

If in fact the molecule adopts the bridging configuration, it is likely that the unbridged configuration is a metastable precursor to the more stable geometry. Unlike the case of benzene/Si(100), where a slow conversion from a single dimer to bridging geometry was observed at room temperature (37,38), no time dependent change in the adsorbate bonding geometry has been observed here. Direct adsorption into the bridge state at room temperature would imply that the barrier for breaking the cyclopropyl ring and forming two additional Si-C bonds is less than 0.85eV. The cyclopropyl ring opening is somewhat surprising as cyclopropane has been reported to not react with the Si(100) surface (39). Evidently, favourable positioning of the single dimer-bound species together with the opportunity to make repeated attempts at cyclopropyl-dimer reaction leads to ring opening in the case of carene. We speculate that similar ring opening reactions may serve to anchor various other organic molecules on Si(100).

We have shown that the addition of 1S(+)-3-carene to the Si(100) surface is enantiospecific, resulting in the formation of a chiral surface. This is a significant step toward the design of surfaces capable of chiral recognition. This work also demonstrates that STM is a useful tool for determining the enantioselectivity of a reaction and, together with calculations of the geometric and electronic structure, can be used to extract details of the bonding geometry for a relatively complex molecule. In the present case, 1S(+)-3-carene is most likely adopting a bridging geometry involving an addition reaction of the double bond followed by opening of the cyclopropyl ring, resulting in the creation of four new asymmetric centers with complete enantiospecificity. This bonding configuration is predicted to be highly stable suggesting that carene saturated Si(100) surfaces may be stable to air exposure.

References

1. Wolkow, R. A. *Ann. Rev. of Phys. Chem.* **1999**, *50*.

2. Linford, M. R.; Fenter, P.; Eisenberger P. M.; Chidsey, C. E. D. *J. Am. Chem. Soc.* **1995**, *117*, 3145-55.

3. Becker, R. S.; Golovchenko, J. A.; Swartzentruber B. A. *Nature* **1987**, *325*, 419-21.

4. Eigler, D. M.; Schweizer, E. K. *Nature* **1990**, *344*, 524-26.

5. Jung, T. A.; Schlitter, R. R.; Gimzewski, J. K. *Nature* **1997**, *386*, 696-8.

6. Lopinski, G. P.; Moffatt, D. J.; Wayner, D. D. M.; Wolkow, R. A. *Nature* **1998**, *392*, p 909.

7. Binnig, G.; Rohrer, H.; Gerber, Ch.; Weibel E. *Phys. Rev. Lett.* *1982, 49*, 57-60.

8. Becker, R.;Wolkow, R. A. In *Scanning Tunneling Microscopy*; Stroscio, J. A.; Kaiser, W. J., Eds.; Academic Press, 1993; vol 27 in "Methods of Experimental Physics".

9. M. J. Bozack, P. A.; Taylor, W. J.; Choyke; Yates, J. T. Jr. *Surf. Sci.* **1986**, *177*, L933-7.

10. Yoshinobu, J.; Tsuda, H.; Onchi, M.; Nishijima, M. *J. Chem. Phys.* **1987**, *87*, 7332-40.

11. Clemen, L. et al., *Surf. Sci.* **1992**, *268*, 205-16.

12. Huang, C.; Widdra, W.; Weinberg, W. H. *Surf. Sci.* **1994**, *315*, L953-8.

13. Fisher, A. J.; Blochl, P. E.; Briggs, G. A. D. *Surf. Sci.* **1997**, *374*, 298-305.

14. Pan, W.; Zhu, T.; Yang, W. *J. Chem. Phys.* **1997**, *107*, 3981-5.

15. Kiskinova, M.; Yates, J. T. *Surf. Sci.* **1995**, *325*, 1-10.

16. Dewar, M. J. S.; Thiel, W. *J. Amer. Chem. Soc.* **1977**, *99*, 4899-907; Davis, L. P. et.al. *J.Comp. Chem.* **1981**, *2*, 433-8; Dewar, M. J. S.; McKee, M. L.; Rzepa, H. S. *J. Amer. Chem. Soc.* **1978**, *100*, 3607; Dewar, M. J. S.; Zoebisch, E. G.; Healy, E. F. *J. Amer. Chem. Soc.* **1985**, *107*, 3902-9; Dewar, M. J. S.; Reynolds, C. H. *J. Comp. Chem.* **1986**, *2*, 140-8.

17. Wolkow, R. A.; Lopinski G. P.; Moffatt, D. J. *Surf. Sci.* **1998**, *416*, L1107.

18. Wolkow, R. A.; Avouris, Ph. *Phys. Rev. Lett.* *1988*, *60*, 1049-52.

19. Boland, J. J. *Surf. Sci.* **1992**, *261*, 17-25.

20. Mayne, A. J.; Avery, A. R.; Knall, J.; Jones, T. S.; Briggs, G. A. D.; Weinberg, H. *Surf. Sci.* **1993**, *284*, 247-56.

21. Liu, H.; Hamers, R.J. *J. Am. Chem. Soc.* **1997**, *119*, 7593-4.

22. Wolkow, R. A. *Phys. Rev. Lett.* **1992**, *68*, 2636-9.

23. Yoshinobu, J.; Tsuda, H.; Onchi, M.; Nishijima, M. *J. Chem Phys.* **1987**, *87*, p 7332.

24. Clemen, L.; Wallace, R. M.; Taylor, P. A.; Choyke,W. J.; Weinberg,W. H.; Yates, J. T. *Surf. Sci.* **1992**, *268*, p 205.

25. Hamers, R. J.; Hovis, J. S.; Lee, S.; Liu, H.; Shan, J. *J. Phys. Chem.* **1997**, *B101*, p 1489.

26. Hovis, J. S.; Hamers, R. J. *J. Phys. Chem.* **1997**, *B101*, p 9581.

27. Lopinski, G. P.; Moffatt, D. J.; Wayner, D. D. M.; Zgierski, M. Z.; Wolkow, R. A. *J. Am. Chem. Soc.* **1999**, *121*, p 4532.

28. Ihm, J.; Cohen, M. L.; Chadi, D. J. *Phys. Rev. B* **1980**, *21*, p 4592.

29. Liu, Q.; Hoffmann, R. *J. Am. Chem. Soc.* **1995**, *117*, p 4082.

30. Konecny, R.; Doren, D. J. *J. Chem. Phys.* **1997**, *106*, p 2426.

31. Tse, J.; Frapper, G. (unpublished).

32. Fang, H.; Giancarlo, L. C.; Flynn, G. W. *J. Phys. Chem.* **1998**, *B102*, p 7311.

33. Frisch, M. J. et al., Gaussian 94 (Revision D.1) (Gaussian Inc., Pittsburgh PA, 1995).

34. Chabal, Y. J.; Raghavachari, K. *Phys. Rev. Lett.* **1984**, *53*, p 282.

35. Weldon, M. K.; Stefanov, B. B.; Raghavachari, K.; Chabal, Y. J. *Phys. Rev. Lett.* **1997**, *79*, p 2851.

36. Tersoff; J.; Hamman, D. R. *Phys. Rev. Lett.* **1983**, *50*, p 1998.

37. Lopinski, G. P.; Fortier, T. M.; Moffatt, D. J.; Wolkow, R. A. *J. Vac. Sci. Technol.* **1998**, *A16*, p 1037.

38. Borovsky, B.; Krueger, M.; Ganz, E. *Phys. Rev. B* **1998**, *57*, R4269.

39. Yates, J. T. *J. Phys.:Condens. Matter* **1991**, *3*, S143-156.

Chapter 21

Chiral Autocatalysis and Selectivity
in the Vicinity of Solid Surfaces

Dilip K. Kondepudi

**Department of Chemistry, Wake Forest University, Box 7486,
Winston-Salem, NC 27109 (email: dilip@wfu.edu)**

Spontaneous chiral symmetry breaking in stirred crystallization is a
general phenomenon. It occurs in crystallization from a solution, as in
the case of $NaClO_3$, and from a melt, as in the case of 1,1'-binaphthyl.
As far as we can tell, it occurs in all crystallizations of achiral
molecules that crystallize in chiral forms and in conglomerate
crystallization of chiral compounds that racemize rapidly compared to
the rate of crystallization. The autocatalysis that drives the observed
chiral symmetry breaking is in the processes of secondary nucleation,
the generation of new crystal nuclei in the vicinity of an existing
crystal. Secondary nucleation is highly chirally selective.

Introduction

Chemical reactions that are symmetric under the interchange of
enantiomers, i.e. reactions that do not favor the production of one enantiomer
over the other can yet generate a large enantiomeric excess in the product. This
is the phenomenon of spontaneous chiral symmetry breaking in which the
product breaks the symmetry of the kinetics that generated it. In mathematical
terms, spontaneous chiral symmetry breaking is said to occur when the solutions
to the kinetic equations do not possess the symmetries of the kinetic equations,

i.e., even though the kinetic equations are invariant under the interchange of the two enantiomers, its solutions do not have the same symmetry. What kind of kinetics exhibit this phenomenon? Spontaneous chiral symmetry may occur if the kinetics have two basic features: chiral autocatalysis and "competition" through which the growth of one enantiomer directly or indirectly inhibits the growth of the opposite enantiomer. When these two features are present, a small random fluctuation that increases the amount of one enantiomer could grow and establish a large enantiomeric excess.

Spontaneous Chiral Symmetry Breaking in Stirred Crystallization

Though several theoretical kinetic models have been proposed for spontaneous chiral symmetry breaking nearly fifty years ago (most of them to explain the possible origin of biomolecular chiral asymmetry(1-3)), laboratory realization and studies of this phenomenon have been relatively recent. It can most easily be observed in stirred crystallization of achiral compounds which crystallize in enantiomeric forms.

Stirred Crystallization of NaClO$_3$

The achiral compound NaClO$_3$ which crystallizes in enantiomeric forms is particularly convenient to study: the optical activity of its crystals can easily be detected because they are isotropic. If crystallization of NaClO$_3$ is performed in a static solution, statistically equal number of l- and d-crystals are found as Kipping and Pope (4) showed a hundred years ago (figure 1(a)). The probability distribution of the crystal enantiomeric excess (CEE = $(N_l-N_d)/(N_l+N_d)$, in which N_l and N_d are the number of l- and d-crystals respectively) is centered at zero. If the crystallization is performed while the solution is continuously stirred, however, one finds that almost all crystals (more than 98% in most cases) are either l or d. Which type, l or d, will dominate in a particular stirred crystallization is found to be random (5). The results of a number of crystallizations are summarized in figure 1(b).

This dramatic difference between CEE generated in stirred and unstirred crystallizations is due to the process of *secondary nucleation* (6,7). In a supersaturated solution, a crystal catalyzes the generation of new crystal nuclei in its vicinity. The nuclei thus generated are called secondary nuclei, and the process, secondary nucleation. In a stirred solution, the secondary nuclei are dispersed by the fluid motion and one sees the formation of a cloud of small crystals. This process can be so effective that most of the excess solute may crystallize out as secondary nuclei in a few minutes. Secondary nucleation, which is clearly autocatalytic, becomes *chirally* autocatalytic in the

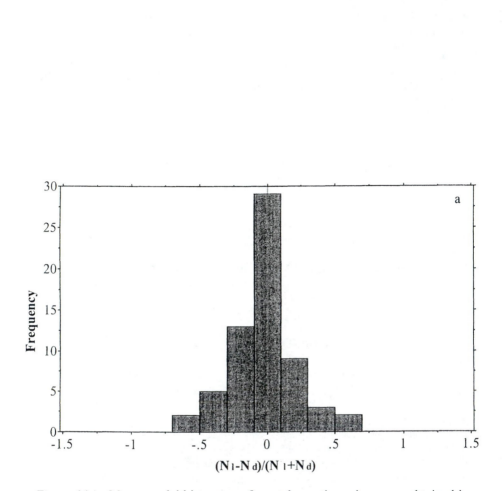

Figure 1(a). Mono-modal histogram of crystal enantiomeric excess obtained in 63 unstirred crystallization of $NaClO_3$. (Reproduced from reference 10. Copyright 1993, American Chemical Society)

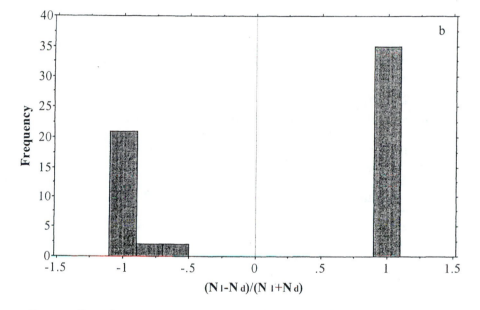

Figure 1(b). Bi-modal histogram of crystal enantiomeric excess obtained in 60 stirred crystallization. (Reproduced from reference 10. Copyright 1993, American Chemical Society)

crystallization of achiral compounds that crystallize in chiral forms. Thus, in stirred crystallization, the "first crystal" that is able to generate secondary nuclei does so with a high degree of chiral selectivity and produces a large number of crystals of the same enantiomorphic form as itself, in a few minutes. This rapid crystallization causes the concentration to drop to a level at which the nucleation rate is virtually zero. Consequently, almost all the crystals generated are overwhelmingly of the same enantiomeric form, viz., that of the "first crystal". In the absence of any chiral influences, the first crystal is l or d with equal probability resulting in the bi-modal probability distribution as figure 1(b) shows.

The transition from a mono-modal probability distribution, shown in figure 1(a), to a bi-modal probability distribution, shown in figure 1(b), as a function of stirring rate has been studied (8). A similar study, in a somewhat different set up, was reported by Martin et al. (9). Though the kinetics of the rates of nucleation and crystal growth are identical for the two enantiomers, because of chiral autocatalysis and rapid drop in concentration (which indirectly inhibits the formation of crystal nuclei that are of opposite enantiomeric form with respect to the "first crystal"), the product is asymmetric with high CEE. The system's kinetic equations which are symmetric under the interchange of l and d enantiomorphs have solutions that do not possess the same symmetry; the large CEE of the product *breaks* the symmetry of the kinetic equations. These qualitative ideas can be used as the basis for a stochastic-kinetic equations that reproduce many of the experimentally observed features well (8,10).

Stochastic Kinetic Equations

Kinetic equations of stirred crystallization, which incorporate the stochastic nature of the process, can be formulated using an empirical rate law for secondary nucleation. There are basically four processes that drive the crystallization: (i) solvent evaporation (ii) homogeneous and heterogeneous primary nucleation (iii) stirring-dependent secondary nucleation and (iv) crystal growth. The kinetic equation can be written in terms of the following variables:

M_W = moles of water in the solution.

M_S = moles of solute in the solution.

$C = M_S/M_W$ the concentration.

C_S = the concentration at saturation.

$E(T)$ = the evaporation rate constant (moles/sec/unit area) which is a function of temperature, T.

A = area of the solution surface at which the solvent evaporates.

N_l = the number of l-crystals.

N_d = the number of d-crystals.

σ_l = the total surface area of the l-crystals.

σ_d = the total surface area of the d-crystals.

$P_x(T,C)$ = stochastic primary nucleation rate which is a function of temperature, T, and concentration, C (x= d or l for the enantioforms).

H = stochastic nucleation rate for nuclei growing at the solution surface
S_x (T,C,s) = secondary nucleation rate which is a function of temperature, T, and concentration, C, and the stirring rate, s (x= d or l for the enantioforms).
$G(T)$ = crystal growth rate constant, a function of temperature.

With the variables thus defined the following kinetic equations are obtained:

$$\frac{dM_w}{dt} = -EA\frac{M_w}{M_w + M_s} \tag{1}$$

$$\sigma_l = \sum_{k=1}^{N_l} 4\pi r_{lk}^2 \qquad \sigma_d = \sum_{k=1}^{N_d} 4\pi r_{dk}^2 \tag{2}$$

Here r_{lk} 'radius' of the kth l-crystal etc.

$$\frac{dM_s}{dt} = -(\sigma_l + \sigma_d)\frac{dr}{dt} = -(\sigma_l + \sigma_d)G(C - C_s) \tag{3}$$

In equation (3), the linear crystal growth rate is assumed to be:

$$\frac{dr}{dt} = G(C - C_s) \tag{4}$$

For the number of l- and d-crystals we have the equations:

$$\frac{dN_l}{dt} = P_l + S_l + H \tag{5}$$

$$\frac{dN_d}{dt} = P_d + S_d + H \tag{6}$$

For primary nucleation the following rate may be used:

$$\text{Rate} = B\ Exp[\frac{-16\pi N_A \sigma_B v^2}{3RT^3[\ln(C/C_S)]^2}] \tag{7}$$

in which B is a pre-exponential factor, N_A = Avogadro number, v = molar volume of solid $NaClO_3$, σ_B= the surface energy constant and R is the gas constant. This expression is approximately the nucleation rate for heterogeneous nucleation.

Since the stochastic nature of nucleation is important in the processes being studied, we define the nucleation rates P_l and P_d in equations (5) and (6) as follows: $P_l = P_d =$ Poisson distributed random function with average value equal to:

$$B \, Exp[\frac{-16\pi N_A \sigma_B v^2}{3RT^3[\ln(C/C_S)]^2}] \qquad (8)$$

For symmetry breaking processes, the most important feature is the dependence of P_l on the supersaturation ratio C/C_S. The constants in this expression are adjusted to give the experimentally observed nucleation rate. To fully explain the observed symmetry breaking, we also found that it was necessary to introduce an additional random source of nucleation, H, as might occur on the surface of the solution (several minutes before nucleation begins in the bulk of the solution). This term is included in equations (5) and (6).

The phenomenon of secondary nucleation has been much studied in the context of industrial crystallization. For the rate of secondary nucleation, the following empirical rate law is often used:

$$S_l = s\sigma_l K_s (C - C_s)^\alpha \qquad (9)$$

Here s is the stirring rate, K_S is a constant that may depend on the temperature, σ_x is the total surface area of the crystals of type x (x being l or d). Empirical values of the exponent α range from 0.5 to 2.5 depending on the compound.

It has also been reported in the literature that secondary nucleation seems to occur at a significant rate only when the crystal reaches a minimum size, R_{min}. Indeed, in our numerical simulation of this process we found that the introduction of a minimum size $R_{min} = 0.09$ cm was necessary to fit the results of the simulation to the experimental data.

The set of equations (1)-(9), which include stochastic nature of the crystallization process, could be used to simulate the crystallization process. A Mathematica computer code based on these equations was written and its predictions were compared with the experimental data. The code is capable of predicting the time-variation of concentrations and the number and size distribution of l- and d-crystals at any time during the crystallization. The comparison between simulation and experimental data is shown in figure 2. The values of the parameters used in the simulation are: $C_S = 0.2099$; $E = 1.38 \times 10^{-5}$ mol/s; $G = 13.13 \times 10^{-3}$ cm/s/unit conc.; $B=Exp[-3.4]$; $\sigma_B = 2.06 \times 10^{-24}$; $v=45.24$ml/mol; $K= 5.0 \times 10^8$; $s =2$; $\alpha = 2.75$; $R_{min} = 0.09$cm. In the simulation data shown in figure 2, a random crystal of size 0.05 cm was found at about 55 min and this crystal became the "seed" for the generation of secondary nuclei.

The relation between the stirring rpm and the probability distribution of the crystal enantiomeric excess is shown in figure 3(a). From this figure it is clear that as the stirring rpm increases from zero to 900, the probability distribution which is mono-modal at zero starts to becomes bi-modal at about 200rpm and becomes sharply bi-modal at 900rpm. These features can also be reproduced in the numerical simulation of the stochastic kinetic equations, as shown in figure

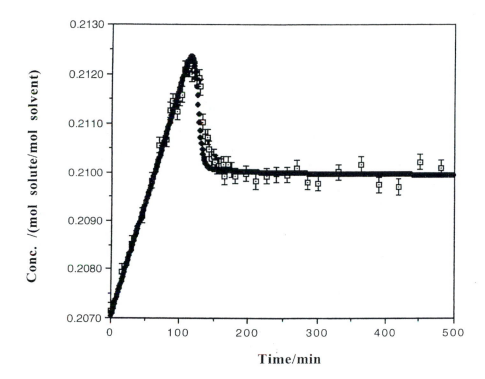

Figure 2. Time variation of concentration of $NaClO_3$ solution in stirred crystallization. Points with error bars are experimental data points and filled points are the results of numerical simulation of the stochastic kinetic equations. (Reproduced from reference 8. Copyright 1995, American Chemical Society)

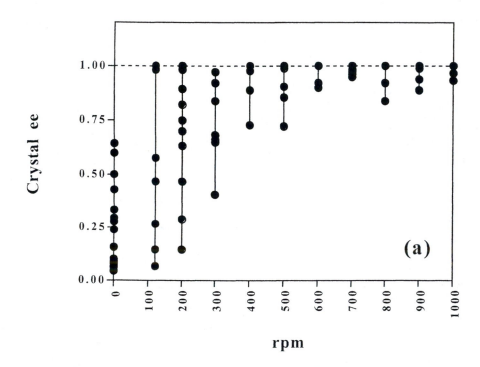

Figure 3(a). Crystal enantiomeric excess obtained in stirred crystallization at various stirring rates. (Reproduced from reference 8. Copyright 1995, American Chemical Society)

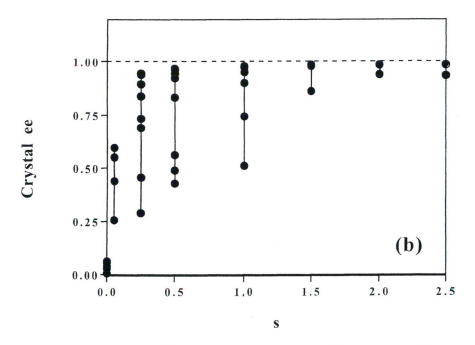

Figure 3(b). Results of computer simulation using the stochastic kinetic equations. Data show crystal enantiomeric excess versus stirring-rate parameter s. Comparison of this data with that shown in figure 3(a) leads to the approximate relation stirring rpm = 600s (Reproduced from reference 8. Copyright 1995, American Chemical Society)

3(b). In the simulation, the parameter s in equation (9) is assumed to be proportional to the stirring rpm. The averages crystal enantiomeric excess of the experimental data at each rpm can be made to fit with that obtained through simulation if rpm = 600s. Thus, an approximate relationship between the parameter s and the stirring rpm can also be obtained through the simulation.

Stirred Crystallization of 1,1'-Binaphthyl Melt.

Chiral symmetry breaking in stirred crystallization can also be realized in crystallization from a melt. 1,1'-binaphthyl is a chiral compound that racemizes rapidly at its melting point(158°C) and it crystallizes as a conglomerate of l- and d-enantiomers. In benzene, at 20°C, the racemization half-life is about 9 hours. Hence, at room temperature, a single crystal (which will consist exclusively of either l- or d-enantiomer) can be dissolved in a solvent, such as benzene, and the optical rotation of the solution can be measured easily. If a sample of 1,1'-binaphthyl is melted and recrystallized, the conglomerate crystals will be randomly made of l- or d-enantiomers. If a large number of independently nucleated crystals are derived in a crystallization, the optical activity of the entire sample would be statistically zero. For small samples that produce only a few crystals, the statistical deviations from zero could be large, but the overall probability distribution of optical activity of a large number of independent crystallizations will be a mono-modal distribution centered around zero, as was shown by Pincock et al. (11). In contrast, if several 2.0g samples of 1,1'-binaphthyl melt are crystallized while the melt is continuously stirred, the optical rotation of each sample could be very large (12). The probability distribution of the specific rotation is bi-modal (figure 4(c)) with EE of most crystallizations being above 80%.

Clearly, secondary nucleation occurs in stirred crystallization from a melt as it does in the crystallization from a solution and, as in the case of crystallization of $NaClO_3$ from a solution, it is a chirally autocatalytic process. There is a strong enantioselectivity for nuclei formed in the vicinity of a crystal. Such strongly enantioselective nucleation also occurs in the vicinity of crystals of other compounds: when seed crystals of l- or d- 1,1'-bi-2-naphthol (whose melting point is much higher than that of 1,1'-binaphthyl) are added to the melt before stirred crystallization was performed, the EE of the resulting 1,1' - binaphthyl crystals was above 80% and correlated strongly to the enantioform of the seeds(12). Thus, the surface of the seed crystals act as highly enantioselective catalysts for nucleation. Furthermore, a small amount of seed crystals are able to produce a much larger amount of crystals with high EE. In such a process, though there is no amplification of EE, there is considerable increase in the amount of material with high EE. The process also demonstrates how EE from one compound can "propagate" to another compound through crystallization.

From Chiral Crystals to Chiral Compounds: Propagation of Chiral Asymmetry

Spontaneous chiral symmetry breaking in NaClO3 results in a large number of crystals of the same enantiomeric form. Since secondary nucleation is a general phenomenon, we can expect similar results for any achiral compound that crystallizes in enantiomeric forms. However, unlike crystals of NaClO3, crystals of most compounds are not isotropic in their optical properties and this makes the identification of l- and d-crystals difficult. In addition, the product of stirred crystallization is often a powder, making it virtually impossible to detect the optical rotation of individual crystals. If the achiral molecules that constitute the enantiomorphic crystals could be converted to chiral molecules through appropriate reactions involving the crystals, then we not only have a way of *propagating* the chirality of the crystals to that of the product molecules but we also have a method for measuring the amount l- and d-forms of the crystals.

Such propagation of chirality from crystal structure to molecular structure can be accomplished in the case of dimethyl-chalcone (1) (figure 5) which contains a carbon-carbon double bond. This achiral compound crystallizes from solution in enantiomeric forms. Bromination in the solid phase using bromine vapor converts this compound to a chiral compound (2) (figure 5). It was shown by Penzien and Smith (13) that, if a single crystal of dimethyl chalcone is made to react with bromine vapor, the chirality of the crystal structure is transferred to the product 2, with a specific rotation of about $[\alpha]_D=10°$ corresponding to an EE of 2 of about 6%. Though the chiral selectivity is not very high, this reaction demonstrates how the chirality of a crystal structure can be transferred to the product. If stirred crystallization of 1 produces a large amount of crystal enantiomeric excess, as it did in NaClO3, then bromination of these crystals using bromine vapor should result in the product having a corresponding EE. A specific rotation of about 10° will indicate that nearly 100% of the crystals have the same handedness. Figure 6 summarizes results obtained recently (14). The results clearly show that bromination of crystals of 1 obtained through stirred crystallization gives product 2 with a specific rotation in the vicinity of 10°. Since bromination of a single crystal produces a specific rotation of about 10° (13) we see that the crystal enantiomeric excess of crystals of 1 generated through stirred crystallization was nearly 100% and that this asymmetry is propagated to the chiral product 2. This reaction provides a specific example of how spontaneous chiral symmetry breaking realized in stirred crystallization can result in the generation of EE. In this case, the EE of the product is only about 6%, but in other systems one might find a much larger EE due to a much higher enantioselectivity in the reaction with chiral crystals.

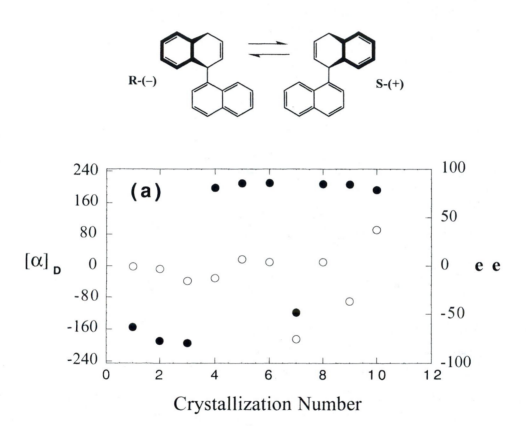

Figure 4. Spontaneous chiral symmetry breaking also occurs in the stirred crystallization of 1,1'-binaphthyl melt. (a) 1,1'-binaphthyl is a chiral molecule that can interconvert rapidly at its melting point. (b) and (c) show the histogram of specific rotation for unstirred and stirred crystallizations respectively. $[\alpha]_D$ = 240° corresponds to 100% EE. (Reproduced from reference 12. Copyright 1999, American Chemical Society)

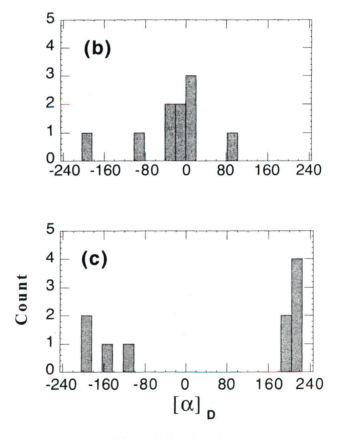

Figure 4. *Continued.*

316

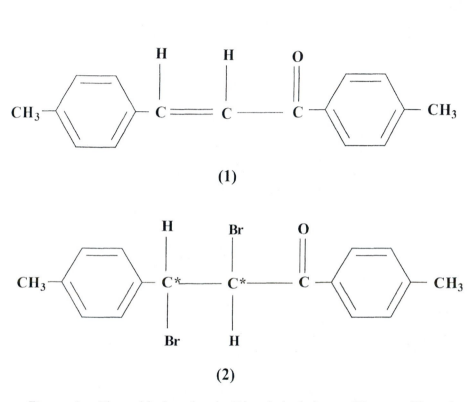

(1)

(2)

Figure 5. The achiral molecule Dimethyl chalcone (**1**) crystallizes in enantiomorphic forms. If the crystals are brominated using Br_2(gas), the chiral compound (**2**) is obtained. The bromination of a single crystal of 1 gives the chiral product with about 6% EE.

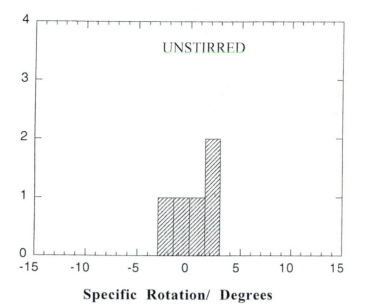

Figure 6(a). Histogram of specific rotations of **2** obtained by bromination of crystalline dimethyl chalcone (**1**) through unstirred crystallization.

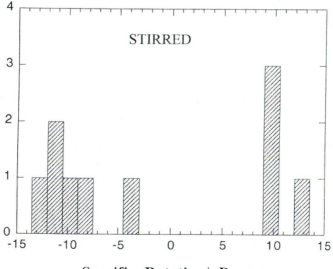

Figure 6(b). Histogram of specific rotations of **2** obtained by bromination of crystalline dimethyl chalcone (**1**) through stirred crystallization. (Bromination of enantiopure single crystal gives a specific rotation of about 10°.)

318

Concluding Remarks

All the above systems show chirally autocatalytic nucleation. They clearly show that nucleation near the surface of a chiral solid can be highly chirally selective. In addition to the above systems, chiral autocatalysis was also reported by Asakura et al. (15-17) in the synthesis of chiral cobalt complexes. However, the exact mechanism that results in a high degree of chiral selectivity is not yet well understood. For the general process of secondary nucleation a basic theory has been proposed by Qian and Botsaris (18). Theories of this kind could act as the basis for an investigation of chiral selectivity in secondary nucleation.

This work was supported by NSF (grant CHE-9527095).

References

1. Frank, F. C.:, *Biochem. Biophys. Acta.*, **1953**, *11*, 459.
2. Bonner, W. A.:, *Origins of Life and Evol. Biosphere, 1991*, *21*, 59.
3. Bonner, W. A.:, *Origins of Life and Evol. Biosphere*, **1994**, *24*, 63.
4. Kipping, F.S. and Pope, W.J.:, *J. Chem. Soc. (London) Trans.*,**1898**, *73*, 606.
5. Kondepudi, D.K., Kaufman, R. and Singh, N:, *Science* **1990**, *250*, 975
6. Jancic, S.J. and Grootscholten, P.A.M.:, *Industrial Crystallization*, D. Reidel Pub. Co., Boston, MA,1984
7. *Separation and purification by crystallization*; Botsaris, G.D.;Toyokura, K. Eds. ACS Symposium Series 667; American Chemical Society, Washington, D.C. 1997
8. Kondepudi, D.K., Bullock, K., Digits, J. et al.:, *J. Am. Chem. Soc.* **1995**, *117*, 401.
9. Martin, B., Tharrington, A. and Wu, X-l :, *Phys. Rev. Lett.* **1996**, *77*, 2826.
10. Kondepudi, D.K., Bullock, K.L., Digits, J.A. et al.:, *J. Am. Chem. Soc.* **1993**, *115*, 10211.
11. Pincock, R. E., Perkins, R. R., Ma, A.S. et al.:, Science, **1971**, *174*, 1081
12. Kondepudi, D.K., Laudadio, L. and Asakura, K.:, *J. Am. Chem. Soc.* **1999**, *121*, 1448.
13. Penzien, K. and Schmidt, G.M. J.:, *Angew. Chem. Int. Ed. Engl* **1969**, *8*, 608.
14. Durand, D., Kondepudi, D.K., Morier, P. and Quina, F.H.: **2000**, Preprint
15. Asakura, K.; Kobayashi, K.; Mizusawa, Y., et al..:, *Physica D*, **1995**, *84*, 72.
16. Asakura, K.; Inoue, K., Osanai, S.; et al.:, *J. Coord.* **1998**, *46*, 159.
17. Asakura, K; Kondepudi, D.K.; and Martin, R.:, *Chirality*, **1998**, *10*, 343.
18. Qian, R-Y. and Botsaris, D.; *Chem. Eng. Sci.* **1997**, *52*, 3429, *Chem. Eng. Sci.* **1998**, *53*, 1745;.

Chapter 22

Chirality in Giant Phospholipid Tubule Formation

Britt N. Thomas[1], Janet E. Kirsch[2], Chris M. Lindemann[1],
Robert C. Corcoran[2], Casey L. Cotant[2], and Phillip J. Persichini[3]

[1]Department of Chemistry, Louisiana State University,
Baton Rouge, LA 70803
[2]Department of Chemistry, University of Wyoming, Laramie, WY 82071
[3]Department of Chemistry, Allegheny College, Meadville, PA 16335

The role molecular chirality plays in tubule formation is probed through the development of a new class of tubule-forming lipids. While the new compounds produce tubules under the same mild conditions as does the prototypical tubule-forming compound, DC(8,9)PC, several important morphological differences appear in tubules formed by these compounds. Atomic force- and optical microscopy show the new tubules' diameters to be about twice as great as those made with DC(8,9)PC, which can be interpreted in accordance with a "chiral packing" class of tubule structure theory. As with DC(8,9)PC, a helical trace is found upon the new compounds' tubule exteriors, imparting a sense of chirality to these microscopic cylinders. Surprisingly, enantiomerically-pure preparations of the new compounds contain helices of both chiral senses, contrary to the heretofore-observed relationship between tubule helix handedness and phospholipid chirality, and inconsistent with the chiral packing theory. Another unexpected result is that one of the new tubule-forming molecules can generate tubules with diameters some ten times of that of those previously reported.

Introduction

Chirality can emerge in surprisingly simple systems, such as close-packed spherical balls confined in cylinders (*1*), and is found at all lengthscales in nature, from the macroscopic helical vine to the chiral seashell down to the single molecule. A topic of intense theoretical and experimental interest that spans the vast range of lengthscales from the molecule to the macroscopic is the self-assembled cylindrical structure known as the "tubule" (*2-4*). These structures possess micron-dimensioned chiral features that reflect the chirality of the molecules from which they self-assemble, and thus present an extraordinary opportunity to elucidate how the Angstrom-scaled property of molecular chirality may be expressed as a macroscopic feature. Attendant with this opportunity, however, are a number of fundamental challenges to the experimentalist and the theorist alike.

The primary challenge confronting the experimentalist is that a molecule's chirality is not an easily accessible, continuously-adjustable parameter. Sometimes the regulation of a system's *net* chirality, through the stoichiometric control of enantiomeric excess, is sufficient to modulate the system's chiral behavior. For example, the degree of chiral twist found in multilayer ribbons made from a gemini surfactant was recently found to be continuously adjustable by varying the amounts of chiral counter-ions present in the surrounding solution (*5*). Such straightforward behavior is not the rule, however, and in general the experimental control of chiral expression is difficult to achieve.

The challenges theorists face in describing these complex structures over lengthscales extending through several orders of magnitude are formidable. For the tubule system we focus upon here, these difficulties are compounded by the recently emerging picture that quite different formation and growth mechanisms dominate at different times in a tubule's growth. Most importantly, the first of these mechanisms appears to be insensitive to the molecule's chirality (*6*). These recent results require re-evaluation of the assumption that tubules are equilibrium structures, and impose the heavy demand that a mechanistic theory be developed.

To elaborate further on these experimental and theoretical challenges it is necessary to more thoroughly describe phospholipid tubules.

Phospholipid tubules

Saturated ethanolic / water solutions of the synthetic diacetylenic phospholipid, 1,2-bis(10,12-tricosadiynoyl)-sn-glycero-3-phosphocholine (hereafter "DC(8,9)PC", **1**; see Figure 1) self-assemble upon cooling into hollow cylindrical tubes, highly unusual for lipids, of typical length 10 μm < L < 100 μm and diameter D ≈ 0.6 μm (*2-4*). The first step of tubule formation under these conditions is the eruption of helically-wound ribbons directly from the cooling spherical vesicle, as shown in Figure 2. These helices grow axially at about 1 μm per second, and over the course of seconds, the ribbons widen to form closed cylinder, leaving a helical trace reminiscent of a

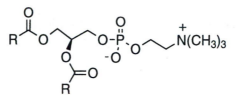

<u>1</u>

<u>2</u>

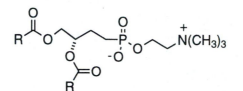

<u>3</u>

R= (CH₂)₈-C≡C-C≡C-(CH₂)₉CH₃

*Figure 1. DC(8,9)PC (top, compound, **1**), its "C4" derivative (**2**) and its "C3" derivative, (**3**).*

322

paper drinking straw. This process draws all of the material from the spherical vesicle. Minutes later, these cylinders are ensheathed by lipid from the cooling saturated lipid solution; the axial growth rate for this process is an order of magnitude slower than that of the inner cylinder growth (6). This ensheathment is repeated, and the result is a set of coaxially-nested cylinders formed by very different mechanisms.

A variety of novel potential applications derive from tubule structural features: Tubules' potential utility in nanofabrication, purification, medical, and encapsulation applications (2, 7, 8) fuels much of the interest in current tubule research. Tubules may be aligned with magnetic or electric fields (9, 10) and their capacity to be metal-plated (11) suggests tubules' potential in microelectronics as well as magnetic template applications. The diacetylene

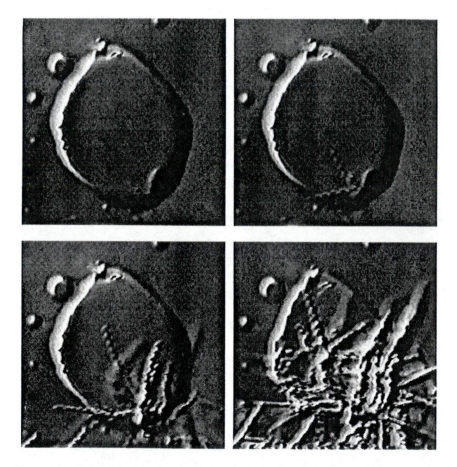

Figure 2. Videomicroscopic sequence of tubules erupting from a large, cooling DC(8,9)PC vesicle. The frames are evenly spaced over 12.4 seconds.

groups within the hydrocarbon tails are easily polymerized and may thereby enhance the mechanical, electrical and optical properties of tubules. Furthermore, tubules' hollowness suggests medicinal (*12*) and industrial encapsulation applications (7, *8*), as well as filtration and purification applications (*13*).

Realization of the full technological potential of tubules requires the ability to optimize their morphology for a given application. Tubule length is controllable over a range of a few μm to hundreds of μm through solvent composition (*14*) or through the precise control of the rate at which the L_α-phase spherical vesicles are cooled to form the L_β'-phase cylindrical tubule (*15*).

However, tubule diameter has proven to be insensitive to regulation through control of solvent composition, lipid concentration, co-surfactants or kinetics (*15*). The regulation of tubule diameter is crucial in determining their suitability for technological applications, *e.g.*, if tubules are to be used as encapsulation agents, the diameter directly affects their "payload" capacity, the diffusion rate of encapsulated material, and the spatial dispersion of tubules in aerosols (*12*).

Some guidance regarding how tubule diameter may be varied is found in a theory in which the chiral shape of the tubule-forming molecule influences the packing of the molecules as they form bilayer ribbons (16, *17*). This chiral packing produces a twist in the ribbon that ultimately leads to the helical winding the ribbon assumes, and it also determines the tubule diameter. A key prediction is that mirror-image enantiomers yield mirror-image helices, which was confirmed experimentally. When chiral packing effects are reduced to zero in this theory, the tubule diameter is predicted to diverge to infinity, that is, membrane curvature goes to zero, and so flat sheets rather than tubules are expected to form.

A test of the chiral-packing tubule structure theory illustrates the difficulties faced in addressing molecular chirality as an experimental parameter. To nullify the effects of chirality, a net zero-chiral environment was created by mixing equal amounts of the *R*- and *S*- enantiomers of the chiral tubule-forming molecule. Surprisingly, however, rather than producing flat sheets as predicted by theory, the racemic mixture yielded left- and right-handed tubules of unchanged diameter, with left- and right-handed tubules found in the product. It was inferred from this outcome that a spontaneous enantiomeric separation preceded tubule formation in the racemate (*18*). That is, despite the experimenter's efforts to negate the effects of molecular chirality, its apparent robustness produced chiral microenvironments in which tubules formed through efficient chiral packing.

Violations of the correspondence between molecular chirality and helix handedness have been reported recently in enantiopure DC(8,9)PC preparations (*6*), in an unrelated chiral cholesteric bile salt systems (*19*) and in the new molecules described in this chapter. These results require a reinterpretation of the enantiomer-mixing results and lend some support to a competing class of chiral symmetry breaking theories.

324

In chiral symmetry-breaking models, the origin of membrane chirality, and hence membrane twist and helix handedness, does not lie in chiral intermolecular interactions, but rather in a collective tilt of the ensemble of molecules composing the lamella. How this may occur is shown in Figure 3, which schematically represents a monolayer composed of achiral rods. The

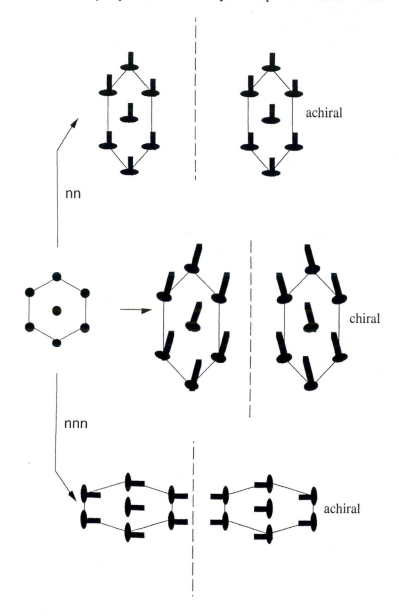

Figure 3. Schematic diagram of chiral symmetry breaking, described in the text

hexagonal arrangement at the center left-hand side of Figure 3 is a view down the membrane normal of close-packed rods that define this monolayer. One may think of these rods as close-packed pencils held normal to a tabletop, with the erasers representing the lipid headgroups in contact with the tabletop. The upward arrow (labeled "nn") leads to a depiction of the results of a collective "nearest neighbors" tilt, the ovals represent the pencil erasers. The downward arrow (labeled "nnn") depicts the outcome of a collective "next nearest neighbors" tilt. These arrangements are distinct and achiral, as shown by their reflections, drawn to the right of the dashed lines. However, for tilt directions lying between the "nn" and "nnn" directions (depicted in the middle of Figure 3 as fundamentally an "nn" tilt with a bit of "nnn" character) a chiral arrangement of rods results. In this way, chiral membranes may result from achiral molecules, and the random tilt direction chosen by the ensemble should produce a 50:50 ratio of left- to right-handed chiral structures.

While the recently-observed violations of the molecular chirality / helical handedness relationship are contrary to the predictions of chiral packing theory, and indeed lend some support to the symmetry-breaking model, the vast preponderance of tubules found in these systems are nevertheless of a single sense of helical handedness. In view of the 50:50 ratio of left- to right-handed helical structures predicted by the collective tilt / symmetry-breaking theories, neither class of tubule structure theory is entirely satisfactory. More solid definition of these systems' remarkable self-assembly behaviors must be experimentally determined before theory can reasonably be expected to advance.

The synthesis project described below began as an attempt to modify tubule diameter in accordance with the predictions of the chiral packing theory. We hoped that by making sufficiently small changes in bond lengths and bond angles near a tubule-forming molecule's chiral center, and hence to its chiral shape, the postulated chiral packing would change and cause a change in tubule morphology. In a sense, this alteration of the molecule's chiral center may be thought of as a refinement of the enantiomer-mixing experiment, but with the changes to chiral packing effected at the molecular level, rather than averaged over the entire system through stoichiometry.

As we shall describe, the hoped-for changes in tubule morphology were indeed obtained, for the tubule diameter was found to nearly double. Large numbers of open helical ribbons, that is, helically-wound ribbons whose helical pitch exceeds that of the ribbon width, were found in preparations made from each of the new molecules. Interestingly, significant numbers of left- and right-handed helices were found in all enantiopure preparations. While the diameter change is consistent with the chiral packing theory, the increased incidence of helices of the unexpected sense of handedness is not. Another surprising result is that under controlled conditions where the rate of formation is very slow, tubules with diameters of about 12 µm, approximately twenty times that of DC(8,9)PC tubules, were found in the enantiopure preparations.

Phosphonate tubule-forming compounds

Phosphonate DC(8,9)PC derivatives

The tubule monomer prototype, DC(8,9)PC, is composed of three structural units: 1) a pair of long hydrocarbon tails, each containing an apparently "enabling" diacetylene group in its midsection; 2) a chiral glycerol-derived backbone; and 3) a polar phosphatidylcholine headgroup. Structural modifications thus far have focused on the first and third units. Modifications of the head group have involved substitution of the choline portion of DC(8,9)PC by other polar groups (20, 21). For the most part, these substitutions resulted in analogs which either failed to form tubules, or which formed tubules closely resembling those formed from DC(8,9)PC. The one exception to this generalization came from the work of Markowitz and coworkers (22) in which choline was replaced by short-chain glycols $HO(CH_2)_nOH$ (n = 2, 3, 4). Tubule formation in the presence of $CuCl_2$ under rigidly defined conditions of pH and ionic strength led to a bimodal population of tubules having 0.1 μm and 0.9 μm external diameters, with the former predominating. Slight variations in tubule formation conditions again led to dramatic decreases in tubule yields or formation of tubule of morphologies very close to those formed from DC(8,9)PC.

Changes to the tail section of the parent structure have mostly centered on repositioning the di-yne moieties along the tails' lengths (23, 24), but have also included removal of the ester carbonyls (to give ether derivatives) (25, 26) as well as addition of polymerizable groups at the tips of the hydrocarbon chains. The results of these structural variations are similar to those found with modifications to the headgroup. While in many cases the modifications seem to have been so subtle as to be unnoticeable (i.e., tubules of virtually identical morphologies to DC(8,9)PC are produced), in other cases the modifications have proven to be sufficiently extreme that tubule formation does not occur; a middle ground of tubule formation with significantly altered morphologies is curiously lacking.

To our knowledge, no modifications of the glycerol backbone of the tubule monomer have been examined. The lack of interest in this portion of the monomer is somewhat surprising given the clear influence which it has on tubule morphology; DC(8,9)PC monomers having an (R)-configuration at the middle carbon of the glycerol backbone form tubules having a right-handed sense of winding, while those derived from monomers having an (S)-configuration are left-handed (27).

We have modified the chiral glycerol backbone of the DC(8,9)PC in two ways, as shown in Figure 1. First, through the replacement of the oxygen linkage to the phosphate group with a methylene group (-CH_2-) to give the phosphonate **2** which, due to the addition of a carbon atom to the three-carbon glycerol backbone, we refer to as the "C4-phosphonate". The second molecule we synthesized is the DC(8,9)PC phosphonate analog 3, in which the polar head group has been moved closer to the stereogenic center contained within the glycerol backbone. This translation is accomplished through the *removal* of the linking phosphate oxygen atom, rather than the (-CH₂-) replacement that led to phosphonate analog 2. While the (-CH₂-) -for- oxygen substitution resulted in the lengthening of the DC(8,9)PC three-carbon glycerol backbone to four contiguous carbon atoms, the linking phosphate oxygen removal does not change the glycerol chain length, and we shall refer to phosphonate analog 3 as the "C3-phosphonate." The synthetic approaches are described elsewhere (*28, 29*).

Physical probes of new tubule forming compounds

Three probes were used to characterize the new phosphonate tubules: the direct imaging probes of optical microscopy and atomic force microscopy (AFM), and the indirect probe small-angle x-ray scattering (SAXS). The direct probes revealed the substantial differences in the new tubule morphologies, that is, the increase in diameter, the presence of significant numbers of left- and right-handed open helices in the new preparations, and finally, the so-called "giant" tubules. In contrast, the SAXS probes revealed that quantities such as interlamellar spacing are very rigorously conserved from one tubule species to the next. We proceed by describing these conserved quantities first.

Small-angle x-ray scattering

High-resolution synchrotron-based small-angle x-ray scattering was conducted at beamline X20A of the National Synchrotron Light Source at Brookhaven National Laboratory. 1.6122 Å x-rays were used with an in-plane resolution of 0.0007 $Å^{-1}$, commensurate with the Ge (1,1,1) analyzer crystal, while out-of-plane resolution was relaxed to 0.008 $Å^{-1}$ to increase signal intensity. Tubule samples were placed in standard 1.5 mm diameter quartz diffraction capillaries, yielding unoriented powder samples.

To minimize tubule polymerization upon exposure to ionizing radiation, 50 mm of the 80 mm capillary was translated continuously through the ≈ 0.5 mm tall collimated x-ray beam during data acquisition, which had the effect of distributing radiation damage throughout an approximately 100-fold larger sample than a stationary sample would present.

The instrument resolution was deconvolved from the tubule interlamellar (00ℓ) peak and fit to the powder average of the scattering from infinite length multilayer tubule structures having outer diameter D, wall thickness T, and layer spacing d. The coaxial cylinders were assumed to have no bending fluctuations, that is, they were assumed to be rigid membranes (*30, 31*). From this model and the measured half-width at half-maximum, we obtained a correlation length ξ.

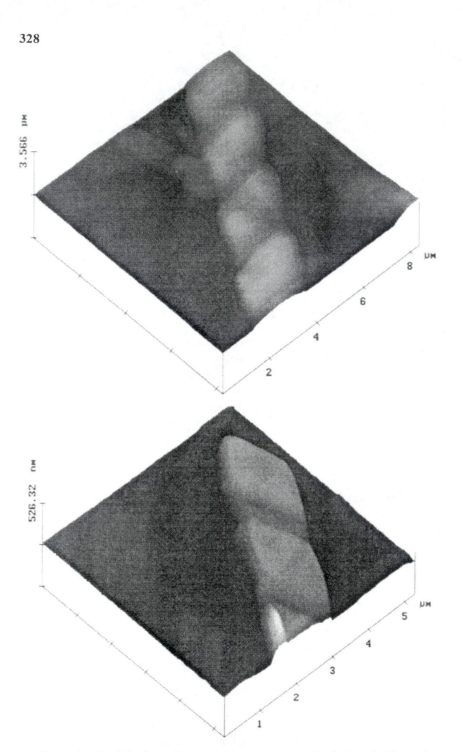

Figure 4. Pseudo-three dimensional perspective of right-handed (top of Figure) and left-handed phosphonate helices (bottom of Figure) in the same phosphonate preparation, obtained with underwater contact-mode AFM. Reproduced from reference 28.

Because the phosphonate and DC(8,9)PC tubule outer diameters D, as determined by AFM, are rather narrowly distributed, the correlation length ξ can be taken to be the mean tubule wall thickness. Once the lamellar spacing d is known, the number of lamellae comprising the tubule wall is known.

Despite a doubling of tubule diameter, both phosphonates' interlamellar spacings are essentially identical to the 62.97 Å spacing found in DC(8,9)PC tubules. In light of this similarity to DC(8,9)PC, it is reasonable to assume the phosphonate intralamellar structure is fundamentally that found previously for DC(8,9)PC; the remarkably small change in interlamellar spacing suggests strongly that the inter-lamellar forces which determine the spacing d are unchanged in the new tubule structure.

Peak-shape analyses for completely rigid membranes leads to a correlation length for the "C3" phosphonate $\xi_{C3} = 197.9$ Å, $\xi_{C4} = 207.0$ Å "C4" phosphonate tubules, while for DC(8,9)PC tubules the correlation length was $\xi_{DC(8,9)PC} = 431.08$ Å. The ratio of tubule wall thickness to interlamellar spacing is the number n of bilayers composing the tubule; we find n to be 6.9 for DC(8,9)PC and 3.2 for both phosphonates. Thus, we find that the two phosphonates produce tubules having nearly identical wall thickness, about half that of tubules made from DC(8,9)PC, and with similarly close cylindrical diameters, about twice that of DC(8,9)PC tubules.

Optical and atomic force microscopy

While the change in the number of lamellae composing phosphonate tubule walls is an unexpected result, it is not surprising: it is entirely plausible that differences in the membrane's mechanical properties, such as the maximum radius of curvature the phosphonate membrane can withstand, underlie these changes. The conservation of interlamellar spacing is not surprising, either, for it is expected in terms of the solvent-mediated interlamellar forces, which should be unchanged after our modification of the DC(8,9)PC molecule. Nor is the AFM- and optically-determined change in tubule diameter a surprise; indeed this was the sought-for experimental outcome. What is surprising is the comparatively large number of stable open helices of both senses of handedness that were found in enantiopure preparations, as shown in Figure 4.

Another surprising outcome that underscores the need for more solid definition of tubule-forming systems' remarkable self-assembly behaviors are the "giant" tubules found to form in "C3" phosphonate preparations, as shown in Figure 5. These structures have diameters of order 10 μm, and are obtained only under conditions of low lipid concentration and very slow cooling of the heated lipid solution to room temperature. Despite our vigorous efforts, no similar structures have been found with DC(8,9)PC or its "C4" analog, despite the similarities of the three molecules. The abrupt and apparently discontinuous order of magnitude increase in these structures' diameters raises a number of profound questions about the tubule formation mechanism.

330

First, given the nature of the abrupt morphology changes, it is reasonable to consider whether these structures are even fairly considered to be tubules. Regardless of that determination, are these structures chiral? AFM examinations do reveal traces of helical windings (see Figure 5), but due to these structures' very thin, deformable walls, their tendency to flatten and intrinsic surface roughness, unambiguous assignments of helical handedness has proven to be elusive. Next, why are these structures found only in "C3" phosphonate preparations? Is this the consequence of the changes installed in the prototype DC(8,9)PC molecule, or is this the result of some as-yet undiscovered experimental artifact? Do these structures lend support to either of the tubule structure theories, or will a new class of theory be required?

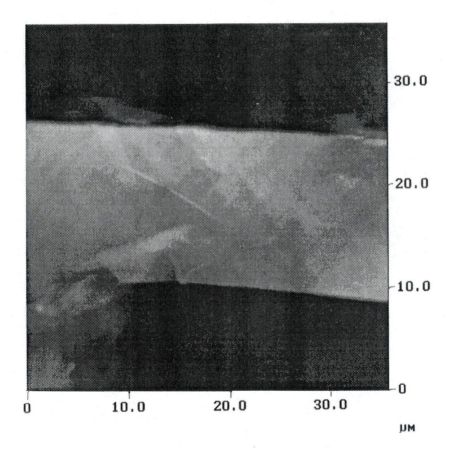

Figure 5. Underwater contact-mode AFM of a giant "C3" phosphonate tubule.

Conclusion

The lengthscale at which chirality appears as a structure-determining factor in phospholipid tubules is an unresolved issue. As the discovery of giant tubules suggests, significant exploration of these systems' remarkable behaviors remains to be done before theory can be expected to advance. To date, the majority of experimental evidence supports the so-called chiral packing theory, in which the chiral shape of the tubule-forming molecule itself is the origin of the tubule's helical winding. Recent observations of violations of the molecular chirality / helix handedness correspondence have lent some weight to a competing chiral symmetry-breaking model, in which the molecule's chirality does not play a role in creating membrane chirality.

Neither theory alone is sufficient to completely explain the experimental results. A symmetry-breaking tilt mechanism should produce a 50:50 ratio of left- to right-handed helical structures, which is not observed, while the recently observed helix-handedness / molecular chirality violations are inconsistent with the chiral-packing theory. Even more confounding is the behavior of DC(8,9)PC systems, where the innermost cylinders' sense of handedness appears to be independent of molecular chirality, and yet the DC(8,9)PC tubule exterior helical trace is of a completely uniform sense of handedness.

If chiral symmetry-breaking is the operative tubule formation mechanism, molecular chirality must, at least in the case of DC(8,9)PC, overwhelmingly bias the otherwise random ensemble tilt, thereby producing tubules with uniformly-handed exterior traces. It is possible that the slower process of tubule exterior growth in the DC(8,9)PC system permits the molecule's chirality to be expressed. The corresponding lack of uniformity in the equilibrium phosphonate tubule preparations indicates that, whatever mechanism is in effect, the modifications we have installed near the DC(8,9)PC stereogenic center have substantially diminished the influence that molecular chirality exerts upon self-assembly product morphology. How this may be related to giant tubule formation is now the central focus of this group's research effort.

References

1   Pickett, G. T., Gross, M. and Okuyama. H., submitted to *Phys. Rev. Lett.*
2   Schnur, J. M. *Science*, **1993**, 262, 1669-76.
3   Yager P. and Schoen, P. E., *Mol. Cryst. Liq. Cryst.* **1984**, 106, 371-81.
4   Yager, P., Schoen, P. E., Davies, C., Price, R. Singh, A., *Biophys. J.* **1985**, 48, 899-906.
5   Oda R, Huc I, Schmutz M, Candau SJ, MacKintosh FC, Nature **1999**, 399 (6736) 566.

6   Thomas, B. N., Lindemann, C. M. and Clark, N. A. , *Phys. Rev. E,* **1999**, 59, (3) 3040-3047.

7   Archibald, D. D.; Mann, S. *Chem. Phys. Lipids* **1994**, 69, 51-64.

8   Baum, R. *Chem. Eng. News*, August 9, 1993, p.19-20.

9   Rosenblatt, C.; Yager, P.; Schoen, P. E. *Biophys. J.* **1987**, 52, 295-301.

10 Li, Z.; Rosenblatt, C.; Yager, P.; Schoen, P. E. *Biophys. J.* **1988**, 54, 289-94.

11 Schnur, J. M.; Schoen, P. E.; Yager, P.; Calvert, J. M.; Georger, J. H.; Price, R. Metal-clad lipid microstructures; U.S. Patent Office, 7 pp., US 4911981 A 900327 US 87-63029 870616, U. S. Navy: USA, 1990.

12 Johnson, D. L.; Polikandritou-Lambros, M.; Martonen, T. B. *Drug Delivery* **1996**, 3, 9-15.

13 Schnur, J. M.; Price, R.; Rudolph, A. S. J. *Controlled Release* **1994**, 28, 3-13.

14 Ratna, B. R.; Baral-Tosh, S.; Kahn, B.; Schnur, J. M.; Rudolph, A. S. *Chem. Phys. Lipids* **1992**, 63, 47-53.

15 Thomas, B. N.; Safinya, C. R.; Plano, R. J.; Clark, N. A. *Science* **1995**, 267, 1635-8.

16 J. V. Selinger and J. M. Schnur, *Phys. Rev. Lett.* **1993**, 71, 4091-4.

17 J. V. Selinger, F. C. MacKintosh, and J. M. Schnur, *Phys. Rev. E* **1996**, 53, 3804-18.

18 Spector, M. S.; Selinger, J. V.; Schnur, J. M. J. Am. Chem. Soc. **1997**, 119, 8533-8539.

19 Zastavker YV, Asherie N, Lomakin A, Pande J, Donovan JM, Schnur JM, Benedek, GB, P*roc. Nat. Acad. Sci.*, **1999**, 96, 14, 7883-7887.

20 Singh, A.; Marchywka, S. *Polym. Mater. Sci. Eng.* **1989**, 675-8.

21 Rhodes, D. G.; Frankel, D. A.; Kuo, T.; O'Brien, D. F. *Langmuir* **1994**, 10, 267-75.

22 Markowitz, M. A.; Schnur, J. M.; Singh, A. *Chem. Phys. Lipids* **1992**, 62, 193-204.

23 Singh, A.; Gaber, B. P. *Polym. Sci. Technol.* **1988**, 38, 239-49.

24 Ruerup, J.; Mannova, M.; Brezesinski, G.; Schmid, R. D. *Chem. Phys. Lipids* **1994**, 70, 187-98.

25 Lee, Y. S.; O'Brien, D. F. *Chem. Phys. Lipids* **1992**, 61, 209-18.

26 Rhodes, D. G.; Xu, Z.; Bittman, R. *Biochim. Biophys. Acta* **1992**, 1128, 93-104.

27 Singh, A.; Burke, T. G.; Calvert, J. M.; Georger, J. H.; Herendeen, B.; Price, R. R.; Schoen, P. E.; Yager, P. *Chem. Phys. Lipids* **1988**, 47, 135-48.

28 B. N. Thomas, R. C. Corcoran, C. L. Cotant, C. M. Lindemann, J. E. Kirsch and P. J. Persichini, *J. Am. Chem. Soc.*, **1999**, 120, 47, 12178-12186.

29 B. N. Thomas, R. C. Corcoran, C. M. Lindemann, J. E. Kirsch and P. J. Persichini, in preparation.

30 Safinya, C. R.; Roux, D.; Smith, G. S.; Sinha, S. K.; Dimon, P.; Clark, N. A.; Bellocq, A. M. *Phys. Rev. Lett.* **1986**, 57, 2718-21.

31 Roux, D.; Safinya, C. R. *J. Phys. France* **1988**, 49, 307-0318.

INDEXES

Author Index

Subject Index

T

Temperature-programmed desorption (TPD)
average peak desorption temperature and average desorption energy, 280*t*
1-chloro-2-methylbutane, 274, 277
experimental, 272
3-methylcyclohexanone, 277, 280
peak desorption temperatures for (R)-3-methylcyclohexanone and (R/S)-3-methylcyclohexanone on Cu(643)^R and Cu(643)^S vs. surface coverage, 279*f*
peak desorption temperatures for (S)-1-chloro-2-methylbutane and (R/S)-1-chloro-2-methylbutane on Cu(643)^R vs. surface coverage, 276*f*
TPD spectra of increasing exposure of (R)-3-methylcyclohexanone to Cu(643)^S surface, 278*f*
TPD spectra of increasing exposures of (S)-1-chloro-2-methylbutane to Cu(643)^R surface, 275*f*
See also Single-crystal surfaces, chiral
Theoretical chemistry, application to chirality, 3–4
Three point model, chiral vapors, 9–10
Three point recognition, model, 4
Trans-2-butene
determining absolute chirality, 194
single chiral molecules and assemblies, 195, 197
Transition current density (TCD). *See* Vibrational transition current density (TCD)
Transition metal ion, X-ray spectroscopy, 159, 160*f*
3,7,7-Trimethyl bicyclo[4.1.0]hept-3-ene. *See* Single molecule at silicon surface
Tripeptides

ab initio simulated IR and VCD spectra for alanine pseudo-tripeptide, 55*f*
computations for model pseudo-tripeptides, 52–53
effects of solvent on α-helical tripeptide VCD simulations, 55–56
See also Chirality in peptide vibrations
Tröger's base
absolute configuration (AC), 24, 28
mid-IR vibrational circular dichroism (VCD), 29*f*
structure, 27
See also Vibrational circular dichroism (VCD)
Tubule formation
1,2-bis(10,12-tricosadiynoyl)-sn-glycero-3-phosphocholine [DC(8,9)PC], 320, 321*f*
changes in tubule morphology, 325
chiral symmetry-breaking models, 324–325
novel potential applications from tubule structural features, 322–323
optical and atomic force microscopy (AFM), 329–330
phosphate DC(8,9)PC derivatives, 326–327
phospholipid tubules, 320–325
phosphonate tubule-forming compounds, 326–330
physical probes of new tubule forming compounds, 327–330
pseudo-three dimensional perspective of right-handed and left-handed phosphonate helices by contact-mode AFM, 328*f*
schematic of chiral symmetry breaking, 324*f*
small-angle X-ray scattering, 327, 329
support to symmetry-breaking model, 325

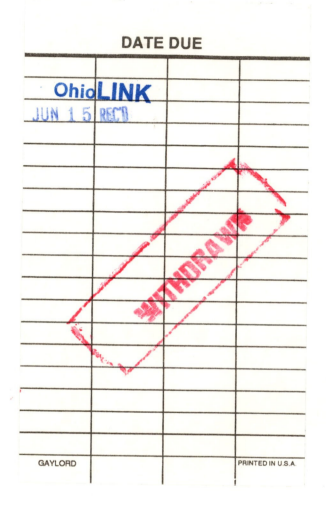
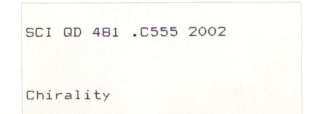